Advances in Intelligent Systems and Computing

Volume 1269

AF832380

The series “Advances in Intelligent Systems and Computing” contains publications on theory, applications, and design methods of Intelligent Systems and Intelligent Computing. Virtually all disciplines such as engineering, natural sciences, computer and information science, ICT, economics, business, e-commerce, environment, healthcare, life science are covered. The list of topics spans all the areas of modern intelligent systems and computing such as: computational intelligence, soft computing including neural networks, fuzzy systems, evolutionary computing and the fusion of these paradigms, social intelligence, ambient intelligence, computational neuroscience, artificial life, virtual worlds and society, cognitive science and systems, Perception and Vision, DNA and immune based systems, self-organizing and adaptive systems, e-Learning and teaching, human-centered and human-centric computing, recommender systems, intelligent control, robotics and mechatronics including human-machine teaming, knowledge-based paradigms, learning paradigms, machine ethics, intelligent data analysis, knowledge management, intelligent agents, intelligent decision making and support, intelligent network security, trust management, interactive entertainment, Web intelligence and multimedia.

The publications within “Advances in Intelligent Systems and Computing” are primarily proceedings of important conferences, symposia and congresses. They cover significant recent developments in the field, both of a foundational and applicable character. An important characteristic feature of the series is the short publication time and world-wide distribution. This permits a rapid and broad dissemination of research results.

**** Indexing: The books of this series are submitted to ISI Proceedings, EI-Compendex, DBLP, SCOPUS, Google Scholar and Springerlink ****

More information about this series at http://www.springer.com/series/11156

Waldemar Karwowski · Tareq Ahram ·
Darko Etinger · Nikola Tanković · Redha Taiar
Editors

Human Systems Engineering and Design III

Proceedings of the 3rd International Conference on Human Systems Engineering and Design (IHSED2020): Future Trends and Applications, September 22–24, 2020, Juraj Dobrila University of Pula, Croatia

IHSED
International
Human Systems
Engineering and
Design Conference
2020
www.ihsed.org

Springer

Editors
Waldemar Karwowski
University of Central Florida
Winter Park, FL, USA

Tareq Ahram
Institute for Advanced Systems Engineering
University of Central Florida
Orlando, FL, USA

Darko Etinger
Juraj Dobrila University of Pula
Pula, Croatia

Nikola Tanković
Juraj Dobrila University of Pula
Pula, Croatia

Redha Taiar
Campus du Moulin de la Housse
Université de Reims Champagne
Ardenne, GRESPI
Reims Cedex 2, France

ISSN 2194-5357 ISSN 2194-5365 (electronic)
Advances in Intelligent Systems and Computing
ISBN 978-3-030-58281-4 ISBN 978-3-030-58282-1 (eBook)
https://doi.org/10.1007/978-3-030-58282-1

This Springer imprint is published by the registered company Springer Nature Switzerland AG
The registered company address is: Gewerbestrasse 11, 6330 Cham, Switzerland

Preface

As noted by the National Academy of Engineering (NAE, 2004) in the report entitled *The Engineer of 2020: Visions of Engineering in the New Century*, in the future, engineering will *"expand toward tighter connections between technology and the human experience, including... the ergonomic design of engineered products"*. Today, human factors discipline promotes a human-centered approach to systems design, with due consideration of a great variety of relevant factors, such as physical, cognitive, neural, social, emotional, organizational, developmental, ecological, environmental that are essential for the socioeconomic development and well-being of the global society.

Recent advances in cognitive computing, machine learning, and artificial intelligence, modeling, and simulation, as well as smart sensor technology, create new opportunities for the field of human systems engineering to contribute to the human-centered approach to the design of products and systems. These new developments also allow expanding the current boundaries of the state of the art of the human factors discipline by investigating the pervasive complexity that underlies the most problems facing contemporary society today.

The present book on *Human Systems Engineering and Design* contains a selected set of novel contributions to the theory and practice of human engineering and design for integrating people and technology in all areas of human endeavor by adopting a human-centered approach, supported by cognitive software and engineering, data analytics, simulation and modeling, and next-generation visualization methods.

The papers contained in this volume are organized into four main sections:

Section 1 Human-centered Design
Section 2 Systems Design and Human Diversity
Section 3 Safety Engineering and Systems Complexity
Section 4 Human Cyber-physical Systems Interactions

The presented papers introduce many innovative examples of systems engineering and design, emphasizing the development of technology throughout the lifecycle development process in the areas of advanced digital manufacturing, smart energy, transportation, urbanization, and infrastructure, healthcare, and cybersecurity sectors on business and industry. Some of the presented studies also include the consideration of user experience in the design of human interfaces for virtual augmented and mixed reality applications.

We hope that this book, which presents the current state of the art in human systems engineering and design, will be a valuable source of theoretical and applied knowledge that enables the human-centered design and applications of a variety of products, services, and systems for their safe, effective, and pleasurable use by people around the world.

Finally, we would like to extend our sincere thanks to the Juraj Dobrila University of Pula, FIPU Faculty of Informatics, for the support of the conference's organizational efforts. Our appreciation also goes to the Scientific Program Advisory Board of the IHSED 2020 Conference, who reviewed papers presented in this volume.

September 2020

Waldemar Karwowski
Tareq Ahram
Darko Etinger
Nikola Tanković
Redha Taiar

Contents

Human-Centered Design

User-Centered Detection of Fake News and Misinformation - Design and Prototypical Implementation in the System Contexter

Kurt Englmeier(✉)

Schmalkalden University of Applied Science, 98574 Schmalkalden, Germany
k.englmeier@hs-sm.de

Abstract. Misinformation or fake news may threaten our democracies, societies, and economies, even individual health and well-being. Humans are usually careful about the things they are being told. They check news or tweets against their knowledge or beliefs and estimate to what extent propositions contain information that is bogus. People have abstract representations of facts in mind. That help them to validate propositions and to search for information suitable for their validation. This paper presents design and prototypical implementation of the Contexter system that enables users to define and manage blueprints of facts or fake news. Contexter takes these blueprints as a schema to detect facts or fake news. It also starts to find variants of these blueprints to detect pieces of text that come semantically close to the propositions addressed by the original blueprint.

Keywords: Information extraction · Fake news detection · Named Entity Recognition · Bag of Words · Information summarization · Text mining

1 Introduction

This paper presents design and prototypical implementation of Contexter, a user-driven text mining tool that enables definition of blueprints for facts or fake news. Contexter takes these blueprints as a schema to detect pieces of text addressing facts or fake news. The system also applies a learning process to find variants of these blueprints to detect pieces of text that come semantically close to the propositions addressed by the original blueprint. User involvement results in a more active role of the users in designing, controlling, and adapting of the learning process that feeds, in this case here, the automatic detection of misinformation and fake news.

Blueprints are abstract representations of the things described in text. At the same time, blueprints represent models of the things the users have in mind, when searching for suitable information or when pursuing fake news. Contexter supports the composition of blueprints by means of Named Entities combined with keywords from Bag of Words (BoW). Contexter takes each set of terms and transforms it into Regular Expressions. The users produce an initial set of a blueprint reflecting facts (in a proposition or fake news) that are subsequently applied to the text collection in order to find all corresponding text instances. In a next step, Contexter modifies this blueprint,

W. Karwowski et al. (Eds.): IHSED 2020, AISC 1269, pp. 3–8, 2021.
https://doi.org/10.1007/978-3-030-58282-1_1

that is, it gradually modifies its elements one by one and tests the resulting variants against the text collection. Contexter presents all encountered new instances for blueprint variants to the users for confirmation or rejection and tries to learn from user feedback. Even though the prototype shown here still represents work in progress, it demonstrates its potential in the detection of fake news and misinformation.

2 Approach and Related Work

The design of Contexter for fake news and misinformation detection bases on a combination of Named Entity Recognition (NER), Bag of Words (BoW), and Word N-Grams [1, 2]. The idea behind the design is that each fact has its individual pattern of words and numeric data. We can assume that a specific combination of keywords and annotated numeric expressions uniquely reflects a particular fact.

The identification of facts starts with information extraction [3] and the annotation of the extracted text pieces according to the meaning they express [4]. Annotation has two roles: first, it adds a meaningful term to the extracted text, in particular the numeric data. Such patterns, for example, represent dates, percentages, numerical data, distances, and the like. Second, the annotations (and keywords) from the first annotation are further annotated. This process (if iteratively performed) produces an increasingly more abstract representation of the text and numeric data in the text piece under consideration. Semantic markers [5] are the smallest fraction of a text covering a certain meaning discernable from the other fractions. Together they mark the meaning of a particular piece of text.

Fake news detection may also include methods and techniques of event mining [6] when it comes to analyze statements over a discourse (in social media) with events that triggered this discourse [7]. developed a framework for narratives of a therapist-patient discourse that is valuable in our context. His work has been summarized and discussed in [8].

3 Recognition and Representation of Facts

The Contexter system enables the definition blueprints or patterns of facts (or fake news). They consist of Named Entities combined with keywords from Bag of Words (BoW). The users describe their blueprints by arranging terms from the BoWs and Named Entities. Contexter takes each set of terms and transforms it into Regular Expressions. The users produce an initial set of blueprints reflecting facts that are subsequently applied to the text collection in order to find all corresponding text instances. In a next step, Contexter gradually modifies each blueprint, that is, one of its elements is turned off or slightly modified. This automatic process aims at finding as many representations of facts as possible. Contexter presents all encountered new instances of facts to the users for confirmation and rejection and tries to learn from user feedback. Representations of facts in text have their unique pattern, independently of the source or language. Contexter provides a number of theme-specific BoWs (for locations, names, expressions of aggression etc.) and Named Entities for common patterns in text reflecting time, amounts, distances, and the like. This process usually combines key words and common text patterns. Finally, each pattern is annotated by an appropriate term that summarizes the meaning of the pattern.

Generic named entities help to standardize factual information and to abstract away the different forms of expressions for essentially the same thing. However, it does not suffice just to annotate generic patterns. We can easily imagine that Named Entities may relate to ontologies that serve specific interpretation or calculation purposes.

Named Entity Recognition (NER) in the context described here operates with bags of words (BoW) addressing locations, persons, organizations, or institutions. There are promising approaches to automatically identify names (and other important key expressions) in texts using conditional random fields (CFR) [9].

The meaning of a piece of text is more than just the meaning of one or more specific terms surrounded by further terms. The blueprints as shown in Fig. 1 possible collocations of prominent text elements for a concept like decrease or infection. They serve two purposes:

```
decrease=casualty.location:country."sinken".numeric
fatality="insgesamt".numeric:total."Menschen"."gestorben"
fatality=duration.increase
fatality='muertes?'.duration.decrease
fatality="total"."fallecidos".increases
infection="Infektion(en)?"."steigen".duration."um".numeric:delta."auf".numeric:total
infection="nuevos positivos";"infección".increase.percent:delta
infection="nuevos positivos";"infección".increase
fatality=increase.?duration
```

Fig. 1. Examples of blueprints addressing facts in relationship with Covid-19 infections and fatalities (in German and Spanish texts). The sequence of the elements of the blueprint are only indicative. Contexter handles them flexibly. Terms like "sinken" or "steigen" (German for decrease and increase) stand for their synonyms, too. They are from the BoW for changes in figures

1. Translated into Regular Expressions, blueprint help to locate pieces of text addressing the concept of the blueprint and extract the text elements that match the elements of the blueprint.
2. Annotation helps to identify the overarching theme manifested by these terms. In only a few cases, the meaning of a piece of text can be inferred just by the individual meaning of its terms. The annotation a standardized and contextualized representation of pieces of text.

In many situations, the overarching meaning is not expressed by even a single word in the phrase. For the correct interpretation of text, we need representations on a higher level, that is, on a more abstract level. This does not mean that we abstract away details. On the contrary, by annotation we add essential details.

Named Entity Recognition (NER) presented so far serves the standardization of data, in terms of assigning standardized annotations to the underlying data. It also helps to detect term patterns and keywords in text fragments that constitute conditional dependencies among the terms. Furthermore, it reveals the correct and unambiguous meaning of the text fragment and its components. By the iterative and incremental application of NER we can produce theme-specific hierarchies of named entities that represent the meaning of the content on different levels of granularity. The hierarchies are the building blocks of content representation of text. We may call them content clusters of Named Entities or content schemas. The Named Entities reflect the overarching theme of the pieces of text addressing the increase (or the decrease of the increase) of infections and fatalities. Distilled and structured data are easier to handle by mining processes in order to detect discrepancies in text, in particular if they are manifested in numerical data (Figs. 373 vs. 3046 in the example below).

BoWs play an important role in the definition of these schemas. A keyword embedded in a pattern of named entities is important when it comes to correctly identify the overarching theme of the piece of text. In the example of Fig. 2 we have a Covid-19 scene of a particular day in Spain. The meaning of each of these elements and the act in the scene can be correctly identified by the Named Entities in conjunction with keywords. Terms like "asciende" (Spanish for increase) or "sinken" (German for "decrease) together with their adjacent terms are an essential ingredient for correctly classifying the respective phenomenon. This example shows that keywords and Named Entities take the role of semantic markers for a particular content.

The identification of semantic markers is a process covering a series of phases:

1. At first, we independently identify clusters of Named Entities and keywords. BoWs provide the keywords while Basic Named Entities contribute standardized versions of small and generic fractions of text. Basic Named Entities may even be aggregated to enhance the content representation. Incrementally, we obtain more complex representations (much like in ontologies) that include semantic relationships among the elements.
2. The next step is the manual definition of more complex content. This is usually an iterative process of defining and testing prototypes for these clusters. If prototypes match on more than one occasion over the whole text collection, in our case Covid-19 reports, we consider them as candidates for further analysis.
3. Phase 2 is the starting point for the learning process to identify term hierarchies. The Contexter system takes the confirmed prototypes of blueprints, and continuously applies them to the text collection. Whenever the system identifies a similar pattern of semantic markers it tries to find further patterns that match this pattern. Each instance found is marked as potential candidate for a new content cluster. If we take the blueprint in Fig. 1 as seed, we may find similar text pattern comprising the same Named Entities but differ in keywords (from BoW) or components of Named Entities applied.
4. The results of the automatic NER are controlled and evaluated. Here, we need again human intervention. The users manage theme-specific BoWs and content schemas and classify the output of the system. A candidate proposed by the system can be confirmed or rejected.

```
<fatality>
    <duration>
        en un día
    </duration>
    <increase>
        <delta>
            123
        </delta>
    </increase>
</fatality>
<infection>
    <increase>
        <delta>
            373
        </delta>
    </increase>
</infection>
<infection>
    <increase>
        <total>
            227.436
        </total>
    </increase>
    <delta>
        <percent>
            0,17%
        </percent>
    </delta>
</infection>
<fatality>
    <increase>
        <total>
            26.744
        </total>
    </increase>
</fatality>
<percent>
    0,54%
</percent>
```

```
<fatality>
    <increase>
        <delta>
            123
        </delta>
    </increase>
    <humans>
        <total>
            26.744
        </total>
    </humans>
</fatality>
<infection>
    <increase>
        <duration>
            binnen 24 Stunden
        </duration>
        <delta>
            3046
        </delta>
        <total>
            227.436
        </total>
    </increase>
</infection>
```

Fig. 2. Examples of extracted and annotated news texts matching the blueprints "fatality" and "infection"

4 Conclusion

This paper presented the state of work of Contexter, a prototypical system that operates on Named Entity Recognition and uses theme-specific Bag of Words to identify semantic markers in text that point to the specific meaning of pieces of text.

The application areas of the content schemas are manifold. The main purpose is identifying facts in texts and represent them in a distilled and standardized way along their respective context. This facilitates the comparison of facts in different sources and, thus, supports the detection of fake news and misinformation.

Named entities and terms from BoWs identify the meaning of words (and numerical data) as they appear in a piece of text. Iteratively applying standardization to already extracted and annotated pieces of text yields semantic hierarchies that enable the identification of meaning along generalization or specification aspects. This, in turn, makes text comparisons more precise and versatile.

Contexter is still work in progress, but we already noticed that our content schemas have a certain proximity to ontologies. We use the schemas for text interpretation on a basic level and gradually produce concept hierarchies. However, we clearly see the necessity to add more functionality to schemas, in particular, when parts of the schema address factual (i.e. numerical) information. Quite often calculations can be helpful to check the plausibility of statements based on numerical information. Let us take the situation when a proposition insinuates a ten-fold increase of infections in a particular location in one week. By comparing the increase rate obtained from trustworthy sources with the predicted one we may notice an extraordinary difference leading to the conclusion that the predicted rate in the proposition is not plausible.

References

1. Wynne, H.E., Wint, Z.Z.: Content based fake news detection using N-gram models. In: Proceedings of the 21st International Conference on Information Integration and Web-based Applications & Services (iiWAS2019), pp. 669–673 (2019)
2. Woods, W.A.: Context-sensitive parsing. Commun. ACM **13**(7), 413–445 (1970)
3. Cowie, J., Lehnert, W.: Information extraction. Commun. ACM **39**(1), 80–91 (1996)
4. Salton, G., Allan, J., Buckely, Ch., Singhal, A.: Automatic analysis, theme generation, and summarization of machine-readable texts. In: Jones, K.S., Willett, P. (eds.) Readings in Information Retrieval, San Francisco, pp. 478–483 (1997)
5. Jancsary, J., Neubarth, F., Schreitter, S., Trost, H.: Towards a context-sensitive online newspaper. In: Proceedings of the 2011 Workshop on Context-Awareness in Retrieval and Recommendation, pp. 2–9 (2011)
6. Calvo Martinez, J.: Event mining over distributed text streams. In: Proceedings of the Eleventh ACM International Conference on Web Search and Data Mining, pp. 745–746 (2018)
7. Schneider, P.: Language usage and social action in the psychoanalytic encounter: discourse analysis of a therapy session fragment. Lang. Psychoanal. **2**(1), 4–19 (2013)
8. Murtagh, F.: Mathematical representations of Matte Blanco's bi-logic, based on metric space and ultrametric or hierarchical topology: towards practical application. Lang. Psychoanal. **3** (2), 40–63 (2014)
9. Sha, F., Pereira, F.: Shallow parsing with conditional random fields. In: Proceedings of the HLT-NAACL Conference, pp. 134–141(2003)
10. Dang, V., Croft, B.W.: Query reformulation using anchor text. In: Proceedings of the Third ACM International Conference on Web Search and Data Mining, pp. 41–50 (2010)

Modular Car Seat for Monitoring the Pressure Distribution on Regions of Pan and Backrest

Alberto Vergnano(✉), Andrea Piras, and Francesco Leali

Department of Engineering Enzo Ferrari, University of Modena and Reggio Emilia, Via P. Vivarelli, 10, 41125 Modena, Italy
{alberto.vergnano, francesco.leali}@unimore.it, 209468@studenti.unimore.it

Abstract. Car driver and occupants monitoring is important for safety and comfort. The systems using vision sensors for monitoring the line of sight can be integrated with pressure sensors embedded in the seats to identify critical Out of Position conditions. A new modular car seat is here proposed to monitor the pressure distribution on different significant regions. The solution improves the limitations of existing technologies embedded in seats. The proportional and fast measurement enables online complex evaluations, while the layout reduces the risk of errors. The experiments proved the effectiveness of the prototype.

Keywords: Weight scale system · Driver monitoring · Driver profiling · Modular car seat · Seat regions · Intelligent vehicle · Safety system · Car seat pressure · Occupant classification

1 Introduction

Monitoring the car driver or the other occupants is important for safety and comfort. Research is focused on developing systems for detecting the drowsy, stress, fatigue and monotony states, [1, 2]. Those systems especially monitor the eyes of the driver. They can reduce the possibility of an accident to occur, but their data are poorly usable for detecting the different Out of Position (OP) states for both the driver and occupants, as in those cases they cannot even find the eyes. On the other hand, the OP conditions are one of the major causes of injuries in crash events, [3]. Those states would be better monitored recurring to a sensor fusion, integrating vision cameras with pressure sensors embedded in the seats, [4].

Most pressure sensors embedded in car seats enable only the detection of the presence of an occupant with ON/OFF logics. Those systems simply inform the vehicle, which checks that the seat belts have been fastened.

Some other systems are based on array of sensors below the fabric in order to monitor the position of the occupant while driving, [5–8]. In dynamic tests, the pressure field is continuously compared with the car accelerations, monitored by an Inertial Measurement Unit (IMU), in order to discriminate the movements of the pressure centre due to a wrong position from those due to inertial effects, [9, 10]. The pressure field must be monitored by thin film piezoresistive sensors, which deliver strongly nonlinear measurement, with hysteresis and drift. Also, not all the pressure passes

W. Karwowski et al. (Eds.): IHSED 2020, AISC 1269, pp. 9–14, 2021.
https://doi.org/10.1007/978-3-030-58282-1_2

through the sensors, since not all the surface is covered by them, leading to possible evaluation errors. Such systems proved to be too prone to the irregularities of the surfaces in contact, which generate local pressure peaks. Moreover, pressure detection with finite area sensors is too sensitive to the occupant biometrics. In case of small percentiles, several sensors can fall outside the pressure field. These limits make the measurement qualitative only.

Other systems are based on a pressure mat with high resolution, [11–14]. The maps of pressure distribution can be used to effectively assess the comfort of the occupant on the seat. However, their construction is too delicate for being provided on passenger cars. A mat consists of three superimposed layers. The outers striped layers require a rows by columns scan strategy, which is too slow for an effective interaction between human and vehicle. Moreover, the operating principle is based again on the piezoresistive material of the middle layer, with the same problems of the previous solutions. In order to get meaningful information, the amount of data must be grouped in few significant regions. So, a reduced number of data results sufficient and effective for the evaluations, [15].

In the present research, a seat for car occupant monitoring has been developed, capable of overcoming many of the described limitations. The seat is subdivided in a proper number of regions in order to detect information that are significant even before data processing.

The paper is organized as follows. Sect. 2 presents the sensor equipment and methods. The driving experiments are reported and discussed in Sect. 3, while the concluding remarks are drawn in Sect. 4.

2 Materials and Method

The seat is conceived with a modular architecture, each module being a significant region for the detection of the interactions between occupant and seat, [13], as shown in Fig. 1a. The seat is composed of 13 modules:

1 and 2. leg areas: front part of the pan, where legs rest or act for balancing during driving maneuvers;
3 and 4. sacral area: rear part of the pan, sustaining most of the load;
5 and 6. lower sides of the pan: inclined cushions with containing function;
7 and 8. lumbar area: lower part of the backrest, loaded by the occupant back;
9 and 10. back area: upper section of the backrest for shoulders support;
11 and 12. upper sides of the seat: inclined cushions with containing function;
13. headrest: for neck and head positioning, not always used by the occupant

In each module of the seat, the load on the padding is transferred to the upper structure, made of a 3mm PVC plate. This plate is supported by four load cells, as shown in Fig. 1b. The squared cells have dimensions 34 × 34 × 8 mm and a payload of 50 kg. So, each module has 200 kg payload, which enable safe experiments without reaching the limit. The load cells are constrained onto a bottom plate which allows them to freely flex thanks to a slot below. All the load from the occupant passes

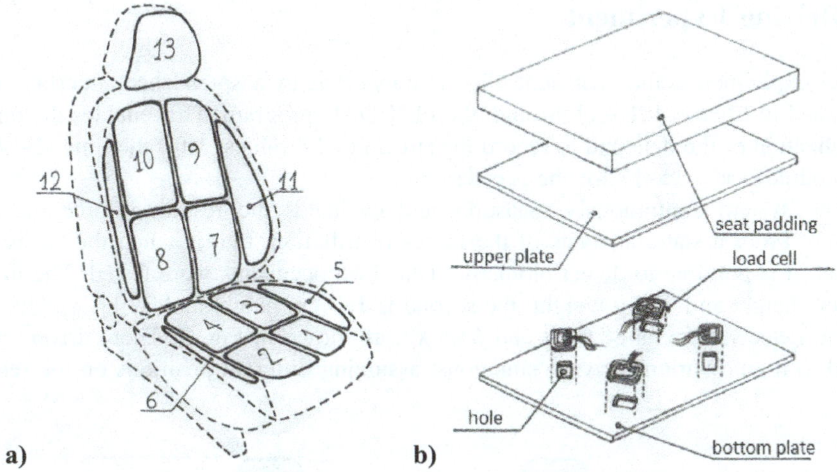

Fig. 1. Layout of a) modules in the seat and b) each module construction.

through the four cells. The four points support is a hyperstatic condition, but the little bending of the plates during working compensates the eventual assembly errors, keeping robust the measurement.

The 4 cells configuration enables the electronic connection as Wheatstone bridge for the power and ground wires. A HX711 amplifier connects the signal wires of all the cells in each module. The HX711 is finally connected with the Arduino Mega 2560 controller through ground, 5V power, digital clock and analogue input. In the end, the modules of the car seat are wired to 13 analogue inputs of the microcontroller. A BNO055 IMU complete the system, monitoring the car as the moving platform, [9, 10]. The installed prototype is shown in Fig. 2.

Fig. 2. Prototype assembled in the car for the experiments.

3 Driving Experiment

In the experiments, the seat controller is connected to a spreadsheet interface programmed in Microsoft Excel through the PLX-DAQ program. This enables the online visualization of the data and to record the stream of 17 values: 1 for the time, 2–14 for the module loads, 15–17 for the accelerations.

The system continuously measures and evaluates the resulting force on each module. From a static analysis of the forces distribution, that is when the car is still parked, it is possible to detect biometric data. Two occupants were tested. The first is 185 cm height and 92 kg weight, the second is 187 cm height and 72 kg weight. The system detects them as 88.9 kg and 68.5 kg, as shown in Fig. 3. Then, drivers were asked to mimic various driving situations, assuming different positions on the seat.

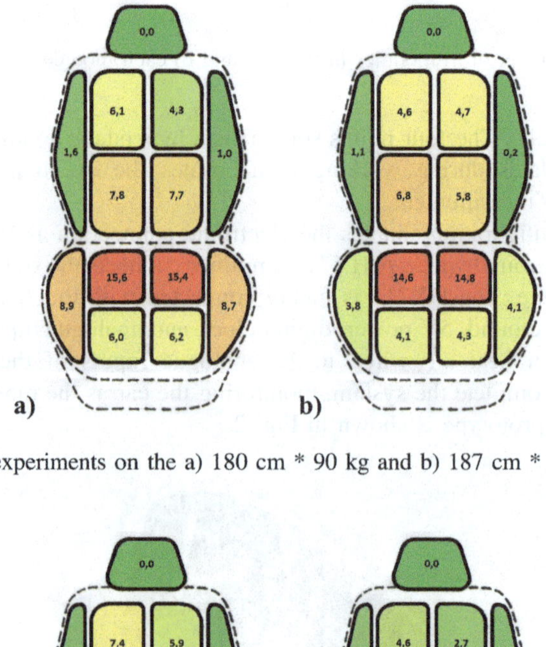

Fig. 3. Static experiments on the a) 180 cm * 90 kg and b) 187 cm * 72 kg occupants.

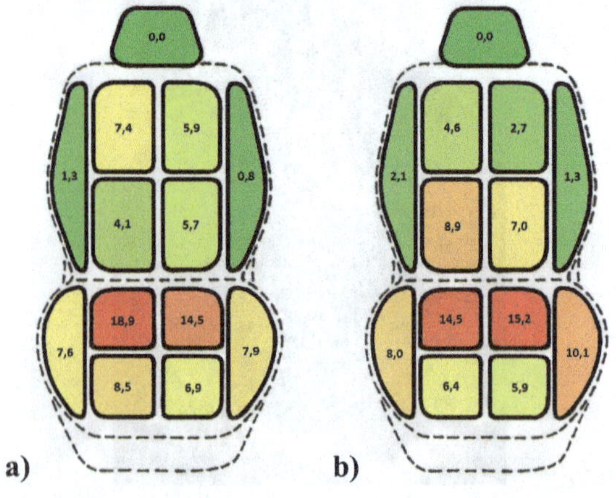

Fig. 4. Experiments on the 180 cm * 90 kg driver in a) relaxed and b) rigid positions.

Fig. 4a shows the experiment on the first occupant in a relaxed position, thus spreading his weight on the modules. The seat detects 89.2 kg. Fig. 4b shows him in a nervous and rigid position, thus with his legs pushing the body against the seat, 85.6 kg detected.

The experiments with the first occupant in a wrong position reveal the OP conditions, as shown in Fig. 5: a) forward reclined against the steering wheel, 83.4kg detected, b) left reclined against the side door, 83.4kg, c) right reclined toward the right seat, 84.7kg. The OP condition is clearly detected by the weight displacement to the front 1-2-3-4-5-6 modules, to the 1-3-5-7-9-11 left ones or to the 2-4-6-8-10-12 right ones respectively.

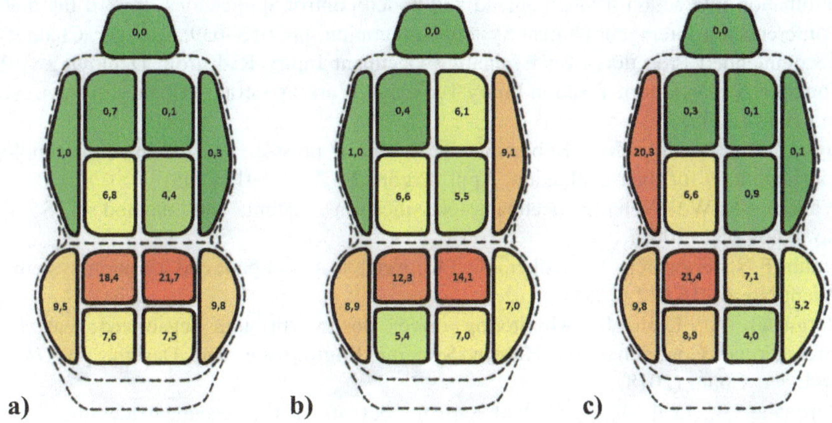

Fig. 5. Experiments on the 180 cm * 90 kg driver in a) forward reclined b) left reclined and c) right reclined out of positions.

4 Conclusion

A novel modular car seat for monitoring the pressure distribution on regions of pan and backrest is developed. Each module is sensorized by four load cells. The load cells are reliable sensors with a metal structure, so reading is fast, proportional to the weight, with no hysteresis, with high resolution and precision, thus enabling complex evaluations. The modular construction is also quite robust. A limited number of analogue inputs enables very fast measurements, thus enabling an effective interaction between human and vehicle, such as described for systems based on a small array of sensors. On the other hand, the layout of the modules delivers effective data, the same as achieved with the pressure mat after data processing. Since all the pressure of the occupant passes through the different modules, the risk of evaluation errors due to surface irregularities is very reduced. The experiments proved the effectiveness of the prototype.

The system can be developed in seats of passenger cars and of industrial vehicles for adjustable ergonomics, according to driving fatigue, personalized settings, different body sizes and driving styles. It can be used in continuous driver monitoring, enabling new safety strategies in intelligent vehicles. Finally, in sport driving, eventually on a simulator, the modular seat would enable to interact with the driver for an augmented drive experience.

References

1. Borghi, G., Venturelli, M., Vezzani, R., Cucchiara, R.: POSEidon: face-from-depth for driver pose estimation. In: 30th IEEE Conference on Computer Vision and Pattern Recognition, pp. 5494–5503 (2017)
2. Marcano, M., Díaz, S., Pérez, J., Castellano, A., Landini, E., Tango, F., Burgio, P.: Human-automation interaction through shared and traded control applications. In: 3rd International Conference on Intelligent Human Systems Integration, pp. 653–659. Springer, Cham (2020)
3. Recommended Procedures for Evaluating Occupant Injury Risk from Deploying Airbags. The Side Airbag Out-of-Position Injury Technical Working Group. Alliance, AIAM, AORC, and IIHS (2003)
4. Gyi, D.E., Porter, J.M., Robertson, N.K.: Seat pressure measurement technologies: considerations for their evaluation. Appl. Ergon. **29**(2), 85–91 (1998)
5. Wallace, M.W.: Vehicle occupant classification system and method. U.S. Patent No. 6,609,054 (2003)
6. Smith, F.N., Forwerck, J., Gauker, B.: Electronic seat occupant classification system. U.S. Patent No. 6,918,612 (2005)
7. Vergnano, A., Leali, F.: Monitoring driver posture through sensorized seat. In: 1st International Conference on Human Systems Engineering and Design, pp. 744–749. Springer, Cham (2018)
8. Vergnano, A., Leali, F.: A methodology for out of position occupant identification from pressure sensors embedded in a vehicle seat. Hum.-Intell. Syst. Integr. (2020)
9. Vergnano, A., Leali, F.: Out of position driver monitoring from seat pressure in dynamic maneuvers. In: 2nd International Conference on Intelligent Human Systems Integration, pp. 76–81. Springer, Cham (2019)
10. Vergnano, A., Pegreffi, F., Leali, F.: Correlation of driver head posture and trapezius muscle activity as comfort assessment of car seat. In: 2nd International Conference on Intelligent Human Systems Integration, pp. 241–247. Springer, Cham (2019)
11. Andreoni, G., Santambrogio, G.C., Rabuffetti, M., Pedotti, A.: Method for the analysis of posture and interface pressure of car drivers. Appl. Ergon. **33**(6), 511–522 (2002)
12. Kyung, G., Nussbaum, M.A.: Driver sitting comfort and discomfort (part II): relation-ships with and prediction from interface pressure. Int. J. Ind. Ergon. **38**(5–6), 526–538 (2008)
13. Ng, D., Cassar, T., Gross, C.M.: Evaluation of an intelligent seat system. Appl. Ergon. **26**(2), 109–116 (1995)
14. Vergnano, A., Muscio, A., Leali, F.: Sensor matrix robustness for monitoring the interface pressure between car driver and seat. In: 2nd International Conference on Human Systems Engineering and Design, pp. 229–235. Springer, Cham (2019)
15. Farmer, M.E., Jain, A.K.: Occupant classification system for automotive airbag suppression. In: IEEE Computer Society Conference on Computer Vision and Pattern Recognition 1, pp. I-I. Washington (2003)

Technical Challenges to Adopting Large Scale Additive Manufacturing for the Production of Yacht Hulls

Eric Peterson[1,2(✉)]

[1] Florida International University, 11200 SW 8th Street, Miami, FL 33199, USA
eric.peterson@fiu.edu
[2] Universita Degli Studi di Genova, Design Navale e Nautico, Stradone Di Sant' Agostino 37, 16123 Genoa, Italy

Abstract. Recent developments in additive manufacturing technology may initially appear as an emerging opportunity for multiple manufacturing sectors, but there are several serious challenges to the potential application of additive manufacturing with Fiber Reinforced Plastic (FRP) to the yacht design and manufacturing industry. While several boat hulls have been produced using additive manufacturing technologies such as the 3Dirigo demonstration project built at the University of Maine [1], vessels of this type remain experimental prototypes that will require additional testing before the technology proves commercially viable as a construction method. The example of Thermwood's Large Scale Additive Manufacturing (LSAM) system to successfully produce a small yacht hull plug mold is a promising application for 3d printing in yacht manufacturing [2]. Other applications of additive manufacturing technology to yacht components have proven viable, especially for small-scale replacement parts and retrofit components [3]. These initial steps into additive manufacturing are promising advances for the yacht design and manufacturing sector, but serious technical challenges remain before FRP hulls can be produced using additive manufacturing technology.

Keywords: Rapid prototyping · Additive manufacturing · 3D printing · Yacht design · Nautical manufacturing · Fiber reinforced plastic · Composite materials · Material performance · Multi-bias printing · Laminar slicer · Robotics · End effector · Extruder · Print head · Tool path

1 Introduction

Advances in additive manufacturing technology over the past three decades have been revolutionary for the field of industrial design, and some of these advances are beginning to work their way into the yacht design and manufacturing sector. At first, 3d printers were used primarily as visualization tools to produce scale models of yachts for design testing and marketing. More recently, additive manufacturing tools have been successfully used to create replacements parts, limited run and customized direct to market marine components.

W. Karwowski et al. (Eds.): IHSED 2020, AISC 1269, pp. 15–20, 2021.
https://doi.org/10.1007/978-3-030-58282-1_3

In the near term it is likely that additive manufacturing will be deployed to aid in the manufacture of yachts and yacht components rather than to 3d print yacht hulls themselves. In 2017 Thermwood used its LSAM technology to produce a plug mold for a small production series skiff [2]. This application of 3d printing technology greatly reduces the time and cost to produce molds for complex manifold surfaces. Molds are one of the most expensive investments for FRP yacht manufacturers. For the yacht manufacturing sector this time saving production tool offers both the opportunity to produce multiple variations of a hull for little additional cost and substantial reductions to the cost of serial production [4] by creating duplicate molds with identical dimensional characteristics thereby allowing multiple manufacturing lines of single high-volume production models.

While there have been several highly publicized demonstration projects featuring full-scale 3d printed boats such as the University of Maine 3Dirigo project, it is clear that boat hulls built using additive manufacturing are not yet ready for mass production. Despite the fanfare surrounding the 3Dirigo project, it is worth noting that the vessel has remained at an indoor facility for nearly a year to undergo testing for seaworthiness [1]. It remains unclear how the surface of the 3d printed material will behave in a hydrodynamic environment and whether the thermal bond between layers of extruded plastic will continue to be impervious to the intrusion of water under load and subjected to various environmental conditions over time. The yacht design and manufacturing sector looks forward to publication of the results of 3Dirigo performance testing.

A hybrid 3d printing application to yacht manufacturing that mitigates some of these issues is a 6.5 m racing yacht built in 2019 by Italian firm Livrea Yacht. For this project Livrea used additive manufacturing to print the inner core of a small sailing vessel that they subsequently coated with a waterproof stiffening carbon fiber skin [5]. A series of production delays have kept the outcome of the project out of the public eye, but the conceptual approach appears promising. However, it is clear from these few examples that significant technical challenges will need to be overcome before additive manufacturing can be used to directly print yacht hulls ready for market.

2 Challenges

In order to understand the technical challenges for adopting additive manufacturing for producing yachts it is useful to understand both the particularities of yacht hulls and how 3d printing technology is used to make large 3-dimensional forms.

Yachts in the sub-40 m range are typically produced using molded Fiber Reinforced Plastics (FRP). The hull shape is formed with multiple layers of biaxial fiberglass or carbon fiber roving applied with thermosetting resin to either a plug or a cavity mold. The outer layer of the hull is a highly polished water impervious plastic surface that is optimized both in form and texture to move through the water with minimal resistance. On the interior, yacht hulls are composite structural systems designed to respond to the dynamic loading conditions of the marine environment. They are composed of two primary parts: a lower FRP hull chemically and mechanically bonded to an internal structural system of metal, wooden, or lightweight foam stringers, ribs, floors, and bulkheads; and an upper FRP deck that is mechanically and/or chemically bonded to the hull and its attendant structural components. Each of the subcomponents is optimized for minimal weight and size to resist loads while maximizing useable interior volume within the hull. All of the subcomponents work together to define, support, and maintain the shape of the hull as it supports its own weight and resists the forces that act upon it in the water.

Additive manufacturing systems applied to large scale projects typically use the Fused Filament Fabrication (FFF) method. They rely on slicing a 3-dimensional computer model of an object into a series of horizontal layers. The 3d printer builds the physical object by tracing the 2-dimensional profiles or slices of the 3-dimensional object with an extruded bead of molten thermoplastic. Each successive layer of molten plastic bonds thermally to the layer below. As the 3-dimensional form begins to emerge, sacrificial support material called scaffolding may be printed to strategically buttress overhanging surfaces and keep the 3-dimensional form from collapsing or deforming under its own weight.

It is clear that FRP yacht hulls have specific requirements that laminar FFF additive manufacturing is poorly suited to address. There are several serious challenges to solve before additive manufacturing can be used by the yacht manufacturing industry to print hulls.

Materials: Traditional FRP molding allows the production of durable, light weight, and highly polished finished hull surfaces integral to the manufacturing process. The finished surface of 3d printed objects is often quite rough, suggesting that additional surfacing applications or polishing procedures may be required to achieve surface qualities required for boat hulls. To achieve similar strength current large-scale additive manufacturing tools use an excessively wide extrusion bead of un-reinforced thermoplastic material that is heavier, thicker, and may be less durable than traditional FRP. The ideal material for 3d printing hulls will require the following characteristics: impervious to ultraviolet light and water or ability to easily bond to UV and water resistant coatings; relatively light weight so as to not significantly increase displacement; retain dimensional stability at a broad range of temperatures; exhibit stiffness, shear strength, and puncture resistance comparable to FRP; and ideally lend itself to low waste and end of use recycling. Given that it is unlikely that a uniform wide bead thermoplastic material will be able to meet these exacting requirements, composite materials with either discontinuous (shredded) or continuous stranded fibers in either a thermoset or thermoplastic matrix appear to be the most likely material candidates. Several promising technologies appear ready for experimentation with modest upscaling including 3d printing solutions offered by Markforged and 9T Labs. Both of these material solutions rely on Continuous Fiber Fabrication (CFF) using relatively expensive thermoplastics: Markforged uses a carbon fiber reinforced nylon [6] and 9T Labs uses a novel semi-crystalline polymer polyetherketoneketone (PEKK) [7]. Both of these material options are significantly more expensive than the thermoset plastics such as polyester and epoxy resins that are commonly used in the marine industry.

Scale: Assuming that continuous fiber reinforced composites present the best material solution for deploying additive manufacturing to yacht production, the technology will need to be scaled up to accommodate the relatively large size of yacht hulls. Currently, there are no examples of commercially available large-scale composite additive manufacturing tools that can make an entire full-size hull as one piece using the CFF process. While the University of Maine 3Dirigo project used a moderate temperature thermoplastic, it did not feature stranded fibers, relying instead upon wood-based cellulose fillers [1]. It is likely that large scale additive manufacturing tools that can print continuous strand composite materials will become commercially available within the next decade. MOI Composites, an Italian start-up firm has conducted an interesting series of demonstration projects with multi-axis non-planar CFF manufacturing using a robotic arm and continuous glass fiber strand impregnated with an ultraviolet curing thermoset resin [8]. Meanwhile, a series of pavilions built as proof of concept computer assisted manufacturing projects at the University of Stuttgart Institute of Building Structures and Structural Design demonstrate that architectural-scale objects can be constructed with robotic arms weaving continuous strand fiber reinforced composite structures that span modest distances without appreciable deflection [9].

Fiber Orientation: Traditional FRP hulls typically use multiple layers of woven glass fiber roving which exhibits bi-directional tensile stiffening when bonded within a thermosetting plastic resin matrix. Optimizing the orientation of fibers within a 3d printed hull for structural reinforcement remains a significant challenge for typical slicer-based additive manufacturing applications. 9T Labs has partially solved this problem for smaller scale parts using a componentry approach featuring a secondary heat and pressure activated bonding process [7], but this is not optimal for large compound curved surfaces, nor is it likely that PEKK will prove to be a financially viable replacement for inexpensive polyester and epoxy resins in the near term. However, very promising developments demonstrating the feasibility of Automated Fiber Placement (AFP) of PEKK impregnated carbon fiber tape by the innovative aerospace firm Automated Dynamics suggest that high performance lightweight reinforced thermoplastic composite structures similar to yacht hulls can be manufactured using automated robotic processes [10]. It should be noted, however, that this technology requires a mandrel or molded surface upon which PEKK impregnated carbon fiber tape is laid prior to laser sintering. One might imagine adapting this technology so that a yacht hull might be laid up within or upon a traditional mold using AFP by one or more robotic arms – not exactly 3d printing a hull, but certainly automating yacht manufacturing using additive processes.

Density: In typical FRP construction, autoclaves or vacuum-bags allow a precise regulation of the ratio of plastic resin to glass fiber. Thermoset FRP parts built using CFF processes tend to exhibit a rough stepped appearance with conspicuous voids and irregularities indicating a high plastic/void to fiber reinforcement ratio. It remains unclear how well this ratio can be controlled using CFF with thermosetting plastics. Additional research and fabrication process testing will be required to find solutions to this challenging problem. Ultimately, yacht manufacturers may need to wait until more expensive but more controllable thermoplastic polymers such as PEKK come down in price before they can be feasibly deployed in the production of yachts.

Structural Integration: As previously discussed, yacht hulls are not purely composite FRP, often relying on wood, foam, or metal structural components that may be encased, chemically bonded, or mechanically affixed to the interior portions of the hull.

Integrating these elements in the additive manufacturing process may present serious challenges. One can imagine a hybrid process whereby a large-scale 3d printer builds integrated pockets, channels, or brackets into the hull. Printing might pause while rigid elements are added, after which printing could resume. This would require a low profile or slender extrusion head mounted on an articulated robotic arm. I am currently working with a research team at Florida International University Robotics and Digital Fabrication Lab on the integration of ancillary structural systems in large scale 3d printed objects. We look forward to publishing the results of our fabrication research early next year.

3 Conclusion

Once these technical challenges are successfully addressed, we may see a new era of FRP yacht hulls produced entirely or in part by additive manufacturing. In the meantime, the most urgent areas of applied research fall into 2 primary areas: software and materials. The typical slicer software used to rationalize 3-dimensional forms for 3d printers is wholly inadequate for the manufacture of yachts. Yacht designers and manufacturers will need to control the robotic placement of materials using easy to define 3-dimensional toolpaths. Further, the programming of patterns of material placement in response to static and dynamic load analysis needs to be nominally standardized and automated. At the same time, the yacht design and manufacturing industry may need to re-evaluate its commitment to cheap thermoset plastics such as polyester and epoxy. While initial projections may indicate that new polymer thermoplastics will be more costly to adopt, the labor savings and economies of scale that accompany widescale industrial implementation of a novel material might quickly bring down the overall production costs for the yacht manufacturing sector. In any case, the next several years promise to be an exciting period for applied research in automated fabrication and presage a potential sea change for the yacht manufacturing sector.

References

1. UMaine News: UMaine composites center receives three Guinness world records related to largest 3D printer. UMaine News (2019). https://umaine.edu/news/blog/2019/10/10/umaine-composites-center-receives-three-guinness-world-records-related-to-largest-3d-printer
2. Thermwood Corporation: Marine boat building applications (2018). http://www.thermwood.com/marine_boat_building_home.html
3. Nazzaro, P.: Printing a finished console. Professional Boatbuilder, p. 62, October/November 2019
4. Post, B. Chesser, P., Lind, R., Sallas, M., Love, L.J.: Feasibility of Using Big Area Additive Manufacturing to Directly Manufacture Boat Hulls. US Department of Energy (2018)
5. Jamie, D.: Livrea: crossing the atlantic in a 3D printed yacht. 3dNatives (2017). https://www.3dnatives.com/en/interview-livrea-yachts-031020174
6. Markforged Inc.: 3D printing materials (2017). https://markforged.com/learn/3d-printing-materials

7. 9T Labs: Your all-in-one solution for composite series production (2020). https://www.9tlabs.com/product
8. MOI Composites: Our technology (2020). https://www.moi.am/#technology
9. Institute of Building Structures and Structural Design: The University of Stuttgart (2020). https://www.itke.uni-stuttgart.de
10. Gardiner, G.: Consolidating thermoplastic composite aerostructures in place. Composites world (2018). https://www.compositesworld.com/articles/consolidating-thermoplastic-composite-aerostructures-in-place-part-1

Human-Machine Interactions for on the Fly Free Text Input Processing

Sergius Dyck(✉) and Almuth Hoffmann

Fraunhofer Institute of Optronics, System Technologies Image Exploitation IOSB, Fraunhoferstr. 1, 76131 Karlsruhe, Germany
{Sergius.Dyck,Almuth.Hoffmann}@iosb.fraunhofer.de

Abstract. Information in machine-readable form has to be entered by users of information systems in various situations. In the input process, forms with a number of individual fields are commonly used. For the creator, this input can be cumbersome and potentially counterintuitive. For this reason, free text components are becoming increasingly important. Information in the form of free text is not machine-readable as such, and extracting information is complex. The accuracy of existing approaches used to extract information from free text components depends directly on the quality of the used internal model. These models have to be created by experts or trained in advance using the appropriate machine learning methods. The main objective presented in this paper is extracting information as well as enhancing the underlying model on the fly. This implies that the information extraction algorithms apply directly to the user's input and that the user is enabled to correct possible deficiencies in the model as they arise.

Keywords: Information extraction · Free-text processing · Human-Machine Interaction · Natural Language Processing

1 Introduction

Information in machine-readable form has to be entered in various situations, whether it is the medical record at the docto's office or other official forms, such as police reports or tax declarations [1]. This paper will refer to this kind of documents as reports. Most common in the input process are complete blank pages where free text can be entered or forms with a number of individual fields for specific pieces of information [1]. Information elements structured in single input fields can be stored machine-readable and can then be reused, quality checked or further processed at a later time. Whereas free text documents are difficult to machine process and thus require manual effort. For the creator, however, both extremes are often cumbersome and potentially counterintuitive [2, 3]. For this reason, combinations like free text components in a structured document are becoming increasingly important for users. However, this limits machine processing to the structured parts; the information from the free text parts is likely to be ignored. In order to make these parts machine-readable as well, NLP (Natural Language Processing) can be used.

W. Karwowski et al. (Eds.): IHSED 2020, AISC 1269, pp. 21–26, 2021.
https://doi.org/10.1007/978-3-030-58282-1_4

The research area NLP deals with the ability to interpret free texts automatically. There are different approaches to extract information from free text components [4]. All approaches have one thing in common: An internal model is used to identify and extract textual information. The accuracy of the information extraction depends directly on the quality of the used model. These models have to be created by experts or trained in advance using the appropriate machine learning methods [5, 6].

The concept presented in this paper focuses on the information acquisition process. In particular, the user should be supported in writing his report as quickly, completely, correctly and intuitively as possible. In this context, approaches are also being investigated as to how the reported information can be recorded and displayed when the information is entered in such a way that it can be automatically retrieved and further processed.

We discuss how these methods can be used to create structured, machine-readable information from free text by means of certain paradigms of HMI (Human–Machine Interaction). The main objective here is extracting information as well as enhancing the underlying model on the fly. This implies that the information extraction algorithms have to be applied directly to the user's input and that the user is enabled to correct possible deficiencies in the model as they arise.

2 Natural Language Understanding

The implementation of the above-mentioned objectives requires the use of technologies and processes from the area of NLP.

The NLP research area addresses automated analysis and synthesis of natural language and is a branch of computational linguistics. NLU (Natural Language Understanding) is a subarea of NLP. It deals with the processing of text on a semantic level to understand the meaning of the text. NLU covers a variety of conceivable applications. These range from small, relatively simple tasks, such as understanding short commands by robots, up to highly complex objectives, such as a complete understanding of newspaper articles. Many applications today fall between these two extremes, for example text classification for automatic analysis of e-mails [4, 7].

Regardless of the used approach, most NLU processes share common components for concept and implementation. Usually they use a lexicon of language, a parser and rules (e.g. grammar rules) to convert sentences into an internal representation and to enrich them with semantic knowledge [8]. The parsers are based on a number of successive basic NLP tasks. The Saarbrücken pipeline model, for example, describes a possible sequence for these basic NLP tasks [9].

A pipeline for basic information extraction consists of the following steps: first single sentences are segmented followed by word tokenization and position-of-speech tagging. This fundamental NLP tasks are the foundation for all further steps such as entity recognition and relation recognition [10].

The prerequisite for this is that the essential information elements for the respective domain are already taken into account when defining the underlying information model. Models are always heavily dependent on the domain in which they are to be used. This affects the ease of adaptability to changed situations in the domain, but also the handling of particularly extensive domains.

In the medical field, for example, abbreviations are often used to keep the typing effort small. In some areas, short, grammatically simple sentences are used that do not correspond to the normal use of grammar. Therefore the normal grammar rules cannot be used in the underlying model. To machine learn these special abbreviations and grammar rules, an adequate dataset for training is needed, but rarely available.

This shows that the interaction with the user is becoming more and more important. In order to be able to process the user input accordingly it requires a model that can be expanded on the fly. We are currently not aware of any procedures that deliberately provide for this. A first starting point for specific model extensions are the basic NLP tasks, like tokenization, POS-tagging and noun phrase chunking for entity recognition. Especially unknown abbreviations and the over-simplified grammar used in some domains can cause problems which continue in later process steps. This basic NLP tasks are based on rules and dictionaries [11]. Therefore they offer a direct interface for extensions on the fly.

To enable the user to interact with further text processing steps like the classification of the entity type is more challenging. State of the art (named) entity recognition methods use neural networks and require annotated training data. Those models are unlikely to be changed at runtime. Therefore a more feasible approach could be to retrain the model with the user corrections overnight. In order to be able to automatically generate training data from user corrections, rules must be derived from the user interactions. Although this approach is technically feasible, the effort and benefits have to be evaluated, especially under the specific language constraints. Dedicated NLP tasks are suitable for examining this approach. For example (named) entity recognition methods are promising, since they could extract a large part of relevant information from texts, when adapted and expanded to the specific information entities.

In addition to the technical implementation of the interaction, it is crucial that the ergonomic design of the corresponding user interfaces is carried out according to the dialogue principles defined by [12]. As the user won't necessary be a domain model expert, the HMI needs to support his interaction with the underlying model in an understandable and self-descriptive manner, without distracting the user's main task of writing the report.

3 Human Machine Interaction

As described in the previous chapter, the user can support and correct the NLP model with his knowledge. In addition, the machine can also support the user in entering his information. This ensures a consistently high quality of information, despite the increasing demands on the user and with regard to the amount of information.

There are certain attempts in the literature trying to include users in NLP processes. However, these papers mainly address the user in the role of a system developer or system administrator and not an end-user [6, 13]. Key questions for beneficial end-user machine interaction are: how much support does the user need while entering information and how can the user support the underlying NLP model?

A possible approach to address above questions will be discussed using the example of creating a medical record for a new patient, in the context of which different information about the patient such as name, address, probable medical history, etc. must be entered.

3.1 User and Machine Support

Providing a completely blank page for the input of information gives the user no clues as to the needed content. This may cause the user to forget to enter essential information. On the other hand offering strongly structured forms with many dedicated input fields could overwhelm or limit the user.

Our approach proposes a compromise between these two extremes by combining corresponding fields, for example all the address information needed, such as street, house number, apartment number, zip code, city, province and country, to one free text like field.

During input, the particular elements are being recognized by means of the NLP model and stored in the system. However, what happens when certain tokens are not recognized correctly or not recognized at all, like the zip code from the example above?

In general there are different HMI paradigms that fundamentally describe the scope for human intervention in machine processes and thus the interaction between human and machine. While HITL (Human-in-the-Loop) provides for very close interaction between human and machine, the HOOTL (human-out-of-the-loop) approach allows the machine to work autonomously without interaction with the human. However, according to [14], the high degree of human involvement for HITL could lead to human-caused errors. And HOOTL, however, makes it difficult if the human is needed at a later time after all [15].

The HOTL (human-on-the-loop) approach tries to avoid these disadvantages. HOTL does not provide for continuous human interaction with the machine processes. i.e. the actual core processes are performed by the machine independently. However, the human has a supervisory role, in which he can intervene if necessary, for example by correcting decisions made by the machine or by taking over certain tasks [16].

3.2 Situation-Dependent and Individual Behavior

A beneficial approach of HMI for information entering can be achieved by switching between HITL and HOTL [17]. It is based on a mutual interaction between human and machine, with both sides contributing to the fulfilment of the task according to their own strengths and weaknesses. However, the roles are not defined in advance, but during the problem-solving process. In some applications, the processes are thus handled by the system, while the human acts as a mere supervisor, i.e. the HOTL paradigm applies. In other cases, the system needs more support from the human, so he is therefore more actively involved, i.e. the HITL paradigm applies.

For the input and processing of free text information, this more flexible method makes a lot of sense from our point of view. Which mode is active at a certain time, however, depends on specific conditions like: (a) Given task: A supervising body raises the priority for improving the model. The mode is set to HITL and users are encouraged to correct errors. (b) Given time: For time-critical tasks, HOTL mode is rather used, for non-time-critical tasks, the HITL mode. (c) Given role: Users who for example have no rights to correct errors are put into HOTL mode automatically (d) Given Information: With information that is of particular importance, for example the medication prescribed by a doctor, the user should be able to correct any errors. In such cases, the HITL mode should be mandatory.

Apart from the context, it is important that the user is put in a position to be able to control the extent of the interaction. If the user cannot or does not want to make a corrective action in the given situation, the machine should not demand HITL mode and proceed in HOTL mode automatically.

Figure 1 shows an example of how the HOTL mode might be applied to an address field. The user enters an address (left). The system recognizes all entries except the 9-digit US zip code (right). The user can then either confirm or reject the result. The shown example is a schematic representation only and not a concrete user interface design. A HITL mode corresponding to this is more complex and will therefore not be further elaborated in this paper.

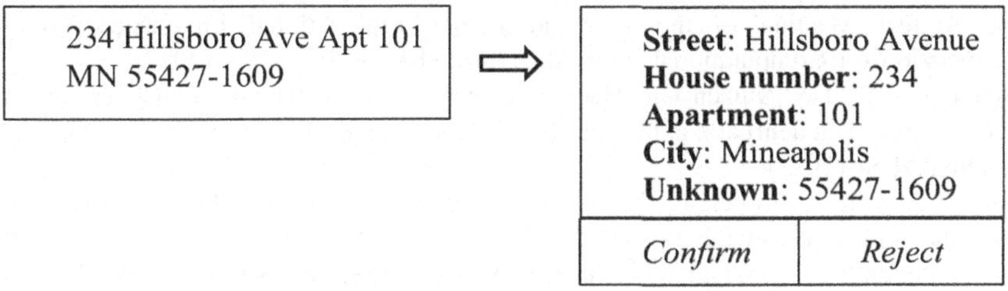

Fig. 1. Example of an address field in HOTL mode

4 Conclusion

In this paper we introduced our basic approach, which allows free text input to be simplified for the user and made machine-readable for the system. The concrete design of the corresponding HMI and user interfaces must be elaborated depending on the domain and the given circumstances. In general, a situation-dependent mode change seems to be appropriate. Since the described approach causes increased effort for the user, especially in HITL mode, it is crucial that the benefits of an improved NLP model are clearly presented to the user. A suitable method for this is, in our perspective, the use of autocomplete methods. Thanks to the improved suggestion lists displayed while typing, the user could immediately see that the extra effort to correct errors is worthwhile.

References

1. Trivedi, G., Handzel, R., Visweswaran, S., et al.: An Interactive NLP tool for signout note preparation. In: ICHI, pp. 426–428, Los Alamitos, CA, USA (2018)
2. Klopfenstein, L.C., Delpriori, S., Malatini, S., et al.: The rise of bots. In: DIS, pp. 555–565. ACM Press, New York, USA (2017)
3. Johnson, C.: Software tools to support incident reporting in safety-critical systems. Saf. Sci. **40**(9), 765–780 (2002). https://doi.org/10.1016/S0925-7535(01)00085-6
4. Gharat, A., Tandel, H., Bagade, K.: Natural language processing theory applications and difficulties. IJTSRD **3**(6), 501–503 (2019)
5. Rizzo, G., Troncy, R.: NERD: a framework for unifying named entity recognition and disambiguation extraction tools. In: EACL, pp. 73–76. European Association for Computational Linguistics, USA (2012)
6. Da Silva, T.C., Magalhães, R., de Macêdo, J., et al.: Improving named entity recognition using deep learning with human in the loop. In: Open Proceedings EDBT. OpenProceedings.org (2019)
7. Baker, K., Bhandari, A., Thotakura, R.: An interactive automatic document classification prototype. In: EuroHCIR Proceedings (2013)
8. Abend, O., Rappoport, A.: The state of the art in semantic representation. In: Proceedings of the Annual Meeting of the Association for Computational Linguistics, pp. 77–89. Association for Computational Linguistics, Stroudsburg, PA, USA (2017)
9. Steinhuber, S.: Das Nutzen von Machine Learning zur Auswertung und Interpretation von schriftlichen Kundenrückmeldungen. FOM Hochschule für Ökonomie & Management, Seminararbeit (2019)
10. Bird, S., Klein, E., Loper, E.: Natural Language Processing with Python: Analyzing Text with the Natural Language Toolkit, 2nd edn., vol. 2. O'Reilly Media, Newton (2009)
11. Kanakaraddi, S.G., Nandyal, S.S.: Survey on parts of speech tagger techniques. In: ICCTCT, pp. 1–6. IEEE (2018)
12. Ergonomics of human-system interaction – Part 110: Dialogue principles. (ISO 9241-110:2006) (2006)
13. Zhang, S., He, L., Dragut, E.C, et al.: How to invest my time: lessons from human-in-the-loop entity extraction. In: ACM SIGKDD, pp. 2305–2313 (2019)
14. Cranor, L.F.: A framework for reasoning about the human in the loop. In: UPSEC, pp. 1–15. USENIX Association, San Francisco, California (2008)
15. Cummings, M.L.: Human supervisory control of swarming networks. In: Annual Swarming: Autonomous Intelligent Networked Systems Conference, pp. 1–9 (2004)
16. Ponsa, P.A., Vilanova, R., Amante, B.G.: Human intervention and interface design in automation systems. Int. J. Comput. Commun. Control **6**(1), 166–174 (2011)
17. Fischer, J.E., Greenhalgh, C., Jiang, W., et al.: In-the-loop or on-the-loop? Interactional arrangements to support team coordination with a planning agent. CPEE (2017). https://doi.org/10.1002/cpe.4082

Cork as a Relevant Material in Fashion: A Study of Socio-Cultural Trends and a Semiotic Reading of the Cork-a-Tex Yarn

Theresa Lobo[1,4](✉), William Afonso Cantú[2,3,4], and Nelson Pinheiro Gomes[2]

[1] Faculdade de Arquitetura, CIAUD, Centro de Investigação em Arquitetura, Urbanismo e Design da, Universidade de Lisboa, Rua Sá Nogueira, Pólo Universitário, Lisbon, Portugal
theresa.lobo@universidadeeuropeia.pt
[2] Faculdade de Letras, ULICES, University of Lisbon Centre for English Studies, Alameda da Universidade, Lisbon, Portugal
[3] IPAM, Instituto Português de Administração de Marketing, Quinta do Bom Nome, Estr. da Correia 54, Lisbon, Portugal
[4] UNIDCOM, Unidade de Investigação em Design e Comunicação, Av. D. Carlos I 4, Lisbon, Portugal

Abstract. The main objective of this research is to understand the symbolic and cultural readings that the Cork-a-Tex yarn can have as a new fashion design textile. The methodology is divided into two parts: literary review and analysis. The literary review contextualizes the main themes of this paper and the analysis approaches both semiotic and trend methods, framing the case study as a 'cool signal' of socio-cultural trends. Both analyses contribute to a cultural reading of meanings behind the case study and promote a better perception of contexts and mindsets. Through this combination of methods, we understand the framework of meanings imprinted on the object. Understanding its cool nature and association with trends reinforces the need for developments of new materials from cork and stimulates research on socio-cultural trends.

Keywords: Cork · Trend Studies · Semiotics · Sustainability · New materials

1 Cork: An Irreverent Material

Cork is a natural material known for the production of cork-stoppers, its most common derived-product. The Portuguese Association of Cork (APCOR) states that "cork is the bark of the cork oak. It has a range of applications to attributes that no technology has managed, until today, to imitate, match or surpass. It is a 100% natural, 100% reusable and 100% recyclable raw material extracted from cork oaks without ever harming the normal development of the species and without damaging the tree" [1].

Cork exports represent 3% of the Portuguese international trading. That makes the country a leader when dealing with this sector. It is also relevant to see all the cultural

W. Karwowski et al. (Eds.): IHSED 2020, AISC 1269, pp. 27–32, 2021.
https://doi.org/10.1007/978-3-030-58282-1_5

and economic environments that cork provides for those who work in this sector and are involved in the management of other derived areas such as product, design, and specific industries [14]. Trough the cellular structure of cork, we can take advantage of its attributes. Its properties are light, waterproof, elastic, compressible, thermal and acoustic isolating, low combustion, hypoallergenic, and resistant to friction [1, 14]. Cork properties make this material an interesting choice when dealing with innovation and new product developments and design [15].

Cork is known as a versatile, unique, and natural material used to develop several cork-materials that allow a significant development of products in different contexts [5, 7, 13]. For this research, we highlight the cork fabric (tecido de cortiça) as an important material development in the fashion industry. This material, also known as "cork skin", is manufactured from rolls of agglutinated and plywood cork that is laminated, forming the sheets of this material. The cork fabric, however, is not a real textile, since cork is a fragile material, some addictive or second layer must be glued on it when laminated. Although it is not a real textile, the cork's agglutination with other fibres makes it possible to get very thin layers of this material. This way, it can be used as a textile-membrane in several fashion-design objects (see, for example, [4]). To solve this question, Sedacor [21], Têxteis Pênedo [23], and CITEVE (University of Porto) developed a new approach to cork and the textile industry. A newly processed material based in cork-powders and leftovers from the industry emerges: the Cork-a-Tex Yarn®. This new yarn is a natural and sustainable material with cork-properties that responds to the need of a cork-thread for sewing and developing new textiles. This new product that takes the form of a natural membrane is a "high-performance textile […] incorporates the properties of textile substrates in terms of comfort, touch and appearance" which takes all the advantages of cork-properties, such as "lightness, thermal insulation, anti-mites, dirt and water repellence, and impermeability, which are traditionally only obtained by the addition of functional chemical agents" [17].

2 Trends: Patterns of Meaning in Society and Culture

Trend Studies, as a scientific approach, has been growing and developing a particular perspective in academia, especially in Portugal [3]. The concept of 'trend', in a socio-cultural approach, is a term that adapts itself in different contexts. It needs to be explained in detail according to the different discourses where it is applied [6, 27]. The concept of 'trend' is generally related to the idea of 'change' as different authors state [10, 19, 27]. We believe that this change is related to the constant and dynamic impulses that stimulate people and, consequently, the environment and culture.

In this research, we are dealing with macro shifts in society and culture. In this sense, socio-cultural trends represent a direction and movements on values on society [6]. Studying them opens new opportunities to recognise the near future through the understanding of cultural manifestations. The study of society in a diachronic perspective, the understanding of present events and the continuous observation of cultural developments underline the need to be aware of new discourses that may be relevant for the identification of future scenarios [10, 12, 18].

Cultural analysis for trends identification is a path for understanding the core movements that enable to recognize the early stage of trends [10, 18]. Tracking upcoming changes in mindsets is particularly important to articulate cultural analysis and brand/product management. For that, operating under the social paradigm of culture highlights the need for the identification of patterns in society, as stated by Williams [26]. Those patterns of meaning are the reflection of trends and enable the creation of better strategies and the crafting of relevant artefacts. When looking for those patterns, it is important to understand the signals that have more chances of indicating a particular movement of change that may become a trend. For that, we must analyse cool signals that are indicators of these changing meanings taking the form of cultural objects.

Looking for cool signals trough a cultural analysis praxis is a relevant tool for discovering the most relevant and prominent signs of change in society. When finding the "coolness" and associated cultural objects that inhabit society, we can interpret meanings and contextualize related practices. When doing that, we can connect the dots and see patterns that lead us to the identification of socio-cultural trends [6, 10], or read the specific object to understand its relevance in present society.

3 Methodology and Case Study

We start by informing the methodological analysis of the case study through a literary review. In this section, defined as non-interventionist, we set the research question that guided this study (What are the symbolic and cultural readings that Cork-a-Tex yarn can have on fashion design?) and contextualized the cork as a natural material and the Cork-a-Tex as a derivative object from cork.

The case study gave clues regarding its relevance, both in Portuguese economy and heritage, but also in the advantages and possibilities of developing new products from this natural material, notably in fashion design. After describing cork as a natural element, we approached Trend Studies to understand the changes that enable us to create relevant and articulated readings for products and brands. Trend Studies' viewpoint is closely linked with tracking culture and instigates the search for layers of meaning in cool objects. In the second part, defined as interventionist [22], we analyse the case study trough both selected methods: (a) semiotic analysis and (b) trend analysis.

Concerning the first method, authors like Barthes [2], Oswald [16], and Rose [20] contribute to the structure of our semiotic analysis. Reading objects permits the identification of meanings behind texts, and semiotics is a method that guides the translation of meanings that exist on the cultural objects [20]. This reading approach of cultural texts helps us to develop new perspectives about the signs that inhabit our society [2, p. 4]. Similarly, Oswald [16, p. 4] underlines that "semiotics adapts linguistic theory to the study of nonverbal signs and symbols and anchors them in the culture", which justifies its relevance in this research. The semiotics system allows a reading focused on the connotative function of signs that contrasts with the denotative function [2, 16, 20]. In this point, it seems relevant to underline both definitions: the denotative reading of the object "resembles the dictionary meaning of a word—it simply indicates the concept as a matter of fact" [16, p. 54]. Meanwhile, "the connotative function of discourse endows signs with nuances and shades of meaning [...] and is highly dependent on the context of the message. Connotations may be culturally based, such as the interpretation of symbols and colors. Connotations may also be highly personal" [16, p. 55].

After this exercise, we develop a trend analysis—with trends identified by Trends Observer—attempting to clarify the symbiosis between the case study as a cool signal and its relation to socio-cultural trends. The concept of cool, although may seem one of the most mystical concepts in Trend Studies, has been discussed by several authors [8, 9, 11]. In this research, we highlighted the definition of the Trends and Culture Management Lab [24], built on the review of several authors like Kerner and Pressman [11], Gloor and Cooper [9] Gladwell [8], among others, which describes cool as something "relevant", "viral", "in the moment", "irreverent", "instigating", and that may reflect a "discontinuity" of the current socio-cultural patterns [24]. We describe the case study as a cool signal and correlate it with socio-cultural trends to review connotations on its symbolic and socio-cultural meaning and possible relation to fashion design.

4 Data Articulation

The possible connotative readings of the Cork-a-Tex yarn call for a symbolic construct associated to the base material: cork. It inhabits Portuguese imaginaries as a well-known reference and natural material. The association with wine, and its bottles, another popular and important Portuguese product, highlights cork's place in a symbolic setting as part of Portuguese narratives and identities. As an essential export product, it also plays a role in many tourism-oriented objects. Now, it takes the role of a textile, an innovative material, and its articulation with fashion represents an imprint of certain narratives and identity elements of Portuguese heritage in fashion creations. It adds depth in symbolic meaning, new layers, which generate positive identifications and associations.

This object, the Cork-a-Tex yarn, encompasses the six cool characteristics [24] already described, as it is: (1) relevant, in tune with emotional identity genes that call for local elements and products with symbolic narratives and associations with emerging socio-cultural trends, as we will see ahead; (2) viral, since the material is expanding to new products and solutions and it highlights the potential of finding new purposes for symbolic materials; (3) in the moment, as it makes sense today and answers to current mindsets; (4) irreverent, as it is not only an avant-garde solution but also a positive and provocative proposal that underlines textile innovation; (5) instigating, thought-provoking, and with a (6) proposal of discontinuity, as it underlines again the potential for new materials with emotional associations that are transferred to new practices and objects that represent lifestyles.

On another note, we can see a clear connection with macro socio-cultural trends, considering the symbolic connotative associations and the described nature of cool. For this analysis, we will consider the "Experienced Narratives and Identities" and "The Full Gaze of Sustainability" trends, identified by Trends Observer [25], a trends platform that operates with macro trends and defends a scientific approach. These trends underline the cool nature of the Cork-a-Tex thread. As the first trend description highlights, it is about identity narratives, the practices, and objects that emphasize identity elements and the symbols behind collective stories. In terms of our study object, now we can experience more symbolic elements in the body. It's the articulation between a symbolic and material culture that calls for our memory and plays nostalgic stories in a new experience and medium, a sustainable thread of fashion. A Portuguese story that can be experienced in the body, giving place to a new textile for fashion that articulates important Portuguese economic and symbolic industries, from cork to textiles and fashion. The second macro trend relates in terms of sustainability. This material, as seen before, can help a growing movement in terms of sustainable fashion and can embrace the front line for new developments in the fashion industry trough natural materials with relevant cultural contexts.

5 Conclusions

This research underlined the methodological potential within the articulation of semiotic readings and trend analysis. The connotative analysis imprinted in texts and fashion objects offers an insight into deep symbolic associations and different meanings. This, in turn, highlights associations with emerging socio-cultural trends, as well as their current symbolic constructions. This way, we can see how objects and texts are in tune with different trends and how they may represent change in mindsets and practices when dealing with fashion design. The Cork-a-Tex thread is one of these objects underlining emerging trends, with a robust cool nature and complex symbolic associations that articulated Portuguese identities, memory, and narratives, as well as sustainability. Trough the association with both trends, previously defined, we confirmed not only the cool characteristics of the thread as an innovative and sustainable solution but also the associations that inserts this object in macro trends that deal with narrated identities, experiences and sustainability. In terms of symbolic pictures within socio-cultural dynamics, this research informs the contexts and meanings in the fashion design sector. Future research with this material may lead to the reaffirmation of Portuguese culture and heritage. We also encourage in-depth research in sustainable materials and practices that can contribute to better alternatives in terms of environment and cultural symbolic heritage, as cork provides.

References

1. Apcor. https://www.apcor.pt/
2. Barthes, R.: Elementos de Semiologia. Cultrix, São Paulo (1964)
3. Cohen, S.: Os Estudos de Tendências Enquanto Disciplina Académica. ModaPalavra **10**(20), 89–93 (2018)
4. Coruche Fashion Cork 2018 (2018). https://youtu.be/vPqMMhJepxQ. Accessed 02 Jun 2020

5. Couceiro, M., Veríssimo, C., Cantú, W., Vieira, A., Figueiredo, A.: Cork International Workshop: Primeiro Ciclo, Exhibition catalogue. Observatório do Sobreiro e da Cortiça, Coruche (2018)
6. Dragt, E.: How to Research Trends. BIS Publishers, Amsterdam (2017)
7. ESAD Matosinhos.: 2nd. Skin – Cork Jewellry. Cultural, Porto (2007)
8. Gladwell, M.: The Coolhunt. The New Yorker, Anals of Style, 17 March 1997
9. Gloor, P., Cooper, S.: Coolhunting: Chasing Down the Next Big Thing. Amacon, New York (2007)
10. Gomes, N., Cohen, S., Flores, A.: Estudos de Tendências: Contributo para uma abordagem de análise e gestão da cultura. Moda Palavra **11**(22), 49–112 (2018)
11. Kerner, N., Pressman, G.: Marcas Cool – Como criar produtos que se destacam e modas que permanecem. Lua de Papel, Portugal (2008)
12. Kjaer, L.: The Trend Management Toolkit. Palgrave MacMillan, London (2014)
13. Mestre, A.: Design cork: for future innovation and sustainability. Susdesign, Lisboa (2008)
14. Mestre, A., Gil, L.: Cork for sustainable product design. Ciência e Tecnologia dos Materiais **23**, 52–63 (2011)
15. Niemeyer, L., Cantú, W.: A cortiça como desafios para novos usos no design. Blucher Des. Proc. **2**(9), 4272–4283 (2016)
16. Oswald, L.: Marketing Semiotics: Signs, Strategies and Brand Value. Oxford University Press, Oxford (2012)
17. Pinto, C.: Cork-a-Tex Yarn: Fio com elevada incorporação de cortiça, 18 April 2017. https://www.compete2020.gov.pt/noticias/detalhe/Proj17714-CorkaTex-Yarn
18. Powers, D.: Thinking in trends: the rise of trend forecasting in the United States. J. Hist. Res. Mark. **1**(1), 2–20 (2018)
19. Raymond, M.: The Trend Forecaster Handbook. Laurence King Publishing, London (2010)
20. Rose, G.: Visual Methodologies. Sage, London (2001)
21. Sedacor. https://www.jpscorkgroup.com
22. Teixeira, P., Neto, J.: A typology proposal for action research. Ciência Edu. **23**(4), 1055–1076 (2017)
23. Têxteis Penedo. https://www.tpenedo.pt
24. Trends and Culture Management Lab: A natureza do Cool, um desafio académico (2020). https://creativecultures.letras.ulisboa.pt/index.php/gtc/
25. Trends Observer. https://trendsobserver.com
26. Williams, R.: The Long Revolution. Penguin Books, Harmondsworth (1961)
27. Vejlgaard, H.: Anatomy of a Trend. McGraw-Hill, New York (2008)

Psychological Interpretation of Human Social Behaviors in the Atypical Architectural Shape

Young Lim Lee[1] and Yun Gil Lee[2(✉)]

[1] Department of Psychology and Psychotherapy, Dankook University, 119 Dandae-ro Dongnam-gu, Cheonan-si, Chungcheongnam-do, Republic of Korea
younglee13@dankook.ac.kr

[2] Department of Architecture, Hoseo University, 20 Hoseo-ro 79beon-gil, Baebang-eup, Asan-si, Chungcheongnam-do, Republic of Korea
yglee@hoseo.edu

Abstract. This study is the continuation of our previous study and concerns the development of a system that simulates human behavior for use in the design process for an atypical space. This paper aims to identify the psychological causes of human social behavior in atypical architectural geometry, which are essential for the development of a human behavior simulation system. Users interact both individually with a building's shape and socially with others within an atypical building, and it is necessary to grasp both the form of social action taking place in an atypical space and identify the cause of that form in order to lay the foundation for computerization. In order to analyze the causes of social behaviors of users in an atypical space, we first looked at the actual social behaviors. A representative case was drawn from the results of a survey. Then, a psychological interpretation was attempted based on the literature review. This paper's primary concern is with the psychological interpretation, and the second half presents a proposal for applying the psychological interpretation to a social behavior simulation system. The results of this study can be used as a foundation for technology that can be integrated into the previously developed ActoViz system to realize a more appropriate simulation for user behavior.

Keyword: Human factors · Human social behavior · Psychological interpretation · Atypical architectural shape · Affordance

1 Introduction

Atypically shaped buildings are becoming increasingly popular in modern society and the ability to design atypical buildings is invaluable for an architect. The design process of an atypical building is somewhat different from a traditional architectural design because the atypical design requires an equally unusual form of creating designs. Thus, when designing an atypical building, the architect could overlook users' convenience and safety, which are significant standards of evaluating the value of buildings. In order to prevent users' safety and convenience from being overlooked, the use of human behavior simulation is incorporated into the design process of creating atypical spaces.

W. Karwowski et al. (Eds.): IHSED 2020, AISC 1269, pp. 33–38, 2021.
https://doi.org/10.1007/978-3-030-58282-1_6

This study is the continuation of our previous study and concerns the development of a system that simulates human behavior for use in the design process for an atypical space. This paper aims to identify the psychological causes of human social behavior in atypical architectural spaces, which is essential for the development of a human behavior simulation system. Users interact both individually with a building's shape and socially with others within an atypical building, and it is necessary to grasp both the form of social action taking place in an atypical space and identify the cause of that form in order to lay the foundation for computerization.

In order to analyze the causes of social behaviors of users in an atypical space, we first looked at the actual social behaviors. A representative case was drawn from the results of a survey. Then, a psychological interpretation was attempted based on the literature review. This paper's primary concern is with the psychological interpretation, and the second half presents a proposal for applying the psychological interpretation to a social behavior simulation system. The results of this study can be used as a foundation for technology that can be integrated into the previously developed ActoViz system to realize a more appropriate user behavior simulation.

2 Atypical Building Design and Human Behavior Simulation

Atypical buildings can embody non-Euclidian building forms rather than typical modern Euclidean forms. It is mainly composed of curved walls, slabs, stairs, and shells that are difficult to see in modern architecture shown in Fig. 1. In the design process of atypical buildings, architects can easily ignore human factors because they normally focus more on the unusual shapes of buildings. Human behavior simulation technology can allow architects to figure out problems related to human factors in the middle of the design process. Hong and Lee (2018) state that the representation of human behavior in atypical architectural design has a positive effect, as it improves architects' problem-detection abilities, related design development, the existing design field, and atypical design [10, 11].

Fig. 1. Aspects of human behaviors in atypical architectural spaces.

In the last study, we studied the development of agent technology in the form of a person acting in response to an atypical building. However, the user's behavior is influenced not only by the physical situation of the atypical building, but also by the social context. In order to develop a more ideal user behavior simulation, the purpose of this study is to investigate the social behaviors of users in atypical buildings and to identify their psychological causes.

3 Psychological Interpretations of Human Social Behaviors in Atypical Architecture

To determine the psychological cause(s) of the social behaviors of users in atypical buildings, the behaviors in atypical spaces were investigated and analyzed. We visited representative atypical buildings, observed user behaviors centered on atypical buildings, and classified typical behaviors. The psychological analysis of representative cases is as follows:

3.1 Self-perception

The 'self-perception theory' proposed by Daryl Bem in 1972 assumes that when we are unsure of our own attitude, we take on the observer's stance. According to the theory of self-perception, when our attitude is weak or ambiguous, by observing our behavior and the environment in which it occurs, we infer our attitude as others observe us. Among the behaviors we observed, we saw users sitting on a chair, sitting on a staircase, or leaning on a pillar. Of course, an individual may be doing this for the purpose of rest, but from a psychosocial point of view, when a person leans or sits somewhere, that person may continue to rest because he or she infers, "Oh, I need a rest now." The fact that one's behavior can correct self-perception can be explained by the "facial feedback effect." that our emotions are affected by imitating the expressions or actions of others. When we observe others resting on walls (or sitting on stairs and chairs), we naturally and unconsciously imitate to match our behaviors and coordinate with others' feelings (Fig. 2).

Fig. 2. Examples of human social behaviors related to self-evaluation in atypical buildings.

3.2 Behavioral Conformity

Since human beings exist in society, they are impacted by the existence of others. One of the effects is conformity. Sympathy is not simply following another person's behavior, but following another person's behavior or belief. Whether we are superficially following or actually resonating, we are influenced by the actions of others. The observation of the majority of individuals leaning forward on the wall, resting, or sitting on a chair can be seen as actions in harmony with the actions of people other than one's self or party, and these actions can be seen as conformity without conscious intention. On the other hand, the action of leaning against the wall and taking pictures can be seen as compliance or cooperative team activity due to resonance with one's party (Fig. 3).

Fig. 3. Examples of human social behaviors related to behavioral conformity in atypical buildings.

3.3 In-Group Bias

It may seem natural for people to sit down on a chair. However, there are no peculiar behaviors such as lying on a chair, climbing on a chair, or holding a chair when a large number of people are sitting. Humans have a desire to be part of a community. In addition, there is a tendency to be advantageous to the group to which one belongs and there is a "group bias" identification, which translates to not wanting to leave one's group by doing something unusual. Therefore, by acting the same as the majority, one will feel supported by a close relationship since one belongs to the same group, consciously or not (Fig. 4).

Fig. 4. Examples of human social behaviors related to in-group bias in atypical buildings.

3.4 Behavior Setting

Although individual behaviors are shaped by the groups they belong to, the role of the environment is also emphasized by assuming that the groups are shaped by the circumstances in which they live. Individuals and the groups to which they belong exist within the physical environment in which they are located, and those places eventually affect the dynamics and outcomes of the group. The environment is also a physical space that allows one to act according to the needs of the situation, that is, a behavioral setting. Fast-food restaurant orders, hospital waiting rooms, university computer labs, company conference rooms, and park benches are all applicable settings. The reason is that people's behavior in these spaces is influenced more by the characteristics of the space than by their personal characteristics. Ecological psychologist Roger Barker concluded that most actions (at least regular and routine) are determined by the environment in which they occur. The behavioral situations unfold in a specific space where group members interact. Barker used the term synomorphy to describe the correspondence between people and situations. High synomorphy allows people to adapt well to the physical environment and to properly utilize the components of the environment. Therefore, the physical environment of stairs or chairs shows a high degree of conformity with the behaviors of walking or sitting down (Fig. 5).

Fig. 5. Examples of human social behaviors related to behavior setting in-group bias in atypical building.

4 Conclusion and Discussion

This is a basic study for the development of a user behavior simulation system for the atypical building design process. In this paper, the causes of social user behaviors in atypical buildings were identified psychologically. Through future research, the contents of this study will be computerized and applied to the simulation system (ActoViz).

Acknowledgments. This work was supported by the National Research Foundation of Korea (NRF) grant funded by the Korea government (MSIT) (NRF-2018R1A2B6005827).

References

1. Gibson, J.J.: Observations on active touch. Psychol. Rev. **69**(6), 477–491 (1962)
2. Goldstrin, B.E.: Sensation and Perception, 9th edn. Cengage Learning (2013)
3. Schönhammer, R.: Einführung in die Wahrnehmungspsychologie: Sinne, Körper, Bewegung. UTB GmbH (2013)
4. Arnheim, R.: A plea for visual thinking. Crit. Inq. **6**(3), 489–497 (1980)
5. Lee, Y., Crabtree, C.E., Norman, J.F., Bingham, G.P.: Poor shape perception is the reason reaches-to-grasp are visually guided online. Percept. Psychophysics **70**(6), 1032–1046 (2008)
6. Gibson, J.J.: The Ecological Approach to Visual Perception: Classic Edition. Psychology Press (2014)
7. Philbeck, J.W., Loomis, J.M.: Comparison of two indicators of perceived egocentric distance under full-cue and reduced-cue conditions. J. Exp. Psychol. Hum. Percept. Perform. **23**(1), 72–85 (1997)
8. Todd, J.T., Norman, J.F.: The visual perception of 3-D shape from multiple cues: are observers capable of perceiving metric shape? Percept. Psychophysics **65**(1), 31–47 (2003)
9. Land, M., Lee, D.: Where we look when we steer. Nature **369**, 742–744 (1994)
10. Hong, S., Lee, Y.: The effects of human behavior simulation on architectural major students' fire egress planning. J. Asian Architect. Build. Eng. **17**(1), 125–132 (2018)
11. Lee, Y.: A study on the effect of human factor for atypical design in the architectural design studio. In: Advances in Intelligent Systems and Computing, vol. 876(1), pp. 130–134 (2018)

Automatic Summarization Method for First-Person-View Video Based on Object Gaze Time

Keita Hamaoka(✉) and Yasuyuki Kono

Kwansei Gakuin University, 2-1, Gakuen, Sanda, Hyogo 669-1337, Japan
{dwd11537,kono}@kwansei.ac.jp

Abstract. Several first-person lifelog videos are lengthy in duration, and often include scenes that are not useful. This can be problematic for users as it requires a considerable amount of time to watch such a video. Therefore, in this study, we propose an automatic video summarization system for first-person-view videos by employing gaze tracking and object detection, based on human gaze time on an object. Because gaze is useful for capturing a user's intention and interest, our approach summarily captures their interest and conscious focal points while watching videos. As a result of the experiment, the evaluation value of the summary video generated by the proposed system exceeded that of the summary video in which important scenes are randomly extracted. From these results, it can be said that our system is useful for rapidly watching videos, and summarizing them to reflect user interest. Our system is applicable in many fields, including behavior recognition, visual diary creation, and support for patients having memory impairment.

Keywords: Video summarization · Gaze tracking · Object detection

1 Introduction

First-person-view video is a method of creating video records of daily life, sporting events, etc. It has become widespread as wearable cameras have become smaller, cheaper, and hands-free, thereby enabling a user to record their natural actions. Unfortunately, several first-person lifelog videos are lengthy in duration, and often include scenes that are not useful to viewers. This can be problematic for users as it requires a considerable amount of time to watch such a video. Several studies have been conducted with a focus on video summarization. For example, Higuchi et al. [1] summarized first-person-view video based on four cues that corresponded to the basic user actions of body movement, stillness, hand movement, and human interaction. The user set the importance of the four cues, and the scenes having high importance were reflected in the video after summarization. The contents of the input video were not considered, because the cues were limited to four. Our system can consider the contents of the input video which is different from previous work. In this study, we propose an automatic video summarization system for lengthy first-person-view-videos by employing gaze tracking and object detection. Because gaze tracking is useful for capturing a user's intention and interest, our approach reflects the interest areas and

W. Karwowski et al. (Eds.): IHSED 2020, AISC 1269, pp. 39–44, 2021.
https://doi.org/10.1007/978-3-030-58282-1_7

conscious focal points while watching a video by summarizing the video with a focus on the abovementioned information.

2 System Overview

Our system extracts a user's gazing point using a gaze tracking device, and object area using object detection function. For handling potential object detection failures, our system performs frame interpolation, i.e., it applies detection information into the preceding frame. Our system compares the distances of each object area between different frames to obtain the time-series information of each object. The object having the smallest Euclidean distance between the current and preceding frames is regarded as the same object. The object gaze time is then calculated based on the object area and the gazing point. When the gazing point is within the detected object area, count the object gaze time. If the gaze time for an arbitrary object exceeds the predefined threshold, the scene is considered important. In the generated summary video, important scenes are played back at normal speed, and others are played back at high speed.

3 Calculation of Object Gaze Time

3.1 Gazing-Point Extraction

Our system extracts a user's gazing point using a gaze-tracking device. The data obtained from the device include noise. Therefore, our system smooths the data. Manu et al. [2] presented a method of smoothing gaze data by employing a weighted average. On the basis of that method, if the gazing point in the nth frame is defined as Pn, the gazing point in the current frame, $P_{fixation}$, is given by Eq. (1).

$$P_{fixation} = \frac{(1P_0 + 2P_1 + \ldots + nP_{n-1})}{(1 + 2 + \ldots + n)} \quad (1)$$

3.2 Object-Area Extraction

Our system extracts the object area using YOLO v2 [3], an object detection algorithm that employs a convolutional neural network. Our system compares the distances between each object area of the current and preceding frames to obtain the time-series information of each object. If the upper-left and lower-right x and y coordinates of the object area are defined as xlt, ylt, xrb, and yrb, then the Euclidean distance between the object areas, P and Q, is given by Eq. (2).

$$d(P,Q) = \sqrt{(xlt_p - xlt_q)^2 + (ylt_p - ylt_q)^2 + (xrb_p - xrb_q)^2 + (yrb_p - yrb_q)^2} \quad (2)$$

An object having the smallest Euclidean distance between the current and preceding frames is regarded as the same object. The output result is shown in Fig. 1.

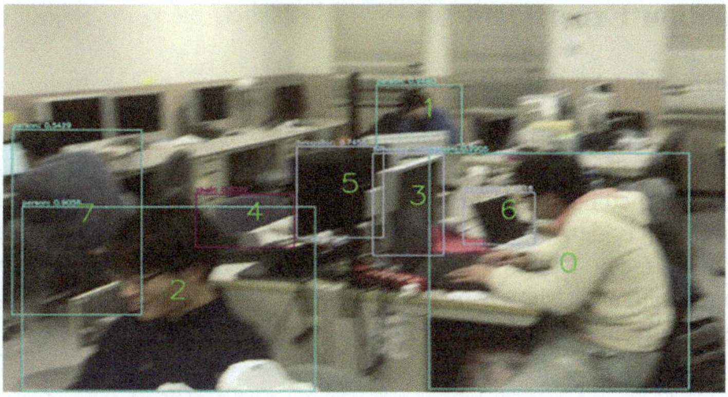

Fig. 1. Each object area is assigned a number that identifies the same object

3.3 Object-Gazing-Time Calculation

The object gaze time is calculated based on the object area and the gazing point. When the gazing point is within the detected object area, count the object gaze time. If the upper-left and lower-right x and y coordinates of the object area are defined as xlt, ylt, xrb, and yrb, the x and y coordinates of the object area are defined as $gazeX$ and $gazeY$, respectively. The conditional expression for adding the object gaze time is given by Eq. 3.

$$\begin{cases} xlt \leq gazeX \leq xrb \\ ylt \leq gazeY \leq yrb \end{cases} \tag{3}$$

4 Summary-Video Generation

4.1 Important-Scene Extraction

Our system determines the importance of scenes based on the object gaze time. If the gaze time for an arbitrary object exceeds a predefined threshold, the scene is considered to be important. The user can adjust the length of the summary video by changing the predefined threshold.

4.2 Generation of Summary Video Based on Importance of the Scenes

Based on scene importance, the first-person-view video is divided into important and non-important segments. The frame rate of important scenes is set to normal speed, and that of the non-important scenes is set to high speed. Subsequently, our system combines them.

5 Evaluation Experiment

5.1 Experimental Method

The subjects included six males and three females. They recorded a 1-h-long first-person-view video using a wearable camera and our system summarized those videos. The subjects watched both the summary video generated by the proposed system and a different summary video in which important scenes were randomly extracted. Questionnaires that used a 5-point Likert scale were then provided to the subjects. The questionnaire content is shown in Fig. 2. We also provided a free-text section for addressing the advantages and disadvantages of our system.

	Questionnaire content
Question 1	Did you feel tired while watching the summary video?
Question 2	Is this system useful for high-speed viewing of videos?
Question 3	Is the interest reflected in the video after the summary? (The summary video generated by the proposed system)
Question 4	Is the interest reflected in the video after the summary (The summary video in which important scenes are randomly extracted)

Fig. 2. Questionnaire contents

5.2 Result

Figure 3 shows the results of the questionnaire. Figure 4 shows the average and standard deviation of the questionnaire results. The standard deviations were calculated to two significant figures.

All summary videos were less than one-third the length of the input video. A high evaluation was obtained in Question 2. The evaluation value of Question 3 exceeded that of Question 4 and considerable difference was confirmed at a significance level of less than 5%. From these results, it was found that our system accurately reflects user interests and their conscious focal points. The evaluation value of Question 1 was low. Since the frame rate of the summary video created by our system is not fixed, it is possible that the user was stressed while watching the video. These problems can be solved by excluding non-important scenes and reflecting only important scenes in the video after summarization.

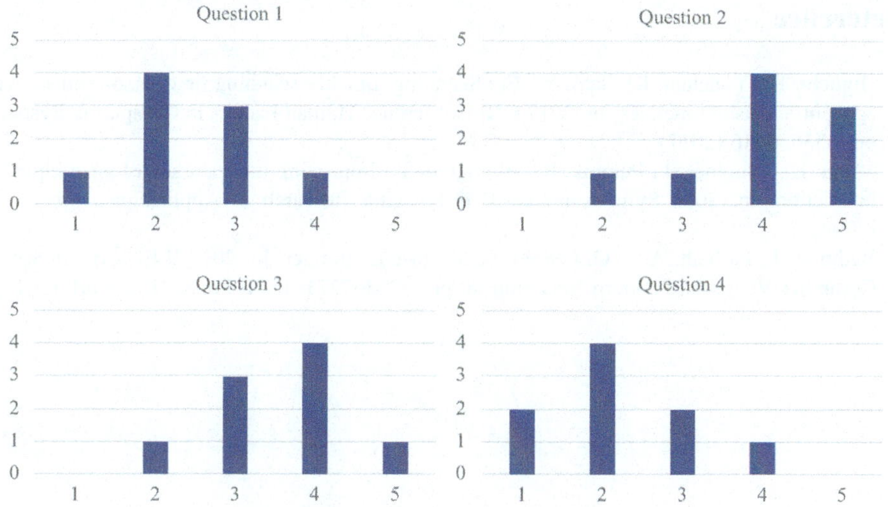

Fig. 3. Questionnaire results

	Average	Standard deviation
Question 1	2.4	0.83
Question 2	4.0	0.94
Question 3	3.7	0.81
Question 4	2.2	0.92

Fig. 4. Average and standard deviation of questionnaire results

6 Concluding Remarks

We proposed an automatic summarization system for first-person-view videos based on object gaze time. From the results of an evaluation experiment, we observed that our system is useful for rapidly watching videos, and summarizing them to accurately reflect user interests. A user opinion was provided in the free-text area that stated, "the scene when I was losing interest was regarded as an important scene and it was reflected in the summary video." Because this system sets important scenes based on the object gaze time, it is not possible to consider the user's mental state. We believe that these problems can be resolved by employing information regarding a user's pupil diameter as there is a correlation between pupil diameter and wakefulness. Thus, our proposed system will be able to accurately reflect mental information by considering the change in pupil diameter. To achieve this, the challenge will be to find the threshold of the amount of change in the pupil diameter during a scene for which the user has a high degree of interest.

Reference

1. Higuchi, K., Yonetani, R., Sato, Y.: EgoScanning: quickly scanning first-person videos with egocentric elastic timelines. In: 2017 CHI Conference Human Factors in Computing Systems, pp. 6536–6546 (2017)
2. Manu, K., Klingner, J., Puranik, R., Winograd, T.: Improving the accuracy of gaze input for interaction. In: 2008 Symposium on Eye Tracking Research & Applications, pp. 65–68 (2008)
3. Redmon, J., Farhadi, A.: YOLO9000: better, faster, stronger. In: 2017 IEEE Conference on Computer Vision and Pattern Recognition, pp. 7263–7271. IEEE Press, New York (2017)

Modelling the Adoption of the Version Control System: An Empirical Study

Tihomir Orehovački[1(✉)], Darko Etinger[1], and Snježana Babić[2]

[1] Faculty of Informatics, Juraj Dobrila University of Pula, Zagrebačka 30, 52100 Pula, Croatia
{tihomir.orehovacki,darko.etinger}@unipu.hr

[2] Polytechnic of Rijeka, Trpimirova 2/V, 51000 Rijeka, Croatia
snjezana.babic@veleri.hr

Abstract. Version control system is a repository of a source code that provides support in all stages of a software development lifecycle. Success of a version control system is affected by its acceptance by end-users. This paper deals with the adoption of the distributed version control system. With an aim to determine antecedents of satisfaction with as well as behavioral intention related to the reuse of the version control system, we proposed a research model composed of constructs that originate from several theories and models including technology acceptance model, theory of reasoned action, and expectation-confirmation theory. An empirical study was then carried out and was composed of two parts. First, study participants had to complete scenario-based interaction with GitHub as a representative sample of distributed version control systems. As a follow up, study respondents had to fill out the post-use questionnaire. Students from two higher education institutions were included in the study as a representative sample of users. The psychometric features of the research model were examined by means of the partial least squares structural equation modelling method. Study findings uncovered implications for both researchers and practitioners.

Keywords: Version control system · GitHub · Technology acceptance model · Theory of reasoned action · Expectation-confirmation theory · PLS-SEM

1 Introduction

Version control systems are applications that are nowadays commonly used in every form of agile software development. Collaborative work on the same project, simplified tracking of changes on snippets of code, repository of all relevant constituent pieces of software being developed, backup that enables restoring all project files to prior or last well working version if something goes wrong with the current one, are just some of many advantages of version control systems. We distinguish centralized version control system that represents only one repository or server containing entire history of versions and different branches of the code and distributed version control system where every single developer or client has their own copy of the history or versions of the code and all of its branches in their local server.

W. Karwowski et al. (Eds.): IHSED 2020, AISC 1269, pp. 45–50, 2021.
https://doi.org/10.1007/978-3-030-58282-1_8

By connecting more than 50 million of developers and almost 3 million of businesses and organization, GitHub is the most popular distributed repository of software artifacts. Apart from being software development platform, GitHub can be also used for educational purposes in terms of hosting course content and student assignments with an objective to monitor their activity on and contribution to projects that are part of software engineering courses. Students who are using GitHub have better learning outcomes, are more prepared for their future career, and have more positive classroom experience [1]. On the other hand, the employment of GitHub in classrooms has its challenges such as students' unfamiliarity with Git and GitHub features, notification overload, and students' unwillingness to share their work publicly [2]. Despite growing presence of version control systems in educational ecosystem, the body of knowledge is lacking studies focused on determining factors that affect students' satisfaction with version control systems and intention to reuse them. With an objective to fill that void, we initiated an empirical study on the adoption in which GitHub was used as a representative of distributed version control system whereas computer science students constituted representative sample of users.

The remainder of the paper is structured as follows. Research model and hypotheses are proposed in the second section. Empirical study findings are reported and discussed in the third section. Conclusions, limitations, and future work directions are provided in the last section.

2 Research Model and Hypotheses

The research model represents an interplay of seven factors where two of them (confirmation and satisfaction) were adopted from the expectation-confirmation theory [3], three of them (attitude towards use, social influence, and intention to reuse) were adopted from the theory of reasoned action [4], and two of them (perceived ease of use and perceived usefulness) were adopted from the technology acceptance model [5].

Confirmation refers to the extent to which version control system has met users' expectations. Satisfaction denotes the degree to which users are pleased with the use of version control system. Attitude towards use indicates the level to which users have a positive perception about the use of version control system. Social influence refers to the extent to which persons which opinion users appreciate think they should use version control system. Intention to reuse denotes the degree to which users are willing to continue to use version control system. Perceived ease of use represents the level to which interaction with version control system is free of effort. Perceived usefulness indicates the degree to which use of a version control system enhances users' performance in completing assignments related to software development. Drawing on the relations among factors in the aforementioned theories and model, we are proposing following hypotheses:

H1. Perceived usefulness is positively influenced by perceived ease of use.
H2. Confirmation is positively influenced by perceived usefulness.
H3a. Attitude towards use is positively influenced by perceived usefulness.
H3b. Attitude towards use is positively influenced by perceived ease of use.
H4a. Satisfaction is positively influenced by confirmation.

H4b. Satisfaction is positively influenced by perceived usefulness.
H5a. Intention to reuse is positively influenced by satisfaction.
H5b. Intention to reuse is positively influenced by attitude towards use.
H5c. Intention to reuse is positively influenced by social influence.

3 Findings

To develop, examine and assess the research model, Partial Least Squares Structural Equation Modeling (PLS-SEM) was employed using the SmartPLS 3.3.2 [6] software package. The analysis procedure includes two steps. First, we assess the validity, reliability and overall quality of the measurement model. After we obtain a valid and reliable measurement model, we proceed to the assessment of the structural model.

To assess the measurement model, Hair et al. [7] suggest the calculation of the following model criteria: indicator reliability (squared standardized outer loadings), convergent validity (average variance extracted - AVE), internal consistency reliability (composite reliability), and discriminant validity (including cross-loadings, the Fornell-Larcker criterion, and the recently introduced heterotrait-monotrait - HTMT ratio of correlations). Confirmatory composite analysis (CCA) was used to establish the reliability of the measurement model items and to test the convergent validity of the proposed constructs [8]. The convergent validity was obtained as the item loadings for each construct were above 0.708 [7]. This means that the average variance extracted – AVE is higher than recommended 0.5, thus we confirm the convergent validity of the proposed constructs. Eight items that have not met the requirements were removed from the measurement model. To explore the internal consistency of the constructs, two indicators were calculated for each construct. First, Cronbach's alpha (CA) in range between 0.868 to 0.919, and second, composite reliability (CR) in range between 0.919 to 0.939 shows that the requirement related to the internal consistency of the constructs was met.

In the next step, we tested the discriminant validity by comparing the AVE of each construct with its squared correlations with other constructs in the model [7]. The comparison results (shown in Table 1) indicate that each construct accounts for more variance in its associated indicator variables than it shares with any other construct in the model.

As a relatively new test for discriminant validity of the constructs, Hair et al. [7] proposed the inclusion of the HTMT ratio of correlations. They indicate that a threshold of 0.85 should not be met. The results of our assessment (in Table 2) shows that this criterion is achieved, thus we can conclude that the discriminant validity of the seven constructs was met. Slight caution should be taken regarding the relationship between confirmation and satisfaction, even though the HTMT ratio of correlations indicates that severe problems weren't detected using the HTMT method.

Table 1. Measurement model assessment and discriminant validity of the constructs

	ATU	CNF	EOU	BI	SAT	SI	PU
Cronbach's Alpha	0.903	0.868	0.919	0.912	1.000	0.903	0.905
Composite Reliability	0.932	0.919	0.939	0.938	1.000	0.939	0.929
AVE	0.775	0.790	0.756	0.791	1.000	0.837	0.724
Attitude towards use (ATU)	**0.880**						
Confirmation (CNF)	0.531	**0.889**					
Ease of use (EOU)	0.459	0.626	**0.869**				
Intention to reuse (BI)	0.568	0.637	0.445	**0.889**			
Satisfaction (SAT)	0.460	0.756	0.609	0.560	**1.000**		
Social influence (SI)	0.380	0.369	0.337	0.400	0.330	**0.915**	
Perceived usefulness (PU)	0.561	0.516	0.370	0.490	0.461	0.381	**0.851**

* square root of AVE on diagonal

Table 2. Heterotrait-Monotrait Ratio (HTMT)

	ATU	CNF	EOU	BI	SAT	SI	PU
Attitude towards use (ATU)	**1**						
Confirmation (CNF)	0.595	**1**					
Ease of use (EOU)	0.499	0.697	**1**				
Intention to reuse (BI)	0.623	0.712	0.487	**1**			
Satisfaction (SAT)	0.480	0.809	0.635	0.587	**1**		
Social influence (SI)	0.420	0.415	0.368	0.439	0.347	**1**	
Perceived usefulness (PU)	0.608	0.575	0.400	0.536	0.478	0.423	**1**

* all values are below threshold of 0.85

The assessment of the structural model is the next step after we established a reliable and valid measurement model. We examined the structural model by employing the PLS analysis, gathered the relevant information about the beta coefficients, tested for the significance of the relationship between constructs through the Bootstrap procedure (with 5000 samples) and extracted the p values for the proposed hypotheses (the results are shown in Fig. 1). The structural model shows a significant positive relationship between all constructs, except for the relationship between perceived usefulness and satisfaction. Therefore, we conclude that all hypotheses are supported, except the H4b hypothesis.

By analyzing the direct, indirect and total effects, it is evident that the strongest relationship is the one between confirmation and satisfaction ($\beta = 0.707$, $p < 0.001$), followed by the relationship between perceived usefulness and confirmation ($\beta = 0.516$, $p < 0.001$) and the relationship between perceived usefulness and attitude towards use ($\beta = 0.453$, $p < 0.001$). Moderate significant relationships are detected as follows: between ease of use and perceived usefulness ($\beta = 0.370$, $p < 0.001$), between ease of use and attitudes towards use ($\beta = 0.291$, $p < 0.001$), between satisfaction and intention to reuse ($\beta = 0.350$, $p < 0.001$), and between attitude towards use and intention to reuse ($\beta = 0.349$, $p < 0.001$). A weak and marginally significant positive relationship is detected between social influence and intention to reuse (β = 0.151, $p < 0.05$). The model explains the variance in the constructs as follows: intention to reuse (45.5%), satisfaction (57.9%), attitude towards use (38.7%), confirmation (26.6%) and perceived usefulness (13.7%).

The effect size f^2 coefficients calculated for the relationships between constructs show strong effect size coefficients between confirmation and satisfaction ($f^2 = 0.870$), and moderate effect size between usefulness and confirmation ($f^2 = 0.362$), between usefulness and attitude towards use ($f^2 = 0.289$), between satisfaction and intention to reuse ($f^2 = 0.170$), between attitude towards use and intention to reuse ($f^2 = 0.163$), and between ease of use and usefulness ($f^2 = 0.159$). According to the provided guidelines [7], we can conclude that all other effect size coefficients can be interpreted as weak.

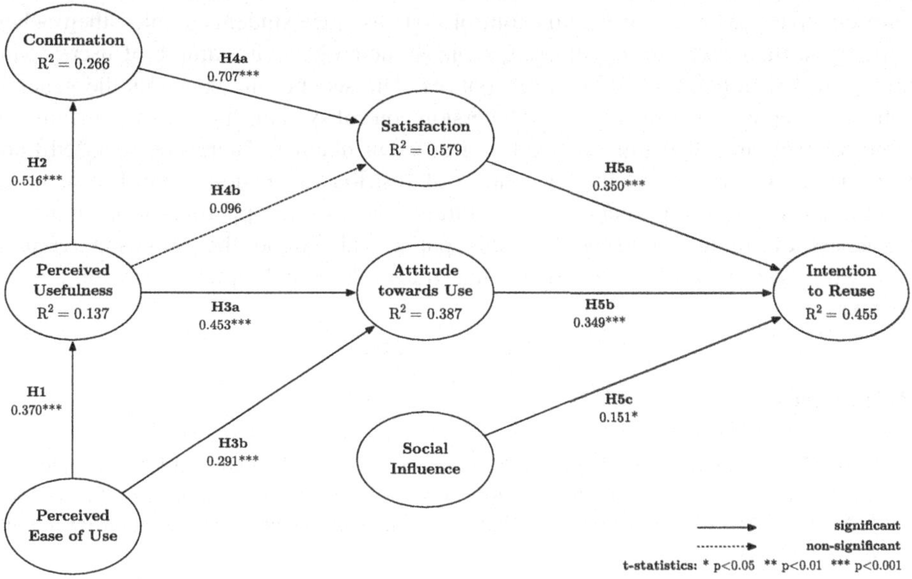

Fig. 1. Research model and PLS results

4 Conclusion

The objective of this paper was to determine the antecedents of satisfaction with and intention to reuse version control system. For that purpose, an empirical study was conducted in which students served as a representative users of GitHub, the most commonly employed version control system. The analysis of the proposed research model revealed that confirmation of expectations strongly contributes to users' satisfaction with the version control system. We also found that the extent to which users perceive version control system as beneficial has medium in size impact on confirmation of their expectations as well as on attitude towards using that system. The same effect appeared to have satisfaction with and attitude toward use of version control system on intention to reuse this system as well as the extent to which the interaction with version control system is effortless on its usefulness as perceived by users. It was also discovered that perceived ease of use has weak impact on attitude towards use of version control system. The same holds for the effect of social influence on intention to reuse version control system. On the other hand, the analysis of the research model uncovered that perceived usefulness does not contribute significantly to the satisfaction when version control systems are considered. Findings reported in this paper can be beneficial to researchers that can employ them as a backbone for future advances in the field as well as for practitioners that can take them into account as guidelines when evaluating existing and designing new version control systems or enhancing them with novel features.

As is the case with other empirical studies, this one also has limitations. The first one is related to the homogeneity of study participants. Although students proved to be representative consumers of version control systems since students employ them when working on their software engineering projects, heterogeneous sample of users could have provided importantly different responses. The second one concerns the generalizability of reported findings. Although version control systems have a lot of features in common, they are often implemented in a different manner. Therefore, reported findings should be interpreted in the context of distributed version control systems in general and GitHub in particular. In our future work, we are planning to examine the robustness of findings presented in this paper and extend the proposed research framework with adoption determinants that reflect the peculiarities of version control systems.

References

1. Hsing, C., Gennarelli, V.: Using GitHub in the classroom predicts student learning outcomes and classroom experiences: findings from a survey of students and teachers. In: Proceedings of the 50th ACM Technical Symposium on Computer Science Education, Minneapolis, pp. 672–678. ACM (2019)
2. Feliciano, J., Storey, M.-A., Zagalsky, A.: Student experiences using GitHub in software engineering courses: a case study. In: Proceedings of the 38th International Conference on Software Engineering Companion, Austin, pp. 422–431. IEEE/ACM (2016)
3. Bhattacherjee, A.: Understanding information systems continuance: an expectation-confirmation model. MIS Q. **25**(3), 351–370 (2001)
4. Fishbein, M., Ajzen, I.: Belief, Attitude, Intention and Behavior: An Introduction to Theory and Research. Addison-Wesley, Reading (1975)
5. Venkatesh, V., Bala, H.: Technology acceptance model 3 and a research agenda on interventions. Decis. Sci. **39**(2), 273–315 (2008)
6. Ringle, C.M., Wende, S., Becker, J.M.: SmartPLS 3, Boenningstedt: SmartPLS GmbH (2015). http://www.smartpls.com
7. Hair, J.F., Hult, G.T.M., Ringle, C.M., Sarstedt, M.: A Primer on Partial Least Squares Structural Equation Modeling (PLS-SEM). Sage Publications Inc., Los Angeles, CA (2017)
8. Hair, J.F., Howard, M.C., Nitzl, C.: Assessing measurement model quality in PLS-SEM using confirmatory composite analysis. J. Bus. Res. **109**, 101–110 (2020)

Evidence on the Use of Gait Analysis - A Review

Afonso Laranjo[1(✉)], Susana Costa[1], Fernando Duarte[2], Miguel Carvalho[3], and Pedro Arezes[1]

[1] ALGORITMI Research Center, School of Engineering, University of Minho, Campus de Azurém, 4800-058 Guimarães, Portugal
alaranjo@hotmail.com,
{susana.costa,parezes}@dps.uminho.pt
[2] Polymer Engineering Department, University of Minho, Campus de Azurém, 4800-058 Guimarães, Portugal
fduarte@dep.uminho.pt
[3] Textile Engineering Department, University of Minho, Campus de Azurém, 4800-058 Guimarães, Portugal
migcar@det.uminho.pt

Abstract. Gait analysis consists of evaluating the individual through kinematic analysis, while walking along a surface. Kinematic analysis relates the relative movement between rigid bodies and finds applications in gait analysis. The purpose of this paper is to find the applications of gait analysis, the methodologies used to perform it and conclude about the different methodologies' uses. A literature search was performed using PRISMA Guidelines. Twenty-two documents fulfilled the inclusion criteria. A total of 15 different countries presented researches in this topic. The areas within which these papers are published include Sports Medicine (7), Pediatric Medicine (1), General Medicine (11), Occupational Medicine (1), Engineering (2). Gait analysis has many different areas of intervention. Some gait parameters are interrelated and there are a few different methodologies available to perform gait analysis. A comprehensive table of results has been developed, where results are presented.

Keywords: Gait analysis · Biomechanics · Kinematic analysis

1 Introduction

Walking is essential for the performance of human tasks and presents itself as a stable movement that can suffer changes as a result of internal or external constraints mostly health constraints.

The biomechanics associated with the human body, specifically in the case of the foot, is an area focused on the perception and understanding of the forces acting on the human body and its respective behavior. The analysis and characterization of the foot is important since it allows the assessment of gait movements and the negative influence of differentiating factors of the foot on normal gait mechanics and on the behavior of the ankle, knee, hip and back [25]. Adequate biomechanics of the foot allows to

W. Karwowski et al. (Eds.): IHSED 2020, AISC 1269, pp. 51–56, 2021.
https://doi.org/10.1007/978-3-030-58282-1_9

maintain a correct posture and naturally a symmetrical distribution of plantar pressure [24]. The walking cycle is divided into two main phases: the support phase and the swing phase. During gait, one foot attends as a support while the other moves to a support position later on with a successive reversal of roles between them. A complete walking cycle goes from the moment that the first foot touches the ground until the moment that the same foot touches the ground again [23].

Gait analysis consists of evaluating the individual through kinematic analysis, while walking along a surface allowing the withdrawal and learning that can be useful in the treatment of pathologies and in the design and idealization of spaces and objects. Kinematic analysis is related to the relative movement between rigid bodies and finds application in gait analysis. In order to analyze the behavior of the human being in situations of static support - when the individual is standing - or dynamic support - when the individual is in movement – some different techniques and systems were used.

This research was aimed at understanding which analysis systems exist and the type of result already registered after these.

2 Methods

The authors resorted to the Preferred Reporting Items for Systematic reviews and Meta-Analyses (PRISMA) Guidelines as a methodological template for this review. Literature searches were conducted between 4th–7th of March 2020 using the ISI Web of Science and Scopus databases. The inclusion criteria were the papers published in Portuguese, Spanish or English language. The following search terms were used in different combinations on both data bases and combined using Boolean operators "AND", "OR" and "*" to allow for words with the same root, including "Gait", "Analysis", "Assess*", "Eval*", "work*", "occupation*", "sport*" and "athlete". Other papers were affected to this review by snowball effect. All duplicates were excluded. Additionally, based on the three-phased search, other papers were excluded according to the titles, abstracts and full-texts which revealed subjects out of the scope of interest for this paper.

3 Results and Discussion

Twenty-two articles were included in the final analysis (Table 1).

Table 1. General characterization of the studies

Citation	Ref.	Area of application
Sadeghisani et al., 2018	[1]	Sports medicine
Oldham et al., 2016	[2]	Sports medicine
Hadizadeh et al., 2016	[3]	Sports medicine
Hortobágyi et al., 2015	[4]	Geriatric medicine
Cippitell et al., 2014	[5]	General medicine

(*continued*)

Table 1. (*continued*)

Citation	Ref.	Area of application
Thakurta et al., 2015	[6]	Occupational medicine
Lee et al., 2012	[7]	Sports medicine
Gouwanda and Senanayake, 2008	[8]	Sports medicine
Aissaoui et al., 1996	[9]	General medicine
LAPHAM and Bartlett, 1995	[10]	Sports medicine
Satoh et al., 2019	[11]	General medicine
Carse et al., 2013	[12]	General medicine
Carson et al., 2001	[13]	General medicine
Dombroski et al., 2014	[14]	General medicine
Kostuj et al., 2018	[15]	General medicine
Nandikolla et al., 2018	[16]	General medicine
Ould-Slimane et al., 2017	[17]	General medicine
Pfister et al., 2014	[18]	Engineering
Piening et al., 2018	[19]	Pediatric medicine
Wright et al., 2010	[20]	General medicine
Yu et al., 2019	[21]	Sports medicine
Zhang et al., 2017	[22]	Engineering

A range of years was not defined in the search, which allowed to verify that the first paper devoted to gait analysis was published was in 1984 and the author is unknown. Then, in the course of the time, the number of studies and the interest increased leading to largest amount of articles in 2017 and 2018 (Fig. 1).

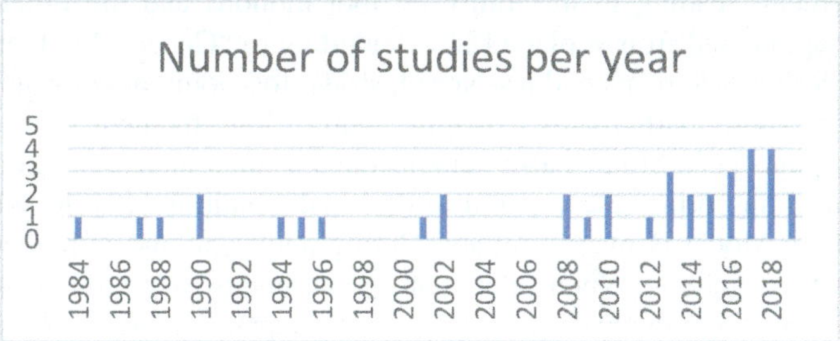

Fig. 1. Number of studies per year.

It is also noteworthy that between years 1996 and 2001 and between years 2002 and 2008 there are no records in the databases consulted of any papers being published in this scientific niche.

Fig. 2. Number of participations in the studies by each Country.

A total of 15 different countries presented researches in this topic, being the USA the one with the highest number of papers published in this niche of research (4), followed by the United Kingdom (3), as seen in Fig. 2. Number of participations in the studies by each Country. Europe is the one continent with more records (13).

The scientific areas within which these papers are published include Sports Medicine (7), Pediatric Medicine (1), General Medicine (11), Occupational Medicine (1), Engineering (2), showing that General medicine and Sports Medicine are the ones more advanced in this theme (Fig. 5). This shows that gait analysis has a very wide application range despite the bigger impact on sports and general medicine. Considering the number of subjects involved in the experiments, the largest amount got found in Hortobágyi et al. (2015) with 2495 people. If looking to sample size intervals, in the final list selection, 8 studies have from 0 to 10 participants, 11 studies have from 10 to 100, 2 studies from 100 to 1000 and just one have more than 1000 participants.

Looking into the documents in the scientific area "Engineering", two conclusions can be made: correlation between Kinect and Vicon stride timing was high and error was small [18]. Yet, in this area results show that unstable shoes with unstable structural elements leading to instability on foot motions and the difference mainly appeared in sagittal and frontal plane [22]. Regarding to "General Medicine", research found that elderly participants had less speed, stride, foot joint movement, moment and power than the young participants. Also, the ratio of joint moment was smaller in the elderly participants [17]. Studies also demonstrated a small increase in arch high index with the 3D printing orthotic compared to the shod condition [14]. Results show that the Oxford Foot Model is overall reliable during adult gait and is higher than previously reported in children. When referencing joint angles to neutral stance decreased error by up to 28 from previous reports [20]. In "Occupational Medicine" papers, it can be observed that women's gait is characterized by greater changes than men [6]. On what concerns to "Geriatric Medicine", studies proved that used exercise can increase the habitual and fast gait speed of healthy old adults in substantial and clinically meaningful ways [4]. When one refers to "Pediatric Medicine", significant relationships between kinematics of Oxford Foot Model and pressure distribution parameters of plantar pressure measurement (PMD) indicate that PMD could give useful supplementary information during a 2D-gait analysis when 3D gait analysis is not obtainable [19]. Concerning to "Sports Medicine", martial arts athletes with low back pain apply symmetrical loads on the lower extremities, similar to healthy people [1]. As well results indicate that gait velocity and range of motion of the center of mass in the coronal plane are sensitive measures of dual-task related changes in concussed patients and should be considered as part of a comprehensive assessment for a sports-related concussion [7].

Researchers focused in this area of research are convinced that there is a clear indication that the research and development of the body-mounted sensor will continue and replace the current technologies in the future [8].

4 Conclusion

Gait analysis has many different areas of intervention with a higher incidence in Sports and in General Medicine. Results show that interest in this theme has been increasing among researchers in the last few years. Some gait parameters are interrelated and there are a few different methodologies available to perform gait analysis, including force plates and motion analysis systems.

Conflict of interest. The authors declare that there is no conflict of interests regarding the publication of this paper.

Acknowledgments. This work has been supported by FCT – Fundação para a Ciência e Tecnologia within the R&D Units Project Scope: UIDB/00319/2020.

References

1. Sadeghisani, M., Sobhani, V., Azimi, H., Karimi, M.T.: Analysis of symmetry of weight-bearing in athletes with and without low back pain during gait. Annals Appl. Sport Sci. **6**(2), 31–36 (2018)
2. Oldham, J.R., DiFabio, M.S., Kaminski, T.W., DeWolf, R.M., Buckley, T.A.: Normative Tandem Gait in Collegiate Student - Athletes: Implications for Clinical Concussion Assessment, pp. 305–311 (2017)
3. Hadizadeh, M., Amri, S., Mohafez, H., Roohi, S.A., Mokhtar, A.H.: Gait analysis of national athletes after anterior cruciate ligament reconstruction following three stages of rehabilitation program: symmetrical perspective. Gait Posture **48**, 152–158 (2016)
4. Hortobágyi, T., Lesinski, M., Gäbler, M., Malatesta, D., Granacher, U.: Erratum to effects of three types of exercise interventions on healthy old adults' gait speed: a systematic review and meta-analysis. Sports Med. **45**(12), 1627–1643 (2015)
5. Cippitelli, E., Gasparrini, S., Spinsante, S., Gambi, E.: Kinect as a tool for gait analysis: validation of a real-time joint extraction algorithm working in side view. J. Sensors **15**, 1417–1434 (2015)
6. Thakurta, A.G., Iqbal, R., Bhasin, H.V.: Comparative analysis of gait kinematics of male and female construction workers while carrying asymmetric loads in different modes. In: International Conference on Industrial Engineering and Operations Management, pp. 1–5 (2015)
7. Lee, H., Sullivan, S.J., Schneiders, A.G.: The use of the dual-task paradigm in detecting gait performance deficits following a sports-related concussion: a systematic review and meta-analysis. J. Sci. Med. Sport **16**, 2–7 (2013)

8. Gouwanda, D., Senanayake, S.M.N.A.: Emerging trends of body-mounted sensors in sports and human gait analysis. In: IFMBE Proceedings, pp. 715–718 (2008)
9. Aissaoui, R., Allard, P., Junqua, A., Frossard, L., Duhaime, M.: Internal work estimation in three-dimensional gait analysis. Med. Biol. Eng. Comput. **34**, 467–471 (1996)
10. Lapham, A.C., Bartlett, R.M.: The use of artificial intelligence in the analysis of sports performance: a review of applications in human gait analysis and future directions for sports biomechanics. J. Sports Sci. **13**, 229–237 (1995)
11. Satoh, Y., Yamada, T., ShimamuRa, R., Ohmi, T.: Comparison of foot kinetics and kinematics during gait initiation between young and elderly participants. J. Phys. Ther. Sci. **31**, 498–503 (2019)
12. Carse, B., Meadows, B., Bowers, R., Rowe, P.: Affordable clinical gait analysis: an assessment of the marker tracking accuracy of a new low-cost optical 3D motion analysis system. Physiotherapy **99**(4), 347–351 (2013). https://doi.org/10.1016/j.physio.2013.03.001
13. Carson, M.C., Harrington, M.E., Thompson, N., O'Connor, J.J., Theologis, T.N.,: Kinematic analysis of a multi-segment foot model for research and clinical applications: a repeatability analysis. J. Biomech. **34**(10), 1299–1307 (2001). https://doi.org/10.1016/s0021-9290(01)00101-4
14. Dombroski, C.E., Balsdon, M.E.R., Froats, A.: The use of a low cost 3D scanning and printing tool in the manufacture of custom-made foot orthoses: a preliminary study. BMC Res. Notes **7**, 443 (2014). https://doi.org/10.1186/1756-0500-7-443
15. Kostuj, T., Stief, F., Hartmann, K.A., Schaper, K., Arabmotlagh, M., Krummenauer, F., Baums, M.H., Lieske, S., Meurer, A.: Using the Oxford Foot Model to determine the association between objective measures of foot function and results of the AOFAS Ankle-Hindfoot Scale and the Foot Function Index: a prospective gait analysis study in Germany (2018)
16. Nandikolla, V.K., Bochen, R., Meza, S., Garcia, A.: Experimental gait analysis to study stress distribution of the human foot. J. Med. Eng. (2017)
17. Ould-Slimane, M., Latrobe, C., Michelin, P., Chastan, N., Roussignol, X., Gauthe, R., Dujardin, F.: Noninvasive optoelectronic assessment of induced sagittal imbalance using the vicon system
18. Pfistes, A., West, A.M., Bronner, S., Noah, J.A.: Comparative abilities of Microsoft Kinect and Vicon 3D motion capture for gait analysis. J. Med. Eng. Technol. **38**(5), 274–280 (2014)
19. Piening, M., Schallig, W., Van der Krogt, M.M.: P 083 - Correlation between Oxford foot model kinematics and plantar pressure during voluntary varus/valgus gait in healthy children. J. Gait Posture **65**, 367–368 (2018)
20. Wright, C.J., Arnold, B.L., Coffey, T.G., Pidcoe, P.E.: Repeatability of the modified Oxford foot model during gait in healthy adults. J. Gait Posture **33**, 108–112 (2010)
21. Yu, C., Gu, Y., Shao, S., Baker, J.S., Awrejcewicz, J.: A comparative biomechanical analysis of the performance level on chasse step in table tennis. Int. J. Sports Sci. Coaching **14**(3), 372–382 (2019)
22. Zhang, Y., Fekete, G., Fernandez, J., Gu, Y.: Analysis of foot kinematics with unstable sole structure using Oxford foot model. J. Biomimetics, Biomaterials Biomed. Eng. **34**, 1–9 (2017)
23. Chaitow, L., DeLany, J.(eds.): Clinical Application of Neuromuscular Techniques, Volume 2, The Lower Body Lower Body, 2nd edn, p. 608. (2011). ISBN 9780443068157
24. Rosário, J.L.P.: A review of the utilization of baropodometry in postural assessment. J. Bodywork Mov. Therapies **18**(2), 215–219 (2014)
25. Rosenbaum, D., Becker, H.-P.: Plantar pressure distribution measurements. Technical background and clinical applications. Foot Ankle Surg. **3**(1), 1–14 (1997)

Task-Technology Fit and Continuance of Use of Web-Based Programming Tool: A Pilot Study

Igor Škorić[1(✉)], Tihomir Orehovački[1], and Marina Ivašić-Kos[2]

[1] Faculty of Informatics, Juraj Dobrila University of Pula, Zagrebačka 30, 52100 Pula, Croatia
{igor.skoric, tihomir.orehovacki}@unipu.hr

[2] Department of Informatics, University of Rijeka, Radmile Matejčić 2, 51000 Rijeka, Croatia
marinai@inf.uniri.hr

Abstract. Web-based coding tools are widely accepted in computer science education. The use of these tools allows us to improve learning, but the requirement is to understand the factors that affect their acceptance. Carefully selecting technology that best suits the needs of the task will allow the optimal use of these tools in education. The purpose of this study is to develop a model that incorporates constructs of the Task-technology fit (TTF) model and the Expectation-confirmation model of IT continuance (ECM) to better understand the impact of the tool's suitability on the user's behavioral intention. The analysis was performed using the partial least squares approach to structural equation modeling. The results show a significant impact of task-technology fit factor on student satisfaction and their continuance intention. Consequently, this demonstrates that the proposed model is appropriate for understanding the acceptance of web-based programming tools in an educational context.

Keywords: Web-based coding tool · Task–technology fit model · Expectation-confirmation model of IT continuance · Introductory programming course · Empirical study · Post-use questionnaire · PLS-SEM

1 Introduction

Web-based coding tools are playing an important role in computer science education. Their importance is further enhanced by the increasing popularity of distance learning and Massive Open Online Courses. These tools offer many advantages over their desktop alternatives (presence on various platforms, availability anywhere and anytime, needlessness for installation and maintenance, suitability for face-to-face and online teaching, and support for collaborative learning), although they have some disadvantages as well. Web-based technology offers opportunities for students to improve their learning, but it is necessary to identify the most effective methods of utilizing those tools in order to improve teaching and learning productivity [1]. Many studies have been conducted to investigate that problem and numerous models have been developed to describe and explain the process of tool adoption and choosing the right tool to solve a

W. Karwowski et al. (Eds.): IHSED 2020, AISC 1269, pp. 57–62, 2021.
https://doi.org/10.1007/978-3-030-58282-1_10

task. Goodhue and Thompson proposed a task-technology fit model and established the linkage between information systems and individual performance [2]. In that model, match between the capabilities of the technology to the demands of the task is expressed by the formal construct known as a task–technology fit (TTF). In the TTF model, task–technology fit is a function of task and technology characteristics. Numerous studies (e.g. [3, 4]) have confirmed that TTF has an influence on user performance as predicted by the TTF model. In the educational context, task–technology fit refers to a learning system's capability to support users in performing their learning tasks. However, while the TTF model integrates technology and task characteristics and connects them through the use of technology with user performance, it does not contain any other factors that may influence technology use. Factors influencing the use of technology are explored by theories of technology acceptance. It has been studied over the years and many models have been developed to describe technology acceptance. Bhattacherjee [5] introduced the unified model of IT continuance in which user's intention to continue using information systems is determined by the confirmation of perceived usefulness, satisfaction with technology, and subjective norm. This model was used as a background for further research dedicated to the acceptance of IT technology by users. Numerous studies confirmed the model in different contexts (e.g. [6, 7]). Also, the model has been validated and expanded in an educational context. Yang [8] analyzed quality factors that influence the continuance intention in the educational ecosystem. Boe [9] integrated IT continuance theory and agency theory (PAT) in higher education settings. Basnet et al. [10] explored students' continuance intention of the automated assessment system in computer science education. Haung [11] identified the importance of hedonic and utilitarian aspects in students' continuance intention toward programming games. Thongmak [12] examined the influence of satisfaction on continued engagement intention in the context of gamifying programming course. Orehovački and Babić [13] examined to what extent pragmatic and hedonic facets of quality affect users' continuance intentions with respect to games designed for learning programming. Current studies (e.g. [14]) have found that task–technology fit plays an important role in explaining users' e-learning continuance intentions. All of the above justifies the use of the IT continuance model in the context of learning programming and its extension with elements of the task-technology fit model. In this paper, we synthesized factors from the TTF model and the expectation-confirmation model of IS continuance to examine the impact of task-technology fit on student acceptance of web-based programming tool and role of satisfaction in that transition. The remainder of the paper is structured as follows. Research model and hypotheses are explained in next section. Research methodology is described in the third section. Findings are reported in the fourth section. Conclusions are drawn in the last section.

2 Research Model and Hypotheses

The proposed research model consists of three constructs adopted from the task-technology fit model: task characteristics, technology characteristics, and task-technology fit. Task characteristics refers to the particularities of a task that can be executed using a web-based coding tool. Technology characteristics denotes the features of a web-based coding tool that may affect its utilization by users. Task-technology fit is the degree to which a web-based programming tool assists students in completing problem-based tasks. Remaining two constructs (satisfaction and continuance intention) were adopted from the expectation-confirmation model of IS continuance. Satisfaction measures the extent to which students are pleased with the use of a web-based programming tool. Continuance intention refers to the degree to which students are willing to remain users of a web-based programming tool and recommend it to their peers. This study proposes that the adoption of a web-based coding tool will depend on how students perceive the fit of this technology to the problem-based tasks they are trying so solve by its means. The mentioned fit will also affect users' satisfaction with the tool, which will in turn contribute to their intention to use the tool. In accordance with the aforementioned, we propose the following five hypotheses:

H1: Task characteristics have a significant positive effect on task–technology fit.

H2: Technology characteristics have a significant positive effect on task–technology fit.

H3: Task–technology fit has a significant positive effect on continuance intention.

H4: Task–technology fit has a significant positive effect on satisfaction.

H5: Satisfaction has a significant positive effect on continuance intention.

3 Methodology

Empirical study was conducted as part of an ongoing project exploring the suitability of various educational tools for acquiring programming skills and their adoption by students. To date, we proposed a taxonomy of tools designed for learning programming [15], identified criteria for selecting the most appropriate tool in that respect [16], and introduced a model in which TAM model was tailored to the context of our research [17]. In this paper, we examined the adoption of a web-based programming tool within an introductory programming course. The sample was composed of undergraduate students enrolled in the Faculty of Informatics and Department of Computer Science of the Juraj Dobrila University of Pula in Croatia. Data were collected using a post-use questionnaire after lab sessions in which students used a web-based programming tool. In the introductory part of the questionnaire, basic demographic data on respondents (gender and age) were collected and the rest of the questionnaire consisted of 17 five-point Likert scale items. These items were adopted from the same literature as their underlying constructs and tailored to the context of our research. The Repl.it tool was used as a representative sample of a web-based programming learning tool. This tool allowed students to work together on the same code, share code and work individually. Students could interact with each other while coding by using the chat feature. At the beginning of the session, features of the tool were briefly presented to students. The lab session included an introduction of a programming problem, a presentation of several solved problems, and a group analysis of the code. Students were also asked to use the aforementioned tool to independently solve several programming problems. The post-use questionnaire was created and distributed to study participants by means of Google Forms. The research model was analyzed with SmartPLS 2.0 M3 tool [18].

4 Findings

A total of 56 students participated in the study. The majority of them (73.2%) were male. At the time study took place, half of the respondents were 19 years old, 19.6% were 20 years of age, 14.3% were 21 years old while remaining 16.07% had between 22 and 25 years of age. The analysis of the research framework followed guidelines suggested by Hair et al. [19] and included the assessment of the measurement and the structural model. First, the outer loadings for all latent variables were examined. All indicators met the criterion according to which their outer loadings should be 0.708 or higher which indicates that latent variables are explaining at least 50% of the variance of each indicator assigned to them. Internal consistency reliability was tested using composite reliability (CR) and Cronbach's alpha (CA). The values for both indicators in the context of all latent variables have exceeded the acceptable threshold of reliability (0.707). Therefore, the internal consistency reliability of items was confirmed. Values of AVE for constructs were in a range from 0.6211 to 0.8427 which is above the cut-off value of 0.50 indicating a satisfactory convergent validity. We tested discriminant validity by extracting the indicator's outer loadings on a construct and its cross-loadings with other constructs. Results indicated that all items' loadings on their respective constructs were above the threshold of 0.707 [20]. Furthermore, each item's factor loading on its respective construct was higher than on any other latent variable in the model. In that respect, this criterion of discriminant validity was met. Discriminant validity was further examined by comparing the square root of the AVE of particular latent variable with the correlation coefficients between this latent variable and remaining ones in the model. Considering the square root of AVE for all latent constructs was superior to the correlation coefficients, we concluded that the Fornell-Larcker criterion of discriminant validity [19] was achieved. Assessment of structural model included the evaluation of latent variables' determination coefficient values, the significance of the path coefficients, the effect size, and predictive relevance [19]. The values of determination coefficient for all dependent variables are above 0.5. Two exogenous variables (task characteristics and technology characteristics) explain 53.2% of variance in task-technology fit which in turn explains 63.9% of variance in satisfaction while 74.3% of the variance in continuance intention is explained by task–technology fit and satisfaction. In order to test the hypotheses, we conducted a bootstrapping analysis with 5000 samples. The strongest relationship was found between task–technology fit and satisfaction ($\beta = 0.7991$, $p < 0.001$) which was followed by the relationship between task–technology fit and continuance intention ($\beta = 0.5588$, $p < 0.001$) and the relationship between technology characteristics and task–technology fit ($\beta = 0.4797$, $p < 0.001$). On the other hand, relationships between task characteristics and task–technology fit ($\beta = 0.3377$, $p < 0.01$) and between satisfaction and continuance intention ($\beta = 0.3472$, $p < 0.05$) proved to be weaker. Since all paths in the structural model were statistically significant, all proposed hypotheses were confirmed. Values of effect size and predictive relevance of exogenous variables were interpreted in accordance with guidance proposed in [19]. We found that task–technology fit strongly affects continuance intention ($f^2 = 0.43$). On the other hand, the effect of satisfaction on continuance intention ($f^2 = 0.15$), task characteristics on task–technology fit ($f^2 = 0.16$)

and technology characteristics on task–technology fit ($f^2 = 0.32$) proved to be moderate in size. The analysis of the research model also revealed that technology characteristics has moderate relevance in predicting task–technology fit ($q^2 = 0.15$). The same holds for task–technology fit in predicting continuance intention ($q^2 = 0.16$). All remaining exogenous latent variables have weak relevance in predicting endogenous latent variables.

5 Conclusion

The aim of this paper was to combine factors from expectation-confirmation model with those of task technology fit model in order to understand to what extent features of a web-based programming tool and particularities of task that can be carried out by its means contribute to behavioral intentions related to the reuse of this tool. Findings presented in this paper are in line with those in current literature since they confirmed positive effects between constructs adopted from the aforementioned two models. More specifically, we found that in the context of the web-based programming tool adoption, degree to which a tool assists students in completing problem-based tasks has a significant positive effect on extent to which students are pleased with the use of that tool. We also found that technology fit has a medium-sized impact on students' willingness to remain users of the tool and to recommend it to their peers. This study has several limitations. It is based on the use of only one tool, and the sample was composed of students from a single university. Therefore, future studies with more participants from different institutions should follow proposed research direction in order to provide more insights into robustness of presented findings.

References

1. Sadaf, A., Newby, T.J., Ertmer, P.A.: An investigation of the factors that influence preservice teachers' intentions and integration of Web 2.0 tools. Educ. Technol. Res. Dev. **64**(1), 37–64 (2015)
2. Goodhue, D.L., Thompson, R.L.: Task-technology fit and individual performance. MIS Q. **19**, 213–236 (1995)
3. Bere, A.: Applying an extended task-technology fit for establishing determinants of mobile learning: an instant messaging initiative. J. Inf. Syst. Educ. **29**(4), 4 (2019)
4. McGill, T.J., Klobas, J.E.: A task–technology fit view of learning management system impact. Comput. Educ. **52**(2), 496–508 (2009)
5. Bhattacherjee, A., Lin, C.P.: A unified model of IT continuance: three complementary perspectives and crossover effects. Eur. J. Inf. Syst. **24**(4), 364–373 (2015)
6. Oghuma, A.P., Libaque-Saenz, C.F., Wong, S.F., Chang, Y.: An expectation-confirmation model of continuance intention to use mobile instant messaging. Telematics Inform. **33**(1), 34–47 (2016)
7. Rahman, M.N.A., Zamri, S.N.A.S., Eu, L.K.: A meta-analysis study of satisfaction and continuance intention to use educational technology. Int. J. Acad. Res. Bus. Soc. Sci. **7**(4), 1059–1072 (2017)

8. Yang, M., Shao, Z., Liu, Q., Liu, C.: Understanding the quality factors that influence the continuance intention of students toward participation in MOOCs. Educ. Tech. Res. Dev. **65** (5), 1195–1214 (2017)
9. Bøe, T., Gulbrandsen, B., Sørebø, Ø.: How to stimulate the continued use of ICT in higher education: integrating information systems continuance theory and agency theory. Comput. Hum. Behav. **50**, 375–384 (2015)
10. Basnet, R.B., Doleck, T., Lemay, D.J., Bazelais, P.: Exploring computer science students' continuance intentions to use Kattis. Educ. Inf. Technol. **23**(3), 1145–1158 (2017)
11. Huang, Y.M.: Students' continuance intention toward programming games: hedonic and utilitarian aspects. Int. J. Hum.-Comput. Inter. **36**(4), 393–402 (2019)
12. Thongmak, M.: Gamifying the first programming class: outcomes and antecedents of continued engagement intention. In: Proceedings of the International Conference on Information Systems Education and Research AIS SIGED (2017)
13. Orehovački, T., Babić, S.: Inspecting quality of games designed for learning programming. In: Zaphiris, P., Ioannou, A. (eds.) LCT, HCII 2015. Lecture Notes in Computer Science, 9192, pp. 620–631. Springer, Heidelberg (2015)
14. Sørebø, Ø., Halvari, H., Gulli, V.F., Kristiansen, R.: The role of self-determination theory in explaining teachers' motivation to continue to use e-learning technology. Comput. Educ. **53** (4), 1177–1187 (2009)
15. Škorić, I., Orehovački, T., Ivašić Kos, M.: Learning to code and collaborate in a web environment. In: Proceedings of the AHFE 2018 International Conference on Human Factors and Systems Interaction, pp. 54–65. Springer, Orlando (2018)
16. Škorić, I., Pein, B., Orehovački, T.I.: Selecting the most appropriate web IDE for learning programming using AHP. In: Proceedings of the 39th International Convention on Information and Communication Technology, Electronics and Microelectronics (MIPRO, pp. 877–882. IEEE, Opatija (2016)
17. Škorić, I., Orehovački, T., Ivašić Kos, M.: Exploring the acceptance of the web-based coding tool in an introductory programming course: a pilot study. In: Proceedings of the 3rd International Conference on Human Interaction and Emerging Technologies (IHIET 2020), Advances In Intelligent Systems and Computing. Springer, Paris (2020)
18. Sarstedt, M., Cheah, J.H.: Partial least squares structural equation modeling using SmartPLS: a software review. J. Mark. Analytics. **7**(3), 196–202 (2019)
19. Hair, J.F.: A Primer on Partial Least Squares Structural Equation Modeling (PLS-SEM). SAGE Publications, Inc., Thousand Oaks (2017)
20. Hulland, J.: Use of partial least squares (PLS) in strategic management research: a review of four recent studies. Strateg. Manag. J. **20**(2), 195–204 (1999)

Impacts of Virtual Communication During Social Isolation of Covid'19

Lilia Raycheva(✉), Neli Velinova, Nadezhda Miteva, and Mariyan Tomov

The St. Kliment Ohridski Sofia University, Sofia, Bulgaria
{lraycheva,hopeace}@abv.bg, nelikdkd@gmail.com, mariyan.d.tomov@gmail.com

Abstract. The COVID'19 pandemic has led to drastic changes in people's lifestyle around the world, incl. those related to the natural way in which individuals interact and communicate. In order to examine some of the effects of virtual communication during the social isolation, an academic research team from the Faculty of Journalism and Mass Communication at The St. Kliment Ohridsky Sofia University of Bulgaria undertook a study in the declared two-month's period (13.03.–13.05.2020) of the state of emergency in the country. The researchers conducted a three-folded study: on the sociological polls and the media coverage during the pandemics; on the impacts of virtual communication during the social isolation, using mixed methodology: quantitative and qualitative survey with three groups of respondents: media users, media professionals and media experts; and on the activities of the telecommunication industry. The results of the effects of virtual communication in social isolation were indicative.

Keywords: Virtual communication · Social isolation · Media · Telecommunications

1 Introduction

The quick spread of COVID'19 found the world unprepared psychologically, economically, medically and socially. It has resulted in a unique and unconventional lockdown of a great percentage of the Earth's population. The World Health Organization has declared the corona virus disease in 2019–2020 to be a pandemic, as well as a critical public health emergency of international concern (PHEIC). Throughout the pandemic period, the WHO has periodically issued a number of recommendations to address the global situation. Thus, in March 2020, at a joint virtual press conference, the UN, the UNICEF and the WHO announced a coordinated global plan to combat COVID'19, with the aim of helping the most financially vulnerable countries [1].

Between early March and the end of May, the European Commission (EC) alone adopted 291 decisions and other acts on COVID'19. They reflected the urgency and the dramatic nature of the situation, as well as the speed at which the EC had to readjust from delivering on long-term objectives to immediate crisis management. The European Commission's revised Work Program for 2020 states, that what started with

W. Karwowski et al. (Eds.): IHSED 2020, AISC 1269, pp. 63–68, 2021.
https://doi.org/10.1007/978-3-030-58282-1_11

isolated cases quickly turned into a public health crisis, an economic shock of unprecedented scale and a pandemic of global and tragic proportions [2].

The aftermath of the gradually easing lockdown in Europe is not problem-free. That is why it makes sense to consider alternative ways to compensate for the limitations of these social challenges through alternative forms of communication. Merit deserves those that technologically enable people to overcome the strain of isolation.

Bulgaria reacted relatively quickly and efficiently at the institutional level to the public health emergency announced by the World Health Organization regarding the disease COVID'19. On February 26, 2020, by order of the Prime Minister, a National Operational Headquarters was established. Among its responsibilities were to organize, coordinate and monitor all actions of the competent authorities in relation to the prevention of the spread of COVID'19, as well as to collect, summarize and analyse all information on the development of the situation in order to inform the media and the public [3]. The National Assembly passed a law on measures and actions during the state of emergency, declared by its decision of March 13, 2020 [4]. The state of emergency in the country lasted exactly two months – until 13.05.2020, when replaced by an emergency until 15.07.2020.

Dealing with the pandemic not only in medical and economic, but also in social and communication terms posed significant challenges to the institutions and the population in Bulgaria during the state of emergency. In order to limit the spread of the corona virus in Bulgaria, the state institutions issued a number of measures and recommendations to citizens, companies and organizations. The traditional understandings of labour, physical communication, working space had to transform ad hoc. The means of virtual communication, including specialized training software, "saved" students from completing the school year on a regular basis. The so-called "home office" became increasingly important and popular. In this new extraordinary situation, people experienced the professional deficits in the field of digital infrastructure and connectivity at national and international level. Although nowadays the health system in the country is already prepared to meet the further challenges of COVID'19, the danger of infection has not yet passed.

2 Aim and Research Methods

Undoubtedly, the COVID'19 pandemic will have a long-lasting influence on how individuals, employees, customers and organizations will work, produce and interact with each other. The transformations of human contacts during the pandemic towards intensive e-communication in a number of social spheres, institutions, communities and even individuals have fundamental scientific importance because it caused and continues to cause a change of paradigms, concepts, social practices and policies.

Though a plethora of research topics would be well fit for examining the pandemics in general, the aim of the undertaken research is to focus on a fundamentally specific societal segment: tracking of the effects of virtual communication in the situation of social isolation and utilizing of the structured and analysed information into patterns for policy measures. The respondents (N=185) were asked 28 questions.

The study is interdisciplinary and it uses mixed research methods, among them:

- Comparative analysis of the derived data from academic sources, sociological surveys, regulatory frameworks and media and telecommunications practices;
- Qualitative research, conducted with three groups of respondents: media ecosystem (traditional media; social media, and social networks) users, media professionals and media experts. The qualitative research techniques included semi-structured interviews; experts' assessments; and analysis of texts and documents, such as government reports, media articles, and websites.
- Quantitative research, conducted via an individual direct poll.

In addition, cross-combining technique for data collection and analysis was applied in order to compensate some of the inherent limitations.

3 Results

3.1 Sociological Polls and Media Coverage During the Pandemics

The national representative surveys of *Alpha Research* [5], Trend [6], and *Gallup International. Center for Public and Political Studies* [7] sociological agencies, conducted during the emergency state (13.03.-13.05.2020) showed an overall acceptance and support by the publics for the health quarantine measures and restrictions. Critical sentiments were present only towards the still unclear economic measures. Most of the participants answered, that increased amount of time spent at home had a negative effect on their mental and physical health. Many Bulgarians realized that the fight against COVID'19 would take a long time until the return to their normal lifestyle.

The declared state of emergency and the accompanying restrictive measures unequivocally displayed that the media environment with its timely and persistent information about the development of the disease was crucial for the normal functioning of the economy and the society. The media also had to readjust ad hoc because they had the responsibility to participate in the national management of the pandemic and in alleviating the stress in society caused by the emergency profound changes.

The point of the public debate in the media about the corona virus, however, was not so much to challenge the constitutional right of the executive power to vital measures during the pandemic, nor to instruct people on how and whether to comply with the undertaken measures. The important issue was to protect people from "infodemia", i.e. from the disproportion in their access to other important information, in order to they do not fall prey to unreliable messages and rumours that spread mostly via social networks.

3.2 Impacts of Virtual Communication During the Social Isolation

In order to examine some of the effects of virtual communication in the conditions of social isolation, an academic team from the Faculty of Journalism and Mass Communication at The St. Kliment Ohridski Sofia University undertook a survey during the two-month's period of the state of emergency in the country (13.03.2020–13.05.2020).

Among the objectives of the survey was to monitor the level of adaptability and the way in which respondents coped socially, psychologically, communicatively and healthily during the isolation associated with the corona virus pandemic. More than 86% of the respondents were satisfied with the opportunity to work and/or study online. 65% of the interviewees gave positive answer to the question whether employers/teachers support distance work. Although 67% claimed that they were not worried of losing their job or of education failure, 59% felt that the pandemic situation negatively affected their work. The prevailing opinion of the respondents was that dealing with the situation, which changed the status quo, requires solidarity, empathy and care for the weaker and riskier groups of our society. Thus they saw the crisis as an opportunity for the Bulgarian society to look at these problems in depth.

Another objective of the survey was to study which channels of information the interviewees had used. 78% of all respondents pointed out the traditional media as the most used information source during the COVID'19. The prevailing part of them stated that they prefer to watch television, then - listen to the radio. Most of the interviewees specified that combine their sources of information, as they want to compare the data with more than one media channel.

The third objective of the survey was to analyze the level of trust in traditional, online, social networks, social media) of the participants in the survey. More than half of the respondents – 65.9% - were adamant that they trusted the traditional media at most. Again, television was the preferred medium, followed by radio – 12.2%, and press - 9.8%. Merit deserved the comparison between consumption (62.2%) and trust (23.7%) in the online news agencies and websites. Similar was the situation with the social networks. Close to half of all respondents (48.1%) were regular users of Facebook, Tweeter, Instagram, Linkdin, etc., but only 12.4% stated that they trusted social networks. Although only 9.2% used as main source of information social media (blogs, YouTube, etc.), the trust in them was higher than in social networks (14.5%).

Most of the respondents were impressed by the level of mobilization of the Bulgarian media and by their responsiveness to the emergency. Among the negative opinions about the media coverage of COVID'19 was the pyramidal paramilitary model of information, as well as the lack of criticism of the official information and the limited debate with various opinions about the corona virus. Some of the respondents explained that the media concentrated mainly on the spread of the virus and did not pay enough attention to other factors following the pandemic, such as the economic challenges in Bulgaria and throughout the world. When asked how they assessed the media coverage of the COVID'19 pandemic, the largest percentage of the respondents claimed that negative news prevailed, which shocked the people (51.6%). For 46.2% of the interviewees, the media paid too much attention to the situation with the corona virus and became intrusive. By doing so media helped to instill fear in society. Regarding the psychological impacts, the largest number of respondents - 59.7% answered that the over saturation with information about COVID'19 depressed them.

3.3 Challenges and Opportunities Facing Telecommunications Industry

For many sectors in the economy (e.g. tourism, restaurant services, retail trade, transport and logistics, media, entertainment industry and film production) the period of forcibly imposed anti-epidemic measures had adverse effects such as temporary or even permanent termination of activities, lay-offs and furloughs, financial losses, re-organization of the supply and distribution chains, filing for bankruptcy. However, for the telecommunications industry in Bulgaria these measures had just the opposite effect. From the very first days of social isolation, mobile operators and internet providers enjoyed peak consumption of their services and got higher income. The explanation for this was connected with the attempts to continue casual activities online (e.g. work and education from home, online shopping, payment of bills, medical consultations, etc.). High levels of distress in people as a result from the abrupt changes in lifestyle, as well as the ban to travel without a valid reason also contributed to overloading the capacity of transmission networks. Their service had to ensure not only the connectivity of their subscribers, but also the ability to operate the entire education system, the public administration, all digital services and the work of a huge number of people from home.

The process of adaptation of the Bulgarian mobile operators to the new situation was speedy and flexible. With this regard, three main reactions were outlined:

- Ensuring the safety of company employees and customers and encouraging the use of existing and the development of new digital services. Providing for additional opportunities for communication, connectivity, entertainment and education was the unifying policy of the three leading mobile operators in Bulgaria (*A1*, *Vivacom* and *Telenor*). Telecommunications companies used the period of forcibly imposed social isolation to promote online purchases of devices, gadgets and accessories by using mobile apps, or discounts, or gifts, or other incentives. Besides, it was a favourable period to promote their own mobile platforms for payment services of all kinds.
- Social solidarity and online security in times of isolation. All mobile operators, in cooperation with the National Operational Headquarters, maintained free of charge the mobile application ViruSafe. Digital services were developed to serve business clients how to run and manage work from home. In April *A1* came up with a free of charge application *A1 Guard* for prevention of telephone fraud – first of a kind in Bulgaria [8]. Its release was in line with the warnings of the Ministry of Interior for activating telephone fraudsters in the conditions of social isolation.
- Donation during quarantine - shared responsibility. Telecommunications companies in Bulgaria started wide-scale donations activity during the state of emergency, built on the principle of shared responsibility. Consumers "invested" funds in the company, which in turn "shared" the gains with the society to overcome the crisis. The telecoms were very active in the direct assistance to health authorities and hospitals to effectively deal with the corona virus infection and support for schools and students in adapting to distance learning from home. The overall amount of the reported donations by the telecommunications companies to fight against COVID'19 exceeded 1 mln BGN (1 320 000 BGN) [9].

During the state of emergency, the telecommunications companies operated under the increased control of the Communications Regulation Commission – the regulatory body for postal services and electronics communications and in the shadow of the legal possibility to use mobile data for uncontrolled monitoring of the publics [10].

4 Conclusion

The study of the virtual communication in the social isolation of the two-month state of emergency regarding COVID'19, led to some major conclusions, namely:

- The restructuring of the communication in the situation of social isolation caused a variety of challenges to the media system and telecommunications sector, which requires new approaches to their management and regulation.
- The effects of virtual communication during the social isolation presupposed the importance of the media system and the telecommunications industry as core determinants in the social management of the processes in extraordinary situations.
- The period of forced social isolation provided an opportunity for media and telecommunications companies to convince society of their responsibility and solidarity.

In crisis situations, such as the one with COVID'19, the effects of virtual communication in social isolation are indicative.

Acknowledgements. The paper has been developed within the framework of COST Action CA16226 *Indoor Living Space Improvement: Smart Habitat for the Elderly* (SHELD-ON) of the European Commission, supported by the research projects KP-06-COST/5-18/06/2019 and KP-06-M35/4-18/12/2019 of the National Scientific Fund of Bulgaria and the Program *Young Scientists and Post Docs* of the Bulgarian Ministry of Education and Science.

References

1. United Nations. https://www.unocha.org/sites/unocha/files/Global-Humanitarian-Response-Plan-COVID-19.pdf
2. European Commission. https://ec.europa.eu/info/sites/info/files/cwp-2020-adjusted_en.pdf
3. COVID'19: Edinen informatsionen portal (in Bulgarian). https://webcache.googleusercontent.com/search?q=cache:WYbmUQ_RSyQJ:https://coronavirus.bg/bg/231+&cd=8&hl=en&ct=clnk&gl=bg
4. State Gazette. https://dv.parliament.bg/DVWeb/showMaterialDV.jsp?idMat=147150
5. Alpha Research. https://alpharesearch.bg/post/969-otrajenie-na-purvia-etap-ot-krizata-s-koronavirusa-vurhu-politicheskata-kartina-v-stranata.html
6. Novini.bg. https://novini.bg/bylgariya/obshtestvo/597366
7. Gallup International: Center for Public and Political Studies. https://www.gallup-international.bg/43323/%d0%b0ttitudes-towards-the-coronavirus-situation/
8. Investor. https://www.investor.bg/telekomunikacii/457/a/tehnologiite-v-borba-s-telefonnite-izmamnici-301664/
9. A1. https://www.telenor.bg/bg/news/; https://www.vivacom.bg/bg/home; https://www.a1.bg/medien-tsentar-informatsia-za-mediite/mc/index/ma/view/infoblock_id/2744
10. Communications Regulation Commission. https://crc.bg/bg/novini/1304/komisiqta-za-regulirane-na-syobshteniqta-preporychva-na-potrebitelite-da-se-syobrazqt-s-tendenciite-v-internet-trafika

Managing Strategic Participation Through Design Principles: A Model for Value Co-Creation in Service-Based Organizations

Rocío Salvatierra(✉)

Pontificia Universidad Católica de Chile, Santiago, Chile
rdsalvatierra@uc.cl,
rocio.salvatierra@alumni.esade.edu

Abstract. Even though it is well-known that co-creation generates opportunities and benefits for organizations, there is still a void in literature concerning participation management practice. This generates mistrust and blurred understandings about it's strategic relevance. This paper presents three models to manage participation through design principles to close the gap between participation as a mindset and participation as a practice of value co-creation. Participation is a scale and each level refers to the depth in which actors are involved in a certain process, initiative or project and, therefore, the level of engagement they reach. It must be strategically managed through time and with the full spectrum of actors that build a service-ecosystem so as to achieve collaboration and multidimensional value.

Keywords: Participation · Management · Strategy · Co-creation · Service design

1 Introduction

We are in an era where management logics have moved from a traditional top-down to a disruptive bottom-up model. Liz Sanders states that *"today it's not "business as usual" anymore"*. The rules have changed and continue to change. The new rules are the rules of networks, not hierarchies" [1]. Public and private organizations are gradually shifting from company-centric or product-centric perspectives to a customer-centric or experience-centric perspective [2]. Also, a paradigm shift is taking place in power structures. Old ways of power are closed and are leader-driven. New power is participatory, open, held by many and promoted by hyperconnectivity and digitalization. The aim is to channel new power and not to hoard it [3]. Companies have spent the last century managing efficiencies, but this century is about managing experiences [4] and ecosystems. To manage both, an "outside-in" view is essential. It implies to move forward from the "inside-out" logic to create real value with and for everyone [5].

These paradigm shifts are happening in a world where most organizations are service-based or are seeking to increase value through servitization [6]. What is a service today? If services are a transformation of the state of reality through the interaction of users and providers [7], then services may be seen as experiences.

W. Karwowski et al. (Eds.): IHSED 2020, AISC 1269, pp. 69–76, 2021.
https://doi.org/10.1007/978-3-030-58282-1_12

Services are co-produced between customers and providers at the moment of delivery. They are perishable and inseparable because they do not exist without all the actors involved [8]. Unlike products, services do not exist without co-production. But this is not limited to a customer-provider perspective. Co-production of services must be understood as a matter of service-ecosystems. A group of actors articulated to deliver multidimensional value. This leads to something obvious but that should not be overlooked, services are intrinsically participative.

Services are highly dependent on human factors such as expectations, motivations, emotions and perceptions. Not only of customers, but of everyone involved in service delivery. It becomes essential for service-based organizations to generate mechanisms that allow them to systematically integrate these human factors to achieve more agile, flexible and permeable organizations.

Managing participation through design principles can be a way to embrace those human factors and turn them into knowledge for decision-making. An outside-in perspective and an experience-centric management logic can be achieved through participation management.

There are three models to close the gap between intention, execution and concrete results of participation management: (1) The Participation Management Scale (IOE-CCS), [9] (2) A Three-variable framework (P.A.L) [9] and (3) An Actionable Model for designing specific participation instances. The three models were applied retro-spectively to a case study in the Pediatric Unit of the Clinical Hospital of Pontificia Universidad Católica de Chile (PUC).

Management and Design have responded to these paradigm shifts through the generation of diverse methodologies, techniques and tools such as Sociocracy, Agile, Lean, user-led innovation, user-centered design and currently, the protagonist is co-design or co-creation. Designers have evolved from involving users as passive agents to partnering with users throughout the design process [10].

In management, value co-creation was promoted by Prahalad and Ramaswamy as *"the process by which products, services and experiences are developed jointly by companies and their stakeholders"* [11] However, there is still a void in literature referring to aspects that enable service co-creation [12]. Value co-creation, co-production, co-design, customer innovation, cooperation, collaboration, co-consumption have been used as synonyms leading to an increase in mistrust and confusion about their benefits. What is clear, is that all concepts hold the intention to open up to participation, involvement and engagement as a common factor [13].

Oertzen, et al., through a systematic literature review of 80 articles detected that: (1) Involvement, engagement and participation are crosswise pre-requisites for all concepts. (2) Few authors employ these concepts referring to multi-actor emphasis, entire service process and emphasis on resource integration [13]. In this paper, Participation will be understood as the global concept that encompasses the other expressions.

Participation management through Design is a way of approaching the entire service process. It focuses on the strategic integration of the full spectrum of actors that build an internal and external service ecosystem. It is a mindset [14], that flows in parallel to the organization's operations and the methodology they decide to use: Design Thinking, Double diamond, Lean, Six Sigma, Scrum, DevOps, among others. This understanding of participation management goes beyond the product and service development phase. It explores participation as an organizational culture and philosophy. It requires constant involvement of partners, suppliers, clients, customers, users, competitors, other companies, communities, government, organizations, foundations, experts, authorities and employees. This point of view shapes participation management as a strategy matter, as an enabler for decision-making, organizational transformation and business impact.

Participation is not an absolute, it is not a dichotomous or binary variable between the existence and the nonexistence of participation. Participation is a scale, and each level refers to the depth in which actors are involved in a certain process and the level of engagement they reach (Fig. 1).

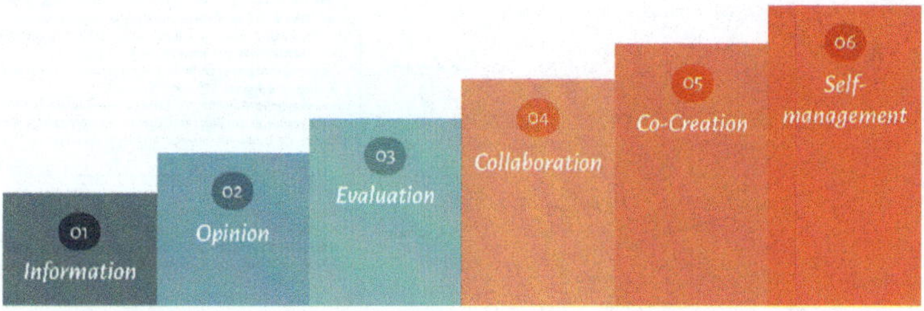

Fig. 1. The Participation Management Scale (IOE-CCS). Based in Arnstein (1969), Weidemann & Femers (1993), Turkucu (2008), Hart (1992), Heimans and Timms (2014)

(1) Information: The minimum level of participation. What has already been decided by someone is merely informed and will be implemented. A hierarchical and unidirectional relationship is generated between the decision-makers and the actors that will be impacted. **(2) Opinion:** It works under the logic of feedback. The participation management team collects the opinions of those involved without necessarily accepting their suggestions or concerns. It includes proposals previously built and imposed by the decision-makers. **(3) Evaluation:** At this level the proposal is submitted to evaluation and criticism by the impacted actors. The team designs mechanisms to integrate those criticism. The focus at this level is to interpret and get as close as possible to the requests and expectations of the actors involved, making the modifications that are possible and pertinent. **(4) Collaboration:** At this level of participation the team works together with the actors impacted. There is still a kind of surveillance or guidance, but the actors are already part of the process. The interaction is bidirectional. The process of collaborating generates value for the team and for the actors impacted as well. **(5) Co-Creation:** At this level the proposal, project or process is managed peer to peer between the team and the actors impacted. The parties are involved at the same level in decision-making. Each one contributes from their experience and expertise. At this level there is a high sense of belonging. Everyone is a builder of the desired future. **(6) Self-Management:** It is the maximum level of participation. Management is completely delegated to the actors involved. The proposal comes exclusively from the other(s) with a minimal intervention of the team. Actors are fully empowered. These levels of participation can be adopted within the same project in different phases, by different actors and at different depths depending on the strategy required to carry out the objectives.

The second model presented is to manage global participation with a strategic approach. Reflections must be guided through 3 variables: (1) Phases and context (2) actors that must participate and (3) levels of involvement and engagement (Fig. 2).

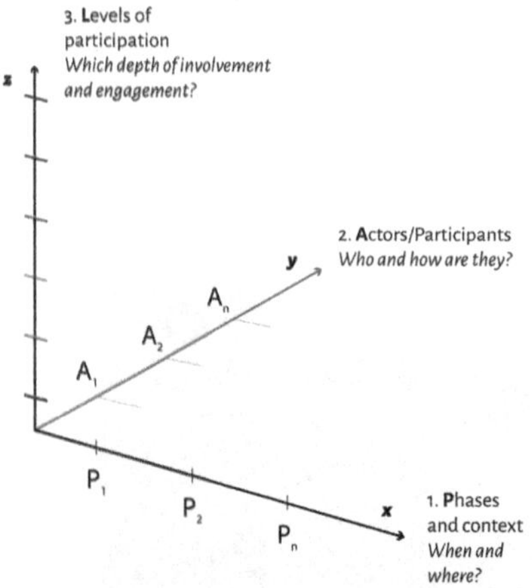

DIMENSION	POSSIBLE REFLECTIONS
1. P - x axis	· What are the **circumstances** in which the project or process is **framed**? · What are the cultural, social, political, intellectual, economic, relational, technical and technological **aspects that surround it**? · What specific **phase** is the **project in**? · Who is **relevant** to participate in that **phase**? · How can participation be generated in **each of the phases of the methodology** used: research, definition, ideation, testing, implementation, continuous improvement?
2. A - y axis	· **Who** are the participants to be involved? (users, end consumers, clients, employees, directors, managers, private companies, competitors, experts, citizens, communities, authorities, public institutions, politicians, NGOs, etc.). · What **motivations and interests** do they have? · Are they **detractors or promoters** of the project or initiative? · Do they have **power**? What kind of power? · How **evangelized** are they in relation to user-centered methodologies ? · What **resources** can they **contribute** to the project with? (intellectual, experiential, emotional, physical, economic, communicational, etc.).
3. L - z axis	· What level of participation will be **sought to get the maximum benefit** from both; the process and the result of the project? · What level of participation will **reach the greatest potential** of each actor? · What level of participation is **relevant for the phase** we are in?

Fig. 2. Three-dimension framework to manage participation

This framework unveils the importance to strategically design, plan, execute and analyze participatory instances. The process requires the orchestration of these instances deliberately over time [15]. To deepen on participatory instances, the third model is proposed to close the gap between participation as a mindset and participation as practice. This Actionable framework (Fig. 3) has seven aspects that need to be designed sequentially and iteratively.

(1) Objectives: the specific objective or objectives of the participatory instance. First of all, specify why participation is needed and the outcomes and/or outputs that are expected to be obtained. They may have to do with socializing certain information, generating a certain atmosphere or mood within the actors, generating specific results like ideas, diagnoses, agendas, projections, products and services, or with only mobilizing dialogue and debate, defying established dynamics, or making some consensual decisions about some issue, among others. **(2) Actors/Participants:** Who is pertinent to participate in that instance according to the defined objective(s) and the global strategy of participation. Foresee their potential contribution, their interests, their areas of expertise or non-expertise, their role, their impact on others behavior, etc. **(3) Dynamics, mechanisms and didactics:** How the objectives will be achieved? It is necessary to think about the general narrative of the participatory instance, the sequence, the work structure, the rhythm, climax, and the overall experience. **(4) Activities, tools and platforms:** What are the best means to accomplish the outcomes and outputs?. Digital platforms, game dynamics, maps, canvases, models, storytelling, mockups, prototypes, frameworks, etc. What resources and materials will be needed?. What will be the roles and tasks of the team and the participants? Leading, outspeaking, accompanying, recording, registering, keeping pace and time and/or taking visual

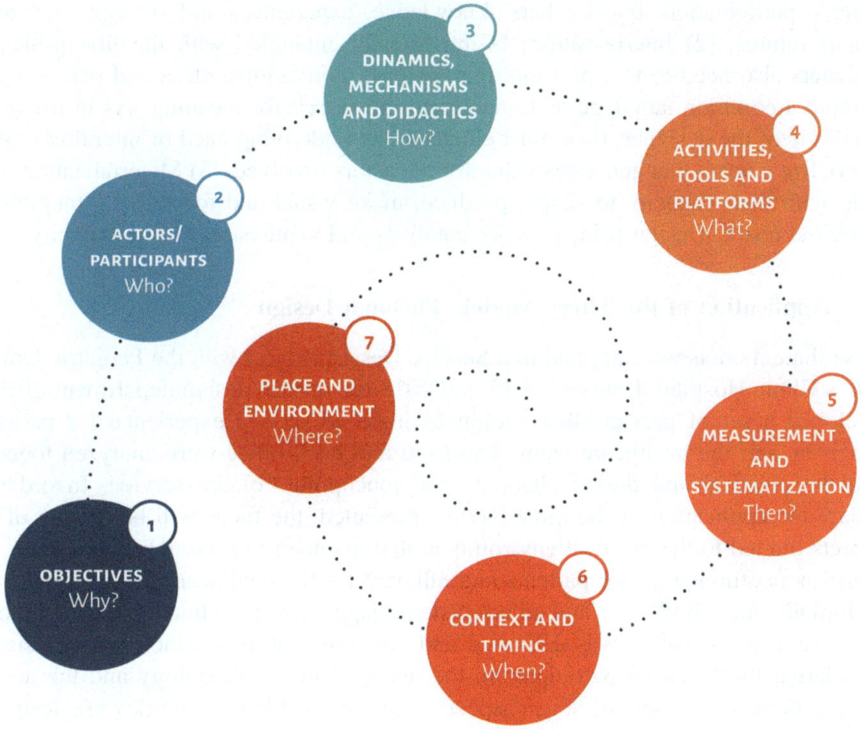

Fig. 3. Accionable framework to manage participation. 7 aspects to design instances

documentation. **(5) Measurement and systematization:** measuring the success or failure of the participation instances is a key aspect to a strategic vision. This involves the previous definition of expected results, KPI's and systematization models. **(6) Context and timing:** The right moment for the instance must be thought. "When" is asking the appropriate day, hour, time, date and the consideration of relevant milestones that can affect the objective of the participation. **(7) Place and environment:** Place and environment: In labor context, in a relaxation place, in neutral or partial location are some of the aspects that must be taken into account.

2 Why Using Design Principles to Manage Participation?

Basically, because of the three principles on which the designer's role is based; Facilitation, Interpretation and Materialization [15]. To design is to project what does not exist using a creative but methodical approach. In this era, we are projecting a world made of digital and physical experiences based in service-ecosystems. Therefore, designers capitalized their strengths to help building that desired future through these three principles. (1) Facilitation: Designers are challenged to be facilitators of change, which means to enable others to design and achieve their objectives [16, 17]. Designers support processes of analysis, discussion, dialogue, reflection, learning, exploration, and sense-making. This is the first principle of design that can be applied to Manage strategic participation; boost others' knowledge, experiences and strengths to create desired futures. (2) Interpretation: Is intrinsically entangled with the first principle. Designers also need to interpret multiple visions, motivations, ideas and perspectives. Promote a common language and encourage the search for meaning. As in the traditional role of the designer, they must elucidate the underlying need or intention to then materialize a result that generates value for the actors involved. (3) Materialization: Has to do with their capacity to shape, produce, make visual and formalize a tangible or intangible result. Design principles are catalysts and synthesizers of complexity.

2.1 Application of the Three Models Through Design

These three models were applied in a Service Design Project with the Pediatric Unit of PUC's Clinic Hospital. Between 2015 and 2018 the service design department of PUC developed a set of projects that sought to improve service experience for patients, caregivers and the healthcare team. The front and back office were analyzed together with the physical and digital channels and touchpoints of the service. In order to explain the application of the three models presented, the focus will be on one of the projects limited to the physical environment that intended to personalize and make the rooms more stimulating for patients and efficient for the healthcare team.

Initially, a methodology and an ecosystem mapping were defined. Then, the global objective of participation was delimited and the participatory instances were designed considering the levels of participation, the phase of the methodology and the actors' profiles. For each session the seven aspects of the actionable framework were designed. After a retrospective analysis about the strategic management of participation, two categories of participation benefits emerged; Outputs and Outcomes.

Outputs: (1) Strategic participation management generates more accurate diagnosis of the problems, challenges and opportunities. (2) Leads to new and better ideas through the integration of diverse views, realities, expectations, motivations and pains. (3) Allows to foresee fastly and cheaply possible technical, functional, formal and definition problems of the products (touchpoints), service or project. Political, technical, social, cultural, economic barriers, among other obstacles, can be seen in advance. (4) Obtains more usable products, services, projects and initiatives because it faithfully adjusts to the real needs of the service-ecosystem and the context where the project lives. (5) Participation management generates more implementable products, services and projects. It increases the sense of belonging and ownership of ideas and proposals. This enables synergy and the continuity of the project.

Outcomes: (1) Managing participation strategically evangelizes in people-centered design methodologies as a way to face challenges in the future. (2) Anticipates the impact on the expanded ecosystem. For example to anticipate the political negotiations that should be carried out with high-level decision-makers. (3) Drives and thrusts change within organizations. Through participatory instances, initiatives within an organization can be socialized and scaled easier and more transparently through a strategic management of participation. (4) It installs participatory practices as a top-of-mind option to develop continuous improvement of the products, services and experiences generating a virtuous cycle.

3 Conclusions

In this paper, three models to implement participation management through design were presented. The Management participation scale (IOE-CCS) promotes the understanding of how to manage the different levels of participation. The Three-dimension framework that helps to envision the global strategy of participation through time considering phases and context, actors of the service ecosystem that must participate and the depth in which they will participate. Finally, the Accionable framework that develops the seven aspects that must be detailed to put into practice participation instances. These models seek to leave behind the blurred limits of strategic participation management. It is imperative to comprehend participation as a mindset that focuses on the entire service process and the resources that the full spectrum of the service-ecosystem can provide to generate multidimensional value.

Considering that the world is shifting to bottom-up logics, experience-centric perspectives, new power structures and outside-in strategies, participation management through design principles might be the way to systematically embrace these changes. Human factors in service ecosystems can be turned into knowledge for decision-making guiding us towards flexibility, adaptation and transformation of organizations.

References

1. Sanders, E.B.-N.: Postdesign and participatory culture. In: Proceedings of Useful and Critical: The Position of Research in Design, pp. 87–92 (1999).
2. Prahalad, C. K. Ramaswamy, V.: The Co-creation connection. Strategy Bus. (27), 50–61 (2002). Second Quarter
3. Heimans, J. Timms, H.: Understanding new power. In: Harvard Business Review (2014).
4. Prahalad, C.K. Ramaswamy, V.: The Co-creation connection. Strategy + Bus. (27), 50–61 (2002). Second Quarter
5. Manning, H.P., Bodine, K.: Outside In: The Power of Putting Customers at the Center of Your Business. Amazon Publishing, Seattle (2012)
6. Vandermerwe, S., Rada, J.: Servitization of business: adding value by adding services. Eur. Manag. J. **6**(4), 314–324 (1988)
7. Gadrey, J.: The characterization of goods and services: an alternative approach. Rev. Income Wealth **46**(3), 369–387 (2000)
8. Kotler, P., Armstrong, G.: Marketing. In Versión para Latinoamérica, 11th edn. Ciudad de México. Pearson Educación, México (2007)
9. Salvatierra, R.: Desencriptar, formalizar y sistematizar la co-creación. Propuesta exploratoria del Diseño Participativo entendido como escala y estrategia metodológica. In: Mollenhauer, K., Soto, M., Salvatierra, R. (eds.) Actas del 1er Congreso de Diseño de Servicios en el III Congreso Chileno de Diseño de Servicios. Pontificia Universidad Católica de Chile, pp. 58–65 (2017)

10. Sanders, E.B.N., Stappers, P.J.: Co-creation and the new landscapes of design. Co-Design **4** (1), 5–18 (2008)
11. Ramaswamy, V.: Co-creation of value: toward an expanded paradigm of value creation. Mark. Rev. St. Gallen **6**, 11–17 (2009)
12. Frow, P., Nenonen, S., Payne, A., Storbacka, K: Managing co-creation design: a strategic approach to innovation. Br. Acad. Manage. 1–21 (2015).
13. Oertzen, A., Odekerken-Schröder, G., Brax, S., Mager, B.: Co-creating services—conceptual clarification, forms and outcomes. J. Serv. Manag. **29**(4), 641–679 (2018)
14. Sanders, E.B.N., Stappers, P.J.: Convivial Toolbox. BIS Publishers, Amsterdam (2012)
15. Aguirre, M., Agudelo, N., Romm, J.: Design facilitation as emerging practice: analyzing how designers support multi-stakeholder co-creation. She ji J. Des. Econ. Innovation **3**(3), 198–209 (2017)
16. Tan, L.: Understanding the different roles of the designer in design for social good. a study of design methodology. In: DOTT 07 (Designs of the Time 2007) Projects. Doctoral thesis, Northumbria University (2012)
17. Thackara, J.: In the Bubble: Designing in a Complex World. The MIT Press, Cambridge (2005)

Mixed Reality Application with MR Glasses in the Interaction Exhibition of Mortise and Tenon Structure of Chinese Traditional Furniture

Dehua Yu(✉)

Beijing Institute of Technology,
No. 5 Zhongguancun South Street, Haidian District, Beijing, China
yudehuabit@163.com

Abstract. The aim of this interaction exhibition of mortise and tenon structure of Chinese traditional furniture is to display the exact and exquisite structure of mortise and tenon and increase the interaction between visitors and the exhibits. All the visitors have chances to participate in the fixing and unfixing process of real mortise and tenon directed by virtual installation instruction. The first plan is to project the installation instruction animation on the wall with the aid of sensor technology. However, the interaction animation can only be displayed on the wall. The second plan is mix real mortise and tenon structure and virtual installation instruction together, while participants would immerse in the real and virtual surroundings with the aid of MR glasses. The experience process is easy and interesting without pressure, although some mortise and tenon is complicated to a certain extent.

Keywords: Mixed reality · Chinese traditional furniture · Mortise and tenon · Interaction

1 Introduction

With the development of VR/AR/MR technology, the traditional exhibition of cultural heritage would have more possibilities to display vivid exhibits and interact with visitors. Take Chinese traditional furniture for example, the exact and artful mortise and tenon is one important part of Chinese traditional furniture, but mortise and tenon structure is hidden inside the furniture [1]. Without modern interaction technology, visitors had no choice to understand the inner structure in deep observation in the past. Picture show and real mortise and tenon structure is the only way to display in past ordinary exhibition.

Nowadays, visitors would have more interactive ways to interact with the real and virtual mortise and tenon structure with the help of new technology. For example, two APPs "Wood Joints" [2] and "Traditional Chinese Furniture" [3] which can be used on phones and pad. People can observe different perspectives of mortise and tenon sliding fingers across the screen, and finish some fixing and unfixing process which is interacting and attractive. The exhibition "the Charm of Mortise and Tenon" which was

W. Karwowski et al. (Eds.): IHSED 2020, AISC 1269, pp. 77–82, 2021.
https://doi.org/10.1007/978-3-030-58282-1_13

held in Beijing Science and Technology Museum in 2018 [4], have more interaction between visitors and furniture. Some interaction use real mortise and tenon structures and some have the aid of virtual reality. One example is the pad screen, which has the sensor capturing the ancient painting on the wall. When the screen capture the painting, a 3D model of Chinese traditional furniture will appear on the screen, and the visitor can rotate the 3D model to observe all the perspectives with fingers (Fig. 1).

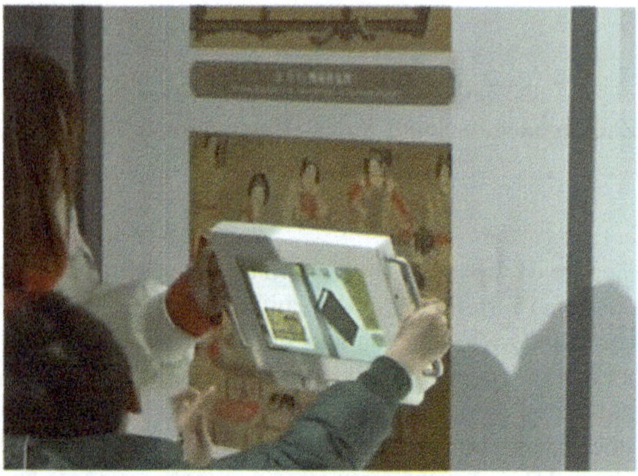

Fig. 1. The interaction with Chinese traditional furniture with the aid of pad

The aim of this study is to increase the interaction experience, and direct visitors to finish the fixing and unfixing process easily with interest without pressure with the help of mixed reality and MR glasses. The real mortise and tenon structure and virtual installation instructions would mix together in front of visitors and participants. In this interaction exhibition, several Chinese traditional high and low dual-purpose tables would be displayed for their exact and unique mortise and tenon structures.

2 Unique Mortise and Tenon Structures of High and Low Dual-Purpose Tables

Chinese traditional high and low dual-purpose tables are unique tables with two functions: high table and low table. The two functions can be transferred freely by fixing and unfixing the high legs with the aid of unique mortise and tenon structures. There are kinds of different mortise and tenon structures to form different high and low tables with similar shapes. That is the charm of mortise and tenon of Chinese traditional furniture.

For example, the poplar high and low dual-purpose table (Fig. 2) is absolutely different from the elm high and low dual-purpose table (Fig. 3) [5]. The two tables have different mortise and tenon structure, although the shapes of them have some similarities. The poplar table's high legs are divided into two pairs, which can be taken away, while the elm table's legs are also divided into two pairs, which can be folded under the short table.

Fig. 2. The mortise and tenon structure of high legs of the poplar high and low dual-purpose table

Fig. 3. The mortise and tenon structure of high legs of the elm high and low dual-purpose table

3 Real Mortise and Tenon and Virtual Installation Instruction Without MR Glasses

The first plan to show the interaction is to project animation of installation instruction on the wall with the aid of sensor technology. When visitors approach the exhibits, the animation would direct them to touch the parts of furniture, then the sensors would detect this action, triggering the play of the installation animation to direct visitors to fix or unfix mortise and tenon structures of Chinese traditional furniture (Fig. 4). No matter how complex the mortise and tenon structure is, visitors can easily install it without pressure with the help of guided animation.

This interaction scheme mixed real mortise and tenon structure and virtual installation instruction together. The weakness is that the installation instruction animation can only be projected on the wall, and the virtual animation is absolutely far from the real mortise and tenon, which is easily distinguished for visitors.

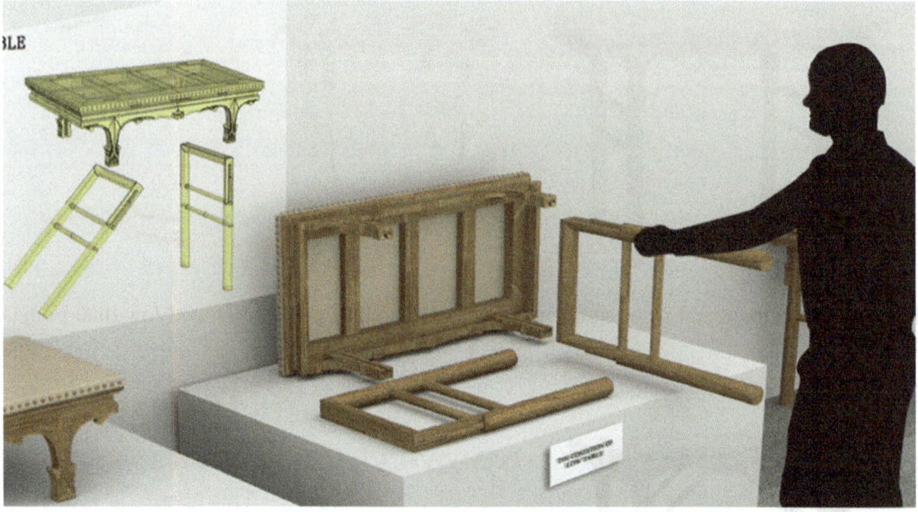

Fig. 4. The fixing and unfixing process directed by the installation animation projected on the wall

4 Real Mortise and Tenon and Virtual Installation Instruction with MR Glasses

The second plan to show the interaction is to mix the real mortise and tenon structure with virtual installation instruction together with the aid of MR glasses. The advantage of MR glasses (for example holoLens 2 [6]) is to help visitors to interact with the exhibits more effectively and interactive. The visitors who want to participate in the fixing and unfixing process of mortise and tenon, would wear MR glasses, after that, the virtual installation instruction would be appeared around (Fig. 5), directing them to experience the installation process. The participants would follow the designed steps to complete the installation step by step with interest without pressure.

Some complicated mortise and tenon structures even have more than five steps, and have more details to operate [7]. However, there is no need to worry about it. The process would be interesting and easy, the interaction guidance make the whole experience comfortable and easy, and all the participants would have good experience wearing the MR Glasses.

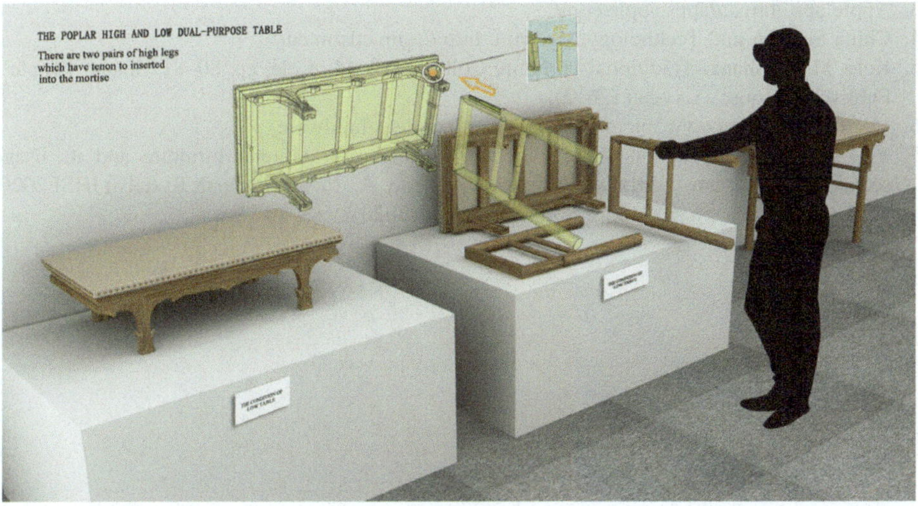

Fig. 5. The fixing and unfixing process directed by virtual installation instruction wearing MR glasses

5 Conclusion

The study focused on the interactive ways and effects with mixed reality technology and MR glasses in the exhibition of mortise and tenon of Chinese traditional furniture. Mixed reality provides more interactive styles between visitors and the exhibits in the exhibition.

Because of the particularity of mortise and tenon of Chinese traditional furniture, which is hidden inside the furniture, mixed reality and MR glasses can give visitors penetrating ability to observe, immerse and even participate in installation process, which breaks the bottleneck of traditional graphic display and gives visitors more vivid and interesting interactive experience.

The interaction exhibition of mortise and tenon structure of Chinese traditional furniture is a positive attempt to introduce mixed reality and MR glasses in the field of cultural heritage display and spread, especially the display of exact and artful mortise and tenon structure of Chinese traditional furniture.

Acknowledgments. This paper is supported financially by MOE (Ministry of Education in China) Project of Humanities and Social Sciences Research Youth Fund (Project No. 19YJC760144).

References

1. Yang, Y.: Study of Ming-style Furniture, p. 40. Beijing Construction Industry Press, Beijing (2002)
2. TagUX Design. https://taguxdesign.com/

3. Apple app. https://apps.apple.com/
4. China Science and Technology Museum. http://cstm.cdstm.cn/
5. Kele, M.A.: Shanxi Traditional Furniture Collected by Ma Kele, pp. 80–81. Shanxi People's Publishing House, Taiyuan (2012)
6. Microsoft. https://www.microsoft.com/zh-cn/hololens/hardware
7. Yu, D.: Virtual reduction and interaction of chinese traditional furniture and its usage scenarios. In: Ahram, T., Karwowski, W., Vergnano, A., Leali, F., Taiar, R. (eds.) IHSI 2020. AISC, vol. 1131, pp. 552–558. Springer, Cham (2020)

Exploring Annotations and Hand Tracking in Augmented Reality for Remote Collaboration

Tiago Madeira(✉), Bernardo Marques, João Alves, Paulo Dias, and Beatriz Sousa Santos

IEETA, DETI, University of Aveiro, Aveiro, Portugal
tiagomadeira@ua.pt

Abstract. Collaboration among remotely distributed professionals is often required in a maintenance context. Professionals need mechanisms with adaptive capabilities to enable knowledge transfer, since the necessary experience and expertise are usually spread among different professionals. To provide a shared understanding, Augmented Reality (AR) has been explored. In this paper, which is part of ongoing research using a human-centered design approach with partners from the industry sector, we describe a framework using annotations to improve the shared perceived realities of different professionals. The framework allows manually freezing the on-site professional context and sharing it with a remote expert to create annotations. Then, the on-site professional can visualize instructions through aligned and anchored annotations, using a see-through Head Mounted Display (HMD). In addition, annotations based on real-time video stream from a remote expert are also available. Hand tracking is used to manipulate the annotations, enabling the adjustment of their position and scale in the real-world according to the context, thus enriching the on-site professional experience and improving visualization of information while conducting maintenance procedures suggested by a remote expert.

Keywords: Remote collaboration · Distributed professionals · CSCW · Maintenance · Framework · Augmented Reality · Annotations · Hand tracking

1 Introduction

Remote collaboration has important value in the industrial domain, since the necessary experience and expertise are usually distributed among different professionals [1, 2]. The field of Computer-Supported Cooperative Work (CSCW) has been concerned with understanding and designing solutions to support remote collaboration [3]. Augmented Reality (AR) can be used to enable knowledge transfer from professionals unavailable on-site [3, 4], since it can provide a common ground environment, i.e., shared understanding for situation mapping, allowing identification of issues, and making assumptions and beliefs visible [5]. A number of earlier studies have explored different annotation mechanisms for remote collaboration [6]. Gauglitz et al. [7] explored how to

W. Karwowski et al. (Eds.): IHSED 2020, AISC 1269, pp. 83–89, 2021.
https://doi.org/10.1007/978-3-030-58282-1_14

stabilize annotations on the real-world using AR tracking in handheld devices. The authors found that users preferred stabilized annotations when compared to the unstabilized alternative.

Kim et al. [8] proposed the use of a manual freeze method to prevent annotations from being anchored to a wrong object, while an on-site user unexpectedly changes the viewpoint. This way, a remote user could manually freeze the live video received from the on-site user and draw on the still video frame, rather than in a live video. Then, s/he can again return to the live video.

Kim et al. [9] introduced a novel auto-freeze method, in which the freezing and unfreezing functions were integrated with the drawing functions to reduce the interactions required. The live video was automatically frozen when a remote user started drawing, and unfrozen when the drawing stopped. Then, a local user could use a Head Mounted Display (HMD) to visualize the annotations, while manipulating physical objects with hand gestures. The authors compared this method with the manual freeze concluding that the auto freeze method was easier to use and speeded up the annotation drawing process.

In this paper, we aim to provide interaction methods to improve cooperation among distributed professionals by using a see-through HMD and hand tracking for visualization and manipulation of different types of AR annotations. Our work, which is part of ongoing research using a human-centered design approach with partners from the industry sector, follows a set of requirements for the design of collaborative solutions described in previous work [reference blind for revision purposes]. The paper makes the following contributions:

- Describes an AR framework developed for remote collaboration;
- Presents different types of annotations as additional layers of information;
- Proposes the use of hand tracking for manipulation of spatial annotations.

2 An Augmented Reality Framework Supporting Annotations

The framework developed within this work is designed to support scenarios that require know-how and additional information from professionals unavailable on-site, as is the case of maintenance scenarios. Therefore, it focuses on two types of users: on-site technicians and remote experts.

Framework Description

Maintenance procedures might be performed in multiple contexts i.e., different equipment and/or environments. Therefore, the framework was designed to function in a diverse set of conditions, using shared 2D annotations that can be placed onto existing equipment, instead of 3D virtual models, thus providing a more generic approach, not depending on the existence of pre-defined 3D models.

Figure 1 presents an overview of the framework. When facing unfamiliar problems, on-site technicians can use the camera of a HMD to manually capture (freeze) the context of the problem and send it to a remote expert for context understanding and to enable the instructions/feedback accordingly i.e., inform where to act and what to do using annotation features (in a laptop or desktop computer), allowing the creation of layers of additional information to illustrate difficulties, identify specific areas of interest or indicate questions. Afterwards, the on-site technician receives the instructions showing the suggestions from the remote expert, which can be visualized using the see-through HMD, overlapping the real-world context, while facilitating the performance of procedures intervention through a handsfree approach. The instructions can be presented using aligned and anchored annotations or through video stream displaying the creation process of the annotations. At any moment, the position of the annotations can be moved using hand tracking with the intent to provide a natural interaction mechanism. The overall collaborative process can be repeated iteratively until the task is successfully accomplished. The proposed framework also supports audio communication despite not being the focus of this research.

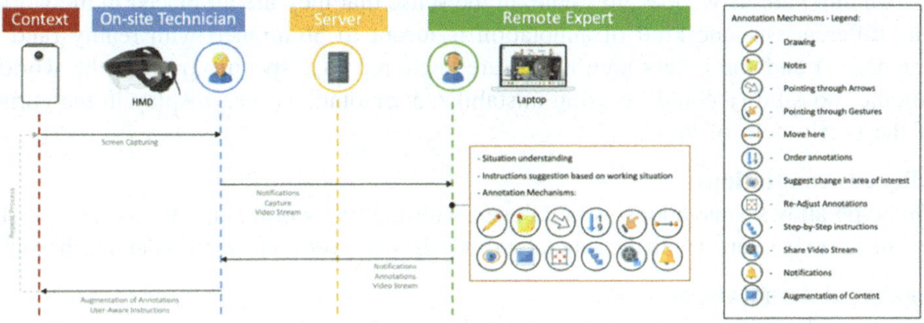

Fig. 1. Framework overview. Goal: Allow an on-site technician to capture the real world using a see-through HMD and share with a remote expert for analysis and to provide instructions through the creation of annotations. Finally, the on-site technician can view the real world augmented with the annotations and perform an intervention

Currently, the framework can be used on devices running Android, MAC OS, or Windows. Since on-site technicians are constantly moving, it seems adequate to equip them with HMD or handheld devices. Regarding the remote expert, we support multiple types of devices, including a laptop or desktop computer, an interactive projector, or an handheld device (Fig. 2). The framework was developed using the Unity 3D game engine, based on C# scripts. To place the AR virtual content in the real-world environment, we used the Vuforia library. Communication between the different devices was performed over Wi-Fi through specific calls to a PHP server responsible for storing and sharing the enhanced pictures accordingly.

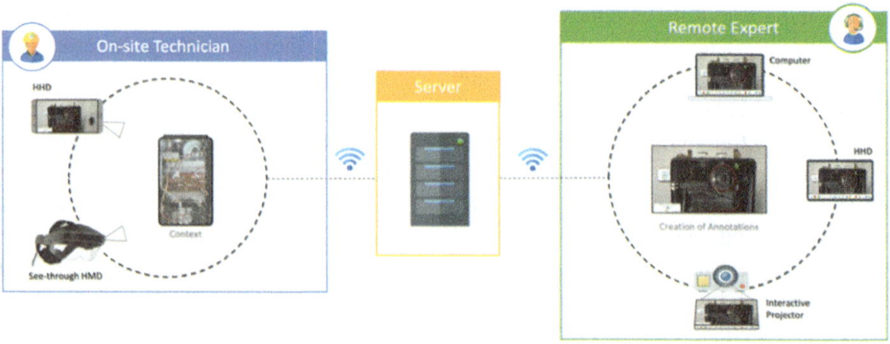

Fig. 2. Framework multiplatform possibilities for the On-site Technician (left: HHD of HMD) and for the Remote Expert (right – Computer, HHD or projection)

All annotations we use are stable, in the sense that they are all placed in the world. The difference is one kind of annotation is meant to be aligned with reality (and is interactive) and the others aren't (and are anchored to a specific pose in the world). Another possibility could be using unstabilized annotations, i.e., display in the corner of the user's field of view.

Aligned Annotations
This type allows annotations to be aligned with the real-world (Fig. 3 - 3). The on-site technician may interact with the annotation and adjust its pose and size to better match reality.

Anchored Annotations
This type of annotation allows an image object to be anchored in a designated pose in the real-world to be consulted by the on-site technician while conducting a given procedure (Fig. 3 - 2). It is also possible for the on-site technician to create one of these annotations by capturing their context through a snapshot saving feature (Fig. 3 - 1).

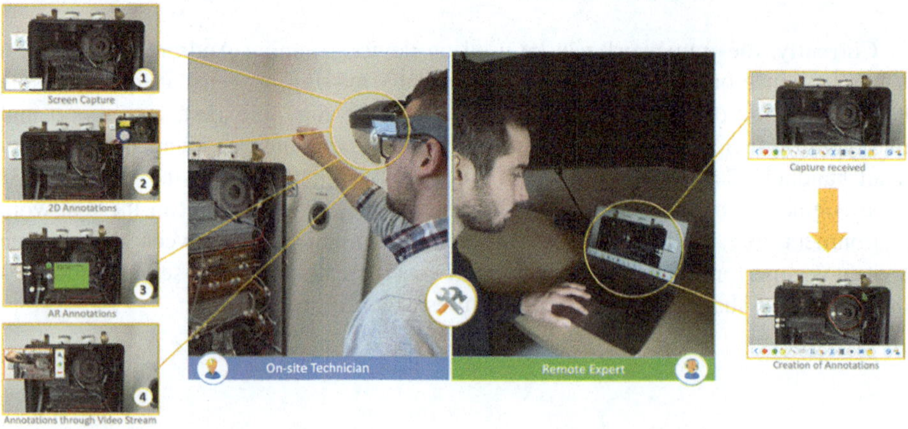

Fig. 3. Example of functions associated to the annotations of the framework for the on-site technician (left – capture and visualization of annotations) and the remote expert (right – situation understanding and creation of annotations)

Anchored Annotations Based on Video Stream
This type of annotation allows a video to be displayed in a designed pose in the world to be consulted by the on-site technician, following the same method previous described (Fig. 3 - 4).

Manipulation of Annotations Using Hand Tracking
We used a see-though HMD featuring gesture-based hands-tracking that enables the technician to interact with the annotations in a natural way. Closing one hand within the collision box of an annotation allows the on-site technician to grab it and move it in 6DOF. If two hands are used, this also enables the rotation and scaling of the annotation, by changing the distance between the user's hands and rotating the hands in relationship to each other, respectively.

3 Discussion

When on-site technicians use a hand-held device to visualize annotations, the need to hold the device with a particular orientation and position for proper observation of the information entails more cognitive and physical demands which may further hinder the ability to perform the task at hand [10]. The handsfree nature of the see-through HMD approach creates a setting where the interaction with the world is facilitated, enabling more complex tasks involving the use of two hands, while not compromising the access to information. Furthermore, the user interface design of the HMD, which leverages gesture-based hand tracking is thought to reduce the learning curve for unexperienced users by using hand interactions and allowing virtual objects to have physical properties of their real-world counterparts.

Allowing the annotations to be aligned with the real-world provides added value by complementing physical objects with information related to them in a straightforward way. The relationship between the target object and the information provided is inferred by enabling the on-site technician to perceive the indications of the remote expert as given directly within the physical context, creating a consistent view of the shared workspace. On the other hand, when using annotations that are not aligned with reality, the relationship between the information provided and the real-world context it refers to must be explicitly expressed, for instance.by presenting the information over a still picture of an object.

This type of remote interaction may also be adequate for the medical field (e.g. particularly in surgery), simulators (e.g. as flight or driving simulators), military operations, and many other scenarios where cooperation or tutoring are required in a set of tasks involving dexterous use of the two hands.

4 Final Remarks and Future Work

In this study, we describe a framework that allows the use of different kinds of spatialized annotations, facilitating remote collaboration, while leveraging the use of hand tracking for manipulation." We followed requirements for the design of collaborative tools using AR in remote scenarios, based on a human-centered design approach with partners from the industry sector. We integrated the use of a See-through HMD with hand tracking capabilities into an existing framework based on sharing of enhanced annotations. Through the use of different types of annotations, the framework allows the customization of the annotations position according to the context, allowing for the enrichment of each user's experience and the merging of the shared perceived realities of different users.

This study is being expanded by conducting a user study to evaluate usability and acceptability and to understand if these new features of the framework can be robust in a real remote setting. Furthermore, we are also planning a focus group to present and discuss the most recent updates to the framework. Then, we intend to conduct formal user studies with domain experts to test our findings and validate the framework.

Acknowledgements. This research was developed in the scope of the PhD grant [SFRH/BD/143276/2019], funded by FCT - Foundation for Science and Technology. It was also supported by IEETA - Institute of Electronics and Informatics Engineering of Aveiro, funded by National Funds through FCT, in the context of the project [UID/CEC/00127/2019]. The context of this research is an industrial partnership collaboration under the Smart Green Homes Project [POCI-01-0247-FEDER-007678], a co-promotion between Bosch Termotecnologia S.A. and the University of Aveiro. It is financed by Portugal 2020 under the Competitiveness and Internationalization Operational Program, and by the European Regional Development Fund.

References

1. Arias, E., Eden, H., Fischer, G., Gorman, A., Scharff, E.: Transcending the individual human mind-creating shared understanding through collaborative design. ACM Trans. Comput.-Hum. Interact. **7**(1), 84–113 (2000)
2. Johson, S., Gibson, M., Mutlu, B.: Handheld or handsfree? Remote collaboration via lightweight head-mounted displays and handheld devices. In: Computer Supported Cooperative Work and Social Computing, pp. 1825–1836 (2015)
3. Ens, B., et al.: Revisiting collaboration through mixed reality: the evolution of groupware. Int. J. Hum Comput Stud. **131**, 81–98 (2019)
4. Jalo, H., Pirkkalainen, H., Torro, O., Kärkkäinen, H., Puhto, J., Kankaanpää, T.: How can collaborative augmented reality support operative work in the facility management industry? In: Conference on Knowledge Discovery, Knowledge Engineering and Knowledge Management 2018, pp. 41–51 (2018)
5. Kangsoo, K., Billinghurst, M., Bruder, G., Duh, H.B.L., Welch, G.F.: revisiting Trends in Augmented Reality Research: A Review of the 2nd Decade of ISMAR. IEEE Trans. Visual. Comput. Graph. **24**, 2947–2962 (2018)
6. Kim, S., Billinghurst, M., Lee, G.: The Effect of Collaboration Styles and View Independence on Video-Mediated Remote Collaboration. Comput. Support. Coop. Work CSCW Int. J. **27**, 569–607 (2018)
7. Gauglitz, S., Nuernberger, B., Turk, M., Hollerer, T.: World-stabilized annotations and virtual scene navigation for remote collaboration. In: ACM Symposium on User Interface Software and Technology, pp. 449–459 (2014)

8. Kim, S., Lee, G., Sakata, Billinghurst, M.: Improving co-presence with augmented visual communication cues for sharing experience through video conference. In: IEEE International Symposium on Mixed and Augmented Reality 2014 (ISMAR 2014), pp. 83–92 (2014)
9. Kim, S., Lee, G.A., Ha, S., Sakata, N., Billinghurst, M.: Automatically freezing live video for annotation during remote collaboration. In: ACM Conference Extended Abstracts on Human Factors in Computing Systems, CHI EA 2015, pp. 1669–1674 (2015)
10. Rehman, U., Cao, S.: Augmented-reality-based indoor navigation: a comparative analysis of handheld devices versus google glass. IEEE Trans. Hum.-Mach. Syst. **47**(1), 140–151 (2017). https://doi.org/10.1109/THMS.2016.2620106

"Homenu": An Interactive Projection Cooking Assistant

Yahong Li, Shuyuan Zhang, and Zhanxun Dong(✉)

School of Design, Shanghai Jiao Tong University, No. 800 Dongchuan Road, Minhang District, Shanghai, China
dongzx@sjtu.edu.cn

Abstract. In recent years, people are increasingly concerned about the safety and health of their diet, which leads to more people preferring to cook for themselves. However, young people do not have complete cooking skills because of poor experience. This paper focuses on their cooking problems and applies user research approach to concluding the pain points of cooking behavior for this cohort of young-adult users. The purpose is to provide users with easy, intuitive, and interesting cooking instructions. Based on the interactive projection technology, we developed a cooking interactive projection system named "Homenu", which has a lightweight and natural interaction.

Keywords: Smart kitchen · Interactive projection · · Interactive design

1 Introduction

According to a study conducted by the United States Department of Agriculture (USDA), millennials spend a larger percentage of their grocery shopping budget on "prepared foods" and they cook "home-prepared food" less than any of the previous generations [1]. But with the improvement of living standards, people have higher requirements for the quality of life. The health, nutrition, and safety of diet have become a public concern [2]. People are more likely to cook for themselves rather than just settle for a takeaway.

With the continuous progress of science and technology, the kitchen has gradually become more intelligent, providing a lot of convenience for cooking. The ways of food purchase, delivery, and storage management have been developed. The smart kitchen has largely eased a lot of preliminary work, but young people still cook less frequently. Based on the previous study, the frequency of cooking is negatively associated with a sense of pleasure (hedonic motivation) [3]. This may occur because they face a lot of difficulties in the process of cooking, which are mainly reflected in the following aspects:

Poor Cooking Experience. Although many people are willing to cook for themselves, most of them are new cooks, especially young people born in the 1980s and 1990s who live and work in big cities. The skill has a positive impact on the likelihood to cook, and attempts to impart skills should also increase the incidence of cooking.

W. Karwowski et al. (Eds.): IHSED 2020, AISC 1269, pp. 90–95, 2021.
https://doi.org/10.1007/978-3-030-58282-1_15

Inefficient Way to Learn Cooking. Currently, the main way people learn to cook is by checking recipe apps on their electronic devices. But this presentation of recipes is not intuitive enough. Users will be busy and get the wrong steps while browsing the tutorial, and the water stains on their hands will also prevent them from touching the electronic screen.

Boring Cooking Process. For many new cooks, the cooking process is boring, especially the cutting and processing of ingredients. The cooking process is perceived to require a considerable amount of manual labor and a need for precise measurements.

Taken together, the purpose of this study is to alleviate the problems encountered by young people who are willing to cook but not experienced in the actual cooking scene and encourage them to engage in healthy cooking habits.

After analyzing the previous research and determining the goals, we carried out the main design process. The stages and behaviors of the cooking process were summarized through observation, user interviews, and other research methods. Then, we analyzed the pain points of the cooking process and obtained the user experience map of the cooking process. In order to achieve the desired interaction effect, we studied the interactive projection technology and determined the development platform and development model. Finally, the cooking interactive projection system was designed and developed.

2 User Research

Firstly, we did field observation. The purpose of this observation is to understand the cooking process of different users, analyze the cooking behavior, and summarize the general process and rules of cooking. Through the observation of the cooking process, it was found that the preparation and production of dishes could be summarized in the following six steps: choose the recipe, prepare ingredients, cut ingredients, process ingredients, heat and cook, decorate and plate.

And then we did the user interviews. The purpose of this interview is to understand the target users' diet status, cooking status, as well as the pain points and needs encountered during cooking learning. And the selection criteria of interviewees in this interview are young people between 20 and 35 years old, with relatively good income and high requirements for quality of life. Six interviewees were selected for the survey, including three females and three males. Three respondents had some cooking experience, while the other three had no cooking experience.

The interview result shows that users cared about life quality and brand experience, and they had a strong sense of healthy diets. They generally believed that cooking at home was healthier than eating take-out food, although they often ate take-out food due to a tight schedule. They expressed a willingness to cook at home if conditions were right.

In terms of their access to culinary instruction, many respondents said they were still too inexperienced to cook independently. In the process of cooking, they often had to look at the recipes on mobile devices like phones, so the operation led to confusion. Also, water-stained and oil-stained on hands during cooking would affect the touch screen, making the whole process terrible. Besides, some respondents said the process of cooking was too boring, especially when cleaning, cutting, and cleaning up afterward.

The cooking behaviors and cooking operations at different stages were summarized in the observation and investigation, as well as the pain points, demands, and typical sayings of users learned in the user interviews. We combined them to create a user experience map (see Fig. 1).

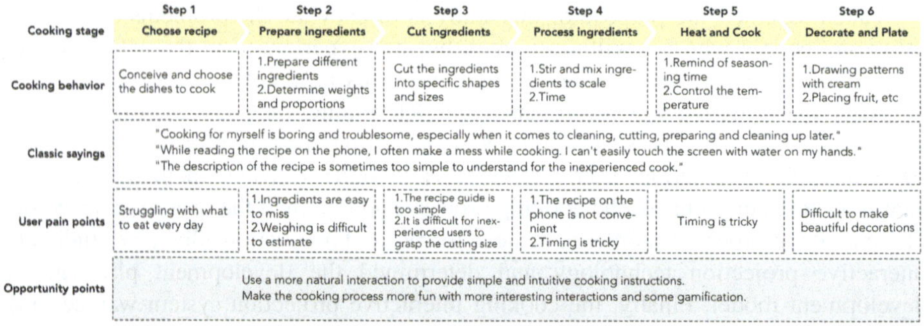

Fig. 1. User experience map and opportunity points

Through the user experience map, we get the opportunity points: Use a more natural interaction to provide simple and intuitive cooking instructions, and make the cooking process more fun with more interesting interactions and gamification.

3 Design and Development of the System

3.1 Hardware and System

Omojola et al. pointed out that with the continuous development of computer technology, we should no longer obtain information only through the old way like mouse, keyboard, and monitor [4]. With the rapid development of sensing technology, more new ways of interaction have emerged.

To demonstrate the design concept and develop the interactive projection system, the carrier used in this design is a technology named interactive projection. It can project information onto a desktop or wall. And it has been developed and widely used in recent years. People can interact by touch, click, and other means. At the same time, the form of projection is free from the constraints of the screen, and its operating environment becomes more diverse. Through interactive projection, ordinary ground, wall, desktop, and other realistic space can become our interactive space.

The interactive projection device we used is "Sony Xperia Touch". It is an android-based interactive projection device. Placed on a desktop, it can project a 23-inch virtual screen. It's agility, interactive, and colorful. Even in a brighter environment, the projected content can be displayed. Users can perform a series of interactive actions on the projection interface, such as playing videos, viewing pictures, downloading and using apps, etc.

Touch is the most basic interaction of "Sony Xperia touch". And because it runs on android, it works just like a normal screen. Manipulation methods such as tapping, touching and holding, pinching in and out, swiping and scrolling, flicking (see Fig. 2).

Current interactive projection technologies can be divided into four types, which are based on trait imaging, infrared sensing, image recognition, and structured light principle [5]. For "Sony Xperia Touch", the interaction is via an infrared sensor and cameras.

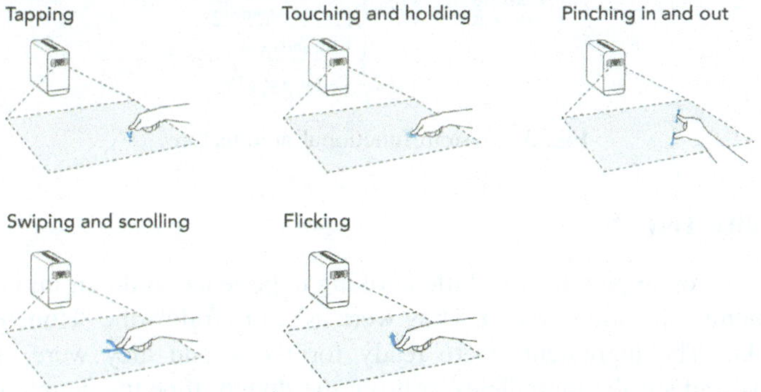

Fig. 2. Touch interaction with "Sony Xperia Touch"

3.2 System Architecture

According to user research and technology exploration, we developed an app called "Homenu". It runs on "Sony Xperia Touch" based on android. "Homenu" has three main functions: personal information, recommended recipe and cooking assistant. Personal information includes personal culinary points, community rankings, and more. Recommended recipe includes today's dishes. After selecting a recipe, users can view the recipe details. After clicking the start button, it will enter the cooking assistant phase. Figure 3 is the functional architecture of the interactive projection system.

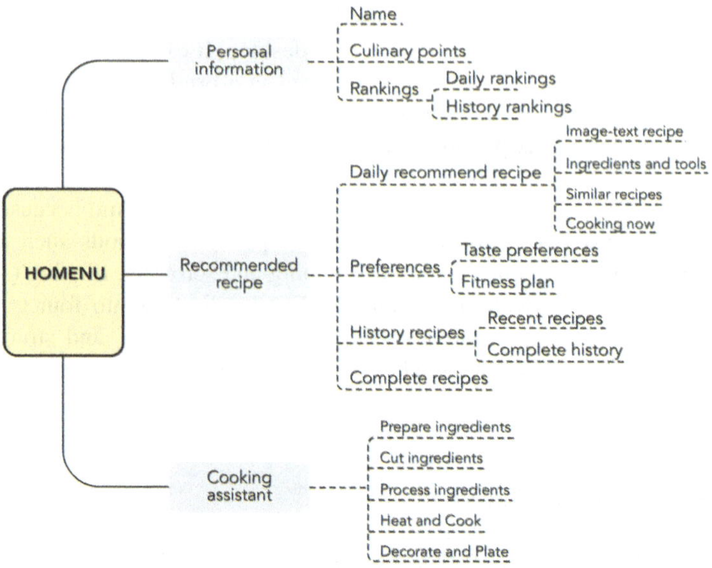

Fig. 3. System functional architecture

3.3 Usability Test

We invited six young people with little cooking experience to do an initial evaluation of the “Homenu” operating system. They were asked to follow the “Homenu” to make a potato cake. The ingredients were ready for them, and they were asked to cut potatoes, mix and stir the ingredients, roll out the dough, time the dough, decorate the cake with cream, and so on (see Fig. 4).

Fig. 4. Usability test

The six participants both thought that this interaction method was enjoyable, and they could feel easier during this cooking process. Owing to the visual guidance, it could help them improve cooking skills. In particular, the projection grid guided cutting of food materials, interactive animation guided decorative production and other functions received unanimous praise from the participants.

And they also expressed worries about long-term effects since cooking for the first time is still fresh and novel for users. Therefore, establishing good feedback is our future efforts to increase users' reliance on interactive cooking assistance systems.

4 Conclusion

This paper focuses on the cooking scene, and conducts research and analysis for young people who are inexperienced in cooking. When cooking for themselves, they encounter many problems like poor cooking experience and the inefficient way to learn cooking. We designed the "Homenu" system based on interactive projection technology to assist young people in cooking. And with the help of Sony Xperia Touch, we completed the development and presentation of the system. The easy and natural interactive way provides users with intuitive and interesting cooking guidance to improve their enthusiasm and participation in cooking. Positive feedback makes cooking more gamified and increases their sense of accomplishment. In addition, the personalized customization of recipes is conducive to the healthy diet of users and can stimulate their creativity.

Smart kitchen is the development trend of future life. It should be a complete process from the garden to the table, involving ingredients recommendation, food purchase, and delivery, cooking guidance, feeling communication, etc. At present, this paper focuses on only one part of the smart kitchen process, the cooking stages. In the future research, we can pay attention to the whole process of smart kitchen, and make good use of the existing network platform data, so that the service of intelligent diet is perfectly integrated into the life of users.

Acknowledgment. This research is supported by Zhejiang Provincial Key Laboratory of Integration of Healthy Smart Kitchen System (Grant No: 2017F02).

References

1. Kuhns, A., Saksena, M.: Millennials devote larger shares of their grocery spending to prepared foods, pasta, and sugar and sweets than other generations. Amber Waves, pp. 1–10 (2017)
2. Wolfson, Julia A., Bleich, Sara N.: Is cooking at home associated with better diet quality or weight-loss intention? Public Health Nutrition **18**(8), 1397–1406 (2015)
3. Namin, A., et al.:. Dine-in or take-out: Modeling millennials' cooking motivation and choice. J. Retailing Consum. Serv. **53**, 101981 (2020)
4. Omojola, O., et al.: An installation of interactive furniture. IBM Syst. J. **39**(3.4), 861–879 (2000)
5. Pool, E., Brosch, T., Delplanque, S., et al.: Stress increases cue-triggered "wanting" for sweet reward in humans. J. Exp. Psychol. Animal Learn. Cogn. **41**(2), 128–136 (2015)

Process Design for Evoking Emotional Response Focusing on Empathy

Akane Matsumae[1(✉)] and Mitsuki Miyahara[2]

[1] Faculty of Design, Kyushu University, 4-9-1 Shiobaru, Minami, Fukuoka, Fukuoka, Japan
matsumae@design.kyushu-u.ac.jp
[2] School of Design, Kyushu University, 4-9-1 Shiobaru, Minami, Fukuoka, Fukuoka, Japan

Abstract. This study attempts to improve video viewing experiences by elucidating the mechanisms of evoking emotional responses focusing not on the video itself, but on the cognitive state of the viewers. The authors focused on empathy, which is considered to be the most important factor in evoking emotional responses, affected by the amount of both knowledge and engagement. The comparative experiment was conducted between two groups: one group was asked just to watch the video; the other group was asked the same but after experiencing a workshop where they were given an opportunity to acquire knowledge of and engage with the video subject matter. As a result, the participants with greater knowledge and engagement before viewing the video displayed a significantly higher level of emotional response on average. The knowledge obtained from this study will contribute to the improvement of the design process to evoke an emotional response in video viewing.

Keywords: Evoking emotional response · Empathy · Engagement

1 Introduction

There is an accumulation of studies and practical knowledge addressing video questions such as: What kind of scenario structures or visual/sound effects can move video viewers? In contrast, video research has barely touched on such issues from the perspective of the cognitive state of the viewers, though it is generally known that even identical videos are perceived differently depending on that cognitive state.

This study seeks to improve the video viewing experience by understanding the cognitive mechanisms which evoke emotional responses.

2 Related Studies

2.1 Evoking Emotional Responses

As a framework for the expression of emotions in emotion research, Tokaji reported that one of the following three is usually adopted:

W. Karwowski et al. (Eds.): IHSED 2020, AISC 1269, pp. 96–101, 2021.
https://doi.org/10.1007/978-3-030-58282-1_16

1. Specific emotional categories typified by joy, sadness, anger, fear, etc.
2. Specific dimensions such as pleasantness-unpleasantness, excitement-sedation, and asleep-awake.
3. The most basic two-dimensional division of positive emotions-negative emotions.

However, evoking emotional response cannot be classified into any of the frameworks used in these studies of emotions, and explained as a special kind of emotional expression. In other words, evoking emotional responses refer to multiple emotions that cannot be captured by a single emotional value. Therefore, Tokaji categorized evoking emotional responses and proposed a theoretical model of its mechanisms centering on empathy [1].

In this study, the authors focused on empathy based on Tokaji's theoretical model of evoking emotional response mechanisms as well as the definition of *feeling deeply and being moved by an object.*

2.2 Empathy

In previous research, Escalas et al. differentiated empathy and sympathy, and found that sympathy is *the concept of a thought-response that is noticed and understood from the perspective of the observer*, whereas empathy is *the concept of an emotional response in which one shares another's feelings from the perspective of a participant* [2]. Ishihara explained that empathy means to enter into the *heart* of another person and to feel the heart of that person, to know the heart of the other person indirectly by becoming *alive* to the physical behavior of that person, who is an external sensory object (for *me*) [3].

Therefore, the authors also considered empathy as a vague and polysemous concept and defined it as *a posture that projects one's own emotions into the object or into others and considers them to be unique to the object.*

Minagawa reported that differences in expression were found depending on the degree of empathy in the context of the learning experience [4]. Furthermore, Oshimi et al. reported that people with greater knowledge and engagement when watching sporting events are more emotionally aroused [5].

Based on these related studies, the authors hypothesized that people with greater knowledge and engagement in the video subject matter would display stronger emotional responses.

3 Methodology

3.1 Experiment

The authors conducted an experiment with two groups to compare how differences in cognitive levels of knowledge and engagement with the video subject matter affect the viewer's emotional responses to the video material. The presence or absence of workshops that enhance engagement and knowledge in the process of video viewing were used to test the hypotheses by means of comparison in a controlled environment.

Participants. Twenty-six healthy participants aged in their twenties participated in the experiment and were divided into two groups, group A and B.

Video Materials. The video materials for the experiment were carefully selected to meet following requirements:

1. Video evoking similar types of emotional responses
2. Video with minimal difference in impressions of participants
3. Video that are unfamiliar in daily life to evoke emotional response
4. Short enough video to keep participants concentration (5–7 min)

Based on these requirements above, the following two videos were selected.
Video 1: A craftsman is demonstrating dragon origami work.
Video 2: A craftsman is demonstrating dragon beard candy work.

Procedure. For Video 1, group A was asked just to watch the video, while group B was asked the same but only after experiencing a workshop where participants were given an opportunity to acquire the knowledge and to become engaged in the video subject matter. After watching each of the videos, the participants were asked to complete a questionnaire. For Video 2, the experiment was conducted with groups A and B switching roles. The workshop for each video took 30 min. The first 10 min comprised a lecture on the video subject matter meant to broaden the viewer's knowledge. The next 20 min were spent with hands-on training to increase engagement with the video subject matter.

Lecture. For Video 1, participants were introduced the following kind of information: This is one of the world's most complex origami works; It is very difficult to create an asymmetrical structure; Introduction of a blueprint for the origami dragon (Fig. 1).

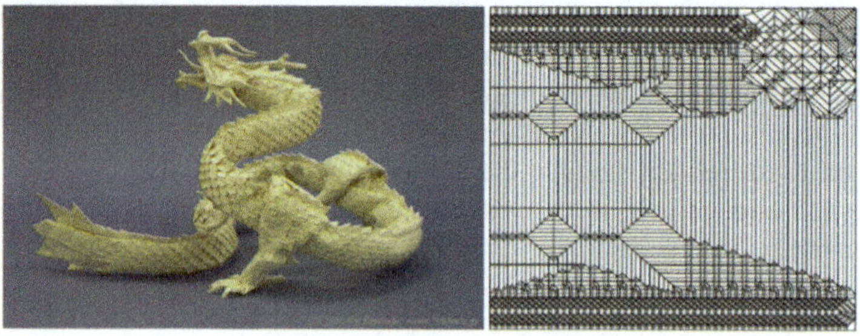

Fig. 1. Dragon origami craft work (*left*) and its blueprint (*right*)

For Video 2, participants were introduced the following kind of information: The malt and other chunks are stretched and folded to make this confection; 16,384 bundles are made at the end; At least 60 kg of grip strength is required to make it (Fig. 2).

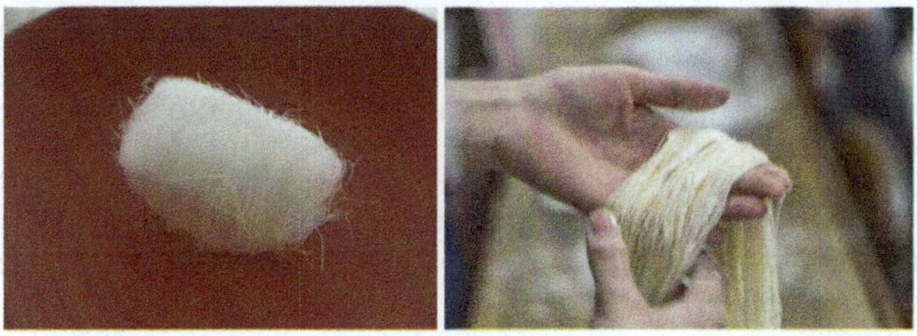

Fig. 2. Dragon beard candy craft work (*left*) and its bundles (*right*)

Hands-on Training. For Video1, participants tried a simple dragon folding exercise. For Video 2, they prepared a lump of syrup and practiced stretching it themselves to make dragon beard candy.

3.2 Evaluation

The authors evaluated each participant in each experimental process: how much emotional response was evoked, as well as how much knowledge and engagement with the video subject matter was acquired before watching the video. The questionnaire items are set as shown in Table 1 and were answered by each of the participants on a 5 point-scale each time they finished viewing the video.

Table 1. Questionnaire items list.

Knowledge			Did you know how to do it?
			Did you know what can be made from it?
Engagement			Have you ever made it?
			Would you like to make it more in the future?
Emotional Rating Scale	Receptive	Overflow	Does it fill your heart? (*love, joy, tears*)
			Does it touch your heart? (tears of gratitude, twilight, loneliness)
		Enjoyment	Does it warm your heart? (*gratitude, peace of mind*)
			Was it captivating? (*beautiful, majestic, stunned silence*)
	Expressive (Positive Emotions)	Fascination	Does it move your heart? (*enthusiastic, deeply touched, stomach butterflies*)
		Excitement	Was it exciting? (*raises me up, makes me want to express myself*)
		Delight	Were you thrilled with it? (*psyched*)
			Were you delighted? (*happy, "yay!"*)
	Expressive (Negative and Neutral Emotions.)	Intense grief	Was it spine-chilling? (*panic, astonishment, tension*)
			Was it unbearable? (*cruelty, trembling, wailing*)
		Awaking	Does it grab your heart (*get goosebumps, trembling, throbbing*)
			Were you enlightened? (*unexpected*)
	General Impression Rating		

Engagement and Knowledge of the Video Subject Matter. Based on the questionnaire used in the experiment by Oshimi et al. [5], the questionnaire items shown in Table 1 are rated on a 5-point scale to evaluate the degree of engagement with and knowledge of the video subject matter.

Evoked Emotional Response. Based on the study by Oode et al. [6], the authors set questionnaire items to evaluate evoked emotional responses on a 5-point scale. Open-ended questions were added at the end.

4 Results and Discussions

An overview of the experiment results is shown below in Table 2 to examine whether there was a significant difference in results between those with and those without the workshop. Mann-Whitney's U test was used with the significance at 5%. Items that showed a significant difference are gray colored in Table 2.

Table 2. An overview of experimental results

WS: workshop w/o: without w/: with			Video 1		Video 2	
			Group A w/o WS	Group B w/WS	Group A w/WS	Group B w/o WS
Knowledge			1.1	2.4	3.8	4.1
			3.4	3.7	4.0	4.1
Engagement			4.0	3.9	2.7	3.2
			2.7	3.6	4.8	4.7
Emotional Rating Scale	Receptive	Overflow	2.9	3.4	3.1	3.3
			2.6	3.0	2.4	2.6
		Enjoyment	2.4	2.2	2.7	2.9
			3.6	4.9	4.3	3.9
	Expressive (Positive Emotions)	Fascination	3.2	4.3	2.8	3.3
		Excitement	3.0	4.1	4.1	4.3
		Delight	3.5	4.2	4.5	4.3
			2.1	2.8	2.9	3.6
	Expressive (Negative and Neutral Emotions)	Intense Grief	1.4	2.1	1.2	1.2
			1.3	1.1	1.0	1.2
		Enlightened	2.8	4.4	3.9	3.3
			2.5	2.6	2.8	2.9
	General Impression Rating		3.6	4.6	4.1	4.2

Looking at the negative and neutral expressive emotions, there was no significant difference between Video 1 and Video 2, all within a low value for all cases of emotions classified as sadness. This is because both videos used in this experiment focused on craftsmanship, and the images were not meant to be heartbreakingly emotional.

The results for Video1 showed that the participants with more knowledge and engagement before viewing the video displayed a significantly higher level of emotional response on average. Four items (fascination, excitement and arousal) improved on the emotion rating scale. The results show that the same images prompted more emotional engagement that came with more emotions. By watching the video based on the knowledge gained in the workshop, some participants were able to experience the artist's technique and compare it to their own images of origami, and to think about the folds and its structure.

The results for Video 2 showed that there were no significant differences among evaluation items. Table 2 shows that knowledge, engagement, and inspirational ratings were all close to 4 for both questionnaires. The participants expressed their interest in the artisanal skills and taste. However, many participants said, "I don't know how to do it, and I can't do it well at all" during the workshop, which indicates that this workshop did not sufficiently deepen their understanding of dragon bearded candy and the artisan. In other words, even after experiencing the workshop, the participants' knowledge was insufficient to be able to compare the actual experience with the skills of the artisans.

5 Conclusion

This study has shown that the process of increasing the viewer's knowledge and engagement can lead to empathy and evoke emotional responses to video, regardless the video itself. The knowledge obtained from this study will contribute to the improvement of the design process to evoke emotional responses in video viewing. The main limitation in this study was the small sample of data. There is a need to examine various kinds of video and viewer positions.

Acknowledgments. We are grateful to Ms. Akiko Fish and Mr. William Fish for their generous and continuous supports to our research activity. This work was supported by JSPS KAKENHI Grant Number JP20K20119.

References

1. Tokaji, A.: Mechanisms for evoking emotional responses of "Kandoh (The state of being emotionally moved)". Cogn. Sci. **8**, 360–368 (2001)
2. Escalas, J.E., Stern, B.B.: Sympathy and empathy: emotional responses to advertising dramas. J. Consum. Res. **29**, 566–578 (2003)
3. Kouji, I.: "Einfühlung" and "Sichversetzen": theory of understanding others in phenomenology and hermeneutics (1). Ann. Rep. Cult. Sci. **48**, 1–19 (1999)
4. Minagawa, N.: A review for the education, bringing intelect and sensibility closer to each other, based on cognitive psychology. Res. Bull. Educ. Sci. **17**, 17–26 (2002)
5. Oshimi, D., Harada, M.: Delight in watching sports: a replication and an examination of moderator effects in the customer delight and satisfaction model. Jpn. J. Sport Manage. **5**, 19–40 (2013)
6. Oode, S., Imai, A., Ando, A., Taniguchi, T.: Evaluation of Kandoh evoked by music-relation between type of Kandoh and affective value of music. NHK STRL R&D **50**, 1111–1121 (2009)

Humanoid Robotics: Guidelines for Usability Testing

Niccolò Casiddu, Francesco Burlando(✉), Claudia Porfirione, and Annapaola Vacanti

Dipartimento Architettura e Design, Università degli Studi di Genova, Stradone di Sant'Agostino 37, 16123 Genova, GE, Italy
{niccolo.casiddu, francesco.burlando, claudia.porfirione}@unige.it, annapaola.vacanti@edu.unige.it

Abstract. In the macrofield of human–computer interaction, human–robot interaction (HRI) is a relatively new and unexplored field of study. As several efforts have been made in robotics to expand the range of functionalities of hardware and software, perhaps the hardest challenge is to ensure good usability and fluid interaction for users. With regard to remote operated robots, these aspects are unrelated and the interaction features may be taken into account at an advanced design stage. Nevertheless, when it comes to humanoid robots, interaction features are the keypoint, therefore the design of the latter should be driven by UX and UCD practices. The actual fact is that too little attention is paid to these issues or, at best, the latter are approached by IT's point of view, who are in charge of the software development. In conjunction, a team of engineers take care of the mechanical parts and, in the most highly developed project, the appearance is curated by industrial designers. Given the importance of the interaction features, the design process should be rearranged in favour of the last-mentioned phase. However, the team in charge of this task should be composed of industrial designers, that we can consider as UI designers, along with UX expert. This study applies UX practices to the design process of a humanoid robot, providing practical guidance and guidelines to put in place usability testing during the various stages of development. Regarding testing, only user satisfaction is often investigated. Instead, the present approach includes an adaptation of typical elements of the web design area, such as learnability, efficiency, memorability and errors.

Keywords: User experience · Human–robot interaction · Usability testing · Humanoid robotics

C. Porfirione, A. Vacanti and F. Burlando—Work developed under the tutorship of Professor Niccolò Casiddu.

W. Karwowski et al. (Eds.): IHSED 2020, AISC 1269, pp. 102–109, 2021.
https://doi.org/10.1007/978-3-030-58282-1_17

1 Human Robot Interaction

In the last decades we have faced a tremendous technological growth in the actions of our everyday life. Almost everything that we were used to do in a certain way, now is being done with the assistance of some technological device. Website and mobile app have become the most important tools in all aspects of our life, from trivial things to life and death matters. As a result, over the years the concept of Interaction Design (IxD) and usability has evolved. If in the past industrial designers were supposed to deal with issues as materiality and ergonomics, the emergence of the Human–Computer Interaction (HCI) scope has brought with it new challenges and nowadays the scope of User Experience (UX) is mainly characterized by digital products.

More recently we have faced the onset of robots in our society with an increasing trend for humanoids (reference) due to the high level of affordances that the latter produce in users (reference). We are now accustomed to see collaborative robots in industries and even if a robot designed to operate in industrial workspace should not be humanlike for any practical purpose, we are seeing an increasing trend of anthropomorphic design of CoBots, as Baxter by Rethink Robotics, as proof of fact that at least some humanoids features are always preferred in the design of a machine bond to work at the side of a human [1].

As for the future, the evolution of the robotics scope and the increasing need of the assistance by machine - especially in areas such as caregiving - suggest that we will see more and more humanoids in our everyday life. The study of Human–Robot Interaction (HRI) and in particular Human-Humanoid Interaction (HHI) will promote the integration of such item in our society. With regard to the above, humanoid robots stand somewhere in the middle between industrial and digital products. The science-fiction imagery sees robots as personal assistants that help us in daily life actions with verbal and physical interaction. A sort of Alexa that can also carry grocery bags and clean the house. Before it comes to anything like that, a lot of challenges in terms of software and hardware must be solved, but we can predict what will happen since same thing happened to other technological sectors as automotive and mobile phone. Let us take the latter as an example: if for a certain time the disparity between products from different companies was made by design and software features, once all the products achieve a good standard in these points, things like the presence of an ecosystem of additional services and the User Interface (UI) started to make a difference. Therefore, even if the robotics scope still has to deal with challenges in terms of software and hardware, it is important to get a head start on the design of the experience that users will have in the interaction with these products.

1.1 End User

The first step for a successful design is to identify who are the end user to which our product or service addresses. When it comes to a humanoid robot, of course it can relate to many different scenarios, but we can identify two main categories:

- **Robots designed directly for end users**
- **Robots that require an intermediate user**

The second group includes those robots that are bought by someone that makes them available for users. Let us take the example of a robot concierge. The end user that interact with it is the customer, but the staff has to take care of the robot for all that concerns the backstage: switching on and off, recharging, storing and in some cases even programming. Moreover, sometimes these steps are carried out by different people, as in the case of an intermediary company that takes care of programming the robot.

Even if also these actions constitute interactions between a human and a robot, for the design of a usability test it should be considered only the interactions that occur between the robot and the end user.

1.2 Humanoid Robots

Another significant distinction must be made between two categories of robot:

- **Directly Interactive Humanoid Robots**
- **Indirectly Interactive Humanoid Robots**

The second group mostly encloses robots that are remote controlled by a mobile app, as Pando by Leju, or by a platform, as T-HR 3 by Toyota. In these cases, the interaction occurs between the user and the tool that allow to control the robot. Certainly, it is necessary to explore this kind of user experience too [2], but the present contribution aims to focus just on the interaction that take place between robots and end users. For this reason, also in this case users that work directly with these types of robot, as maintenance technicians or developers, will not be taken into account since they don't belong to the category of end users.

2 Usability Testing

Given the importance of what stated above, this section provides some practical guidelines to the design of a usability test of a humanoid robot with end users. Not having the presumption of solving all aspects about how to test such a complex product, the scope of usability texting provides useful reference for further insights [3, 4].

2.1 Formative or Validation?

When usability testing is utilized during the design process it is said to be formative, while when used to validate a nearly finished product's efficacy prior to release it is known as summative or validation usability testing.

Since the development of a humanoid robot is a complex and articulated process, it is best to engage with users since the early stage of the design process. At this stage it is advisable to focus on what users expect about interaction features instead of appearance. In fact, the latter can be tested with prototype and mockup at a more advanced stage of the design process. Moreover, in this regard the literature presents many insight [5], while the UX scope in humanoid robotics is HRI can vary widely depending on the specific case [6].

Therefore, at an early stage it is best to organize a focus group in which user can observe a mockup that show the dimensions of the robot. All the interaction features should be presented verbally by the moderator, possibly with the help of video that show pre-existing references. This is very important when the users are not friendly with the technology, as in the case of elderly.

The moderator should present different example of user flow, also with the help of storyboards, asking the users if they agree or if they would act differently. Alternatively, some points of the user flow can be left blank and he may ask to the users to fill them. More elaborate methods as a participatory game design session can be used.

At an advanced stage of the design process a summative usability test can be done to validate the interaction features as well as the appearance of the robot. The following paragraphs address this issue more specifically.

2.2 Lab-Based or Field-Based?

As a result of what stated before, a formative usability test should be lab-based. On the contrary, with regard to summative usability testing, users should interact with the robot in pseudo-real scenarios. With reference to the factors presented by Hass [7] these are the motivation behind this choice:

Participant Access and Accessibility. Let's start with a consideration: every robot should be designed for a precise environment and end-user. If the robot is tested in that precise environment, it's easy to think that the users with which the test had to be done will be there already.

Interaction Fidelity. A user should act in a more natural way if he comes across a robot during his everyday life rather than during a testing session that has been scheduled by time. If participants do not prepare psychologically the event, they will interact with the robot without bias.

Environmental Impact. In order to ensure an interaction fidelity, the environment plays a key role. A sterile environment will not facilitate a truthful and seamless experience and some possible interaction may not be taken into account.

Given the importance of a field-based test, nevertheless most of the research carried out with robot are lab-based. This is because the following problems:

Control. When designing a usability test, the last thing you want is to have distractions that may ruin the regular course of the session and invalidate the results. A field-based study might presents a lot of distractions, so it is of the utmost importance to prevent them. However, real life is chaotic, so a field-based study with distractions better reflect the interaction that will occur between user and robot compared to a lab-based testing. As stated, distractions must be identified in advance in order to remove noise from the data.

Risk. A lab-based study carries less risks for participants and for the robot. Therefore, when designing field-based usability testing every risk must be identified in advance and prevented.

Observability and Data Capture. A lab-based study allows a more complete record of data. It is easier to use cameras, microphones and advanced tools as eye trackers. In a field-based study it can be much more complicated to have a setting that allow to record all those data. The solution can be found in another key role of usability testing.

2.3 Moderate or Autonomously

Once that the correct location and users have been identified, and the scenario of the test has been settled, there is still one key factor that miss. Most of the time, when users first meet a humanoid robot, they tend to attribute to him a high intelligence level and well-developed interaction features. For example, even if the robot does not have vocal recognition, they approach it saying phrases like "hello, what's your name?" expecting to have a conversation with it. If the robot does not answer, they keep talking to it, picturing that it has not hear them or that the question was asked in the wrong way. Maybe the robot does not even speak, maybe it just knows how to play soccer or chess, but users expect to have with him a human-like interaction, since it is a humanoid. This is even more true with non-expert user as children, elderly or people with disabilities.

Therefore, it is necessary to have a moderator that introduces participants to the robot explaining what kind of interactions they can have with it. Given what stated above, the role of the moderator is also to collect all the possible data about the interaction. In order to achieve this result, the best choice is to guide the participant through the session thanks to a pre-designed script. At the same time, the moderator should fill a format with his observations. It is also useful to have microphones and a camera that allow the team to re-observe the session at a later time. However, as has been said it is important to let the participants feel in a comfortable natural environment, therefore these devices should not be invasive.

3 Usability Test Guideline

This section provides directives and guidelines for the design of a validation usability test of a *directly interactive humanoid robot.*

3.1 Sample Size

Different interpretations can be found in the literature about what is the ideal sample size for usability testing. The use of formulas to calculate it [8, 9] is unwise, since the complexity of the interaction elements that are tested. For the same reason, instead of the 5 ± 1 rule [10] it is best to refer to the 10 ± 2 rule [11].

Therefore, the sample size should be of 10 ± 2 participants. After fixing the issues that have arisen during the test, the latter should be repeated at least three time with the same sample size of different participants.

3.2 Location and Recruitment

As stated above, it is best to find participants in a real scenario in which the robot will be employed. If a robot is designed to operate as a museum guide, it would be possible to recruit users while they are visiting the museum. For example, at the end of the exhibition itinerary it may be asked at some customer if they want to visit another room in which they will be guide through the exhibition by a humanoid robot.

3.3 Duration

Since the test is an unexpected event for participants, it is possible that they don't have a lot of time for it. Therefore, it should last preferably less than half an hour thus divided:

- Welcome, consent form administration and study orientation (5 min)
- Pre-Task professional background interview question (5 min)
- 3–5 Interactive tasks (15 min)
- Post-session interview question (5 min)

Through this time schedule every part should have enough time and, if a participant has no hurry, the post-session interview can last longer and be more productive in terms of data collection.

3.4 Invasiveness

To collect data in a closer-to-reality way, it would be better to carry out a contextual inquiry by shadowing rather than a usability test [12, 13]. Nevertheless, as stated above the presence of a moderator is essential to ensure the correct interaction way between participants and the robot. Therefore, the role of the moderator is the key to the success of the test: he has to guide participants and make them comfortable without interfere in any way with their actions, as if he were not there.

3.5 Moderator's Guidelines

Some typical observational goal might be:

- What are participants' first impression of the robot?
- Are participants able to identify the right way to interact whit it or do they need an insight by the moderator?
- Are the participants able to complete key tasks in an efficient and satisfying manner?
- Does the robot stimulate participants to interact with it?
- How do participants interact with the robot? Do they just talk to it or do they also touch it? Where?
- Are participants more engaged by robot's voice or movements?
- Do the participants refer to the robot as if they were talking with a person? Do they show curtesy to it?
- Do the participants understand robot's instructions?

- What are the robot's greatest strengths and weaknesses?

A good way to collect these data is to set some tasks that participants should achieve, both explicitly and implicitly. For example, one task could be: participants should understand that they can touch the tablet on the robot's chest to have a greater interaction with it. The task should be implicitly and, if participants does not achieve it, the moderator should show them the features and observe if they use this way of interaction subsequently.

Other task can be presented in writing or verbally, also by the robot itself. As participants attempt the tasks, they might be asked to think aloud by voicing their impression. Once participants have completed a task, given up on it or when the moderator feels sufficient time has passed to count it as a failure, he should ask some quantitative and qualitative question as "On a scale of 1 to 10 how easy or difficult was to complete this task? Why you gave the rating you did?".

Moreover, the moderator can collect data by filling a pre-built form. In this regard the moderator can rely on typical elements of usability testing as learnability, efficiency, memorability, errors and satisfaction [14].

References

1. Destephe, M., et al.: Walking in the uncanny valley: importance of the attractiveness on the acceptance of a robot as a working partner. Front. Psychol. **6**, 204 (2015). https://doi.org/10.3389/fpsyg.2015.00204
2. Yanco, H.A., et al.: Beyond Usability Evaluation: Analysis of Human-Robot Interaction at a Major Robotics Competition. Human–Computer Interaction, Lawrence Erlbaum Associates, Inc. (2011)
3. Albert, W., Tullis, T.: Measuring the User Experience: Collecting, Analyzing, and Presenting Usability Metrics. Newnes, Oxford (2013)
4. Barnum, C.M.: Usability Testing Essentials: Ready, Set…Test! Elsevier, Amsterdam (2010)
5. Vincent, J., et al.: Social Robots from a Human Perspective. Springer, Heidelberg (2015)
6. Polak, R.F., et al.: Differences between young and older users when interacting with a humanoid robot: a qualitative usability study for rehabilitation. Ann. Phys. Rehabil. Med. **61**, e499 (2018). https://doi.org/10.1016/j.rehab.2018.05.1163 (ScienceDirect)
7. Hass, C.: A practical guide to usability testing'. consumer informatics and digital health: solutions for health and health care. In: Edmunds, M., et al. (eds.), pp. 107–24. Springe, Heidelberg (2019). https://doi.org/10.1007/978-3-319-96906-0_6
8. Turner, C.W., et al.: Determining usability test sample size. Int. Encycl. Ergon. Hum. Factors **3**, 3084–3088 (2006)
9. Schmettow, M.: Sample size in usability studies. Commun. ACM **55**(4), 64–70 (2012). https://doi.org/10.1145/2133806.2133824
10. Nielsen, J., Landauer, T.K.: A mathematical model of the finding of usability problems. In: Proceedings of the SIGCHI Conference on Human Factors in Computing Systems - CHI '93, pp. 206–213. ACM Press (1993). https://doi.org/10.1145/169059.169166
11. Hwang, W., Salvendy, G.: Number of people required for usability evaluation: the 10±2 rule. Commun. ACM **53**(5), 130–133 (2010). https://doi.org/10.1145/1735223.1735255

12. McDonald, S.: Studying actions in context: a qualitative shadowing method for organizational research. Qual. Res. **5**(4), 455–473 (2005). https://doi.org/10.1177/1468794105056923 (SAGE Publications)
13. Holtzblatt, K., Beyer, H.: Contextual Design: Defining Customer-Centered Systems. Elsevier, Amsterdam (1997)
14. Di Pascale, M.: Manuale di sopravvivenza per UX designer: Guida pratica alla progettazione. HOEPLI EDITORE, Milan (2019)

Long-Distance Relationships: Use of Technology Advances in Communication, Idealization and Satisfaction

Pamela Acosta-Rodas[1], Hugo Arias-Flores[2], and Carlos Ramos-Galarza[1,2](✉)

[1] Facultad de Psicología, Pontificia Universidad Católica del Ecuador, Av. 12 de Octubre y Roca, Quito, Ecuador
{mpacosta, caramos}@puce.edu.ec

[2] Centro de Investigación en Mecatrónica y Sistemas Interactivos MIST/Carrera de Psicología/Carrera de Enfermería, Universidad Tecnológica Indoamérica, Av. Machala y Sabanilla, Quito, Ecuador
hugoarias@uti.edu.ec

Abstract. Long distance romantic relationships are becoming more and more common nowadays. There can be various concrete reasons that are the causes for this occurrence, among the most common, the pursuit of career opportunities, educational goals, military service, and migratory issues. The aim of this study is answering the question: how is possible to maintain this kind of relationship? This is explained from two variables: the first one refers to the technology advances which help the communication and the second one studied from the psychological periscope, the romantic idealization. It is approach the perception of this new type of relationships in the general population. Understanding technology influences in human life could help to develop more technological advances to the people to feel closer and to keep healthy relationships with themselves and the others, as well, it could bring lights to counseling and psychological therapy.

Keywords: Long-distance · Romantic relationships · Social networks

1 Introduction

There is an increasing notion that long-distance romantic relationships (LDRs) are becoming more and more common nowadays [1–6]. There can be various concrete reasons that are the causes for this occurrence, such as the pursuit of career opportunities, educational goals, military service, and migratory issues, among the most common.

The literature about long and short-distance romantic relationships' benefits comparison has shown inconsistent results, some studies conducted on college students revealed similar and even higher levels of quality satisfaction when being involved in a long-distance relationship [1, 2, 4, 6–8], while others stated that there are more possibilities of not engaging with the relationship and more chances to break up [3].

W. Karwowski et al. (Eds.): IHSED 2020, AISC 1269, pp. 110–115, 2021.
https://doi.org/10.1007/978-3-030-58282-1_18

Communication and idealization are two variables of great interest in this study as predictors of relationship satisfaction. The development of technology in the present makes possible the interaction between people, even if they are millions of kilometers away, with this tool many couples in long-distance relationships interact as they are beside their partners, studied couples in long-distance relationships who spent extended periods connected by video chat with their partner and concluded that they could feel as being there and having the supporting and the good times as any other couple that sees each other every single day, which is beneficial for the maintenance of the relationship. The use of technological advances for keeping people closer such as WhatsApp, Facebook, Skype, Instagram and others that allow keeping the communication on real-time between the partners will help the relationship, as well as this virtual connection, will help to arise the romantic idealization.

Idealization in romantic relationships, refers to the tendency toward positive distortions that minimize problems, Stafford and Merolla [8] affirm from their study that long-distance partners engage in significantly more positive relationship rumination than short-distance relationships, this may occur in two possible routes such as the cognitive idealization (positive illusions), and the second route termed behavioral idealization, which emanates from blocked communication, indeed, idealization is the decision to view the partner positively and focus on strengths as a way to keep the relationship satisfying and possibly compensate for separation [8], it is important to mention as well the presence of fewer conflicts especially when they are physically together, this time is higher appreciated than in couples involved in the short-distance relationship [3, 4].

For instance, adding long-distance relationships' facts already mentioned: the short time the couple spends together, plus the communication through technological advances, which makes the couple feel like knowing about each other and sharing their daily life through virtual connection, plus arisen romantic idealization of the time when they spent apart wishing being together, bring as a result relationship satisfaction, encouraging them to keep this romantic relationship even kilometers away, as it is represented in Fig. 1.

Fig. 1. Graphic representation of distance romantic relationship through technology devices

2 Method

2.1 Research Design

This is quantitative and descriptive research; it aims to identify people's perception of the usage of technology in long-distance romantic relationships.

2.2 Participants

The sample was composed of 152 participants aged between 17 and 60 years, 30.9% corresponded to females, and 30.9% to males. From the total participants, 55.60% answered that they have had a long-distance relationship. Participation was voluntary and throughout this study ethical standards for human research were respected.

2.3 Measures

To measure the participants' perceptions about long-distance romantic relationships, a questionnaire with 9 items to assess this construct was built and applied.

2.4 Procedure

This study started with a request for voluntary participation to people that would cast their perception about the playing role of technology in long-distance romantic relationships. The application of the questionnaire was possible through the technological application Google Forms and every participant fulfilled it in an anonymously and voluntary way.

2.5 Data Analyses

To analyze the obtained data from the participants, statistical techniques such as central tendency and dispersion were conducted.

3 Results

Following, Table 1 with the values found from the participants' answers in the questionnaire applied, is shown.

Table 1. This table shows the statistical values form the participants' answers.

Item	TD	MD	NA/ND	MA	TA
1. Do you think it is possible to maintain with success a long-distance romantic relationship?	21.1%	25%	18.4%	28.3%	7.2%
2. Do you consider that using WhatsApp application would help to improve couple's communication in the long-distance romantic relationship?	23%	17.1%	13.8%	39.5%	23%
3. Do you consider that Skype's video-call would help to maintain a long-distance romantic relationship?	9.9%	11.9%	13.9%	41.1%	23.2%
4. Do you think usage of Facebook application would help the romantic relationships through the exchange of photographs and living experiences?	21.7%	17.8%	23%	27%	10.5%
5. Do you think that uploading photos in Instagram would help in the long-distance romantic relationship because partners know about each other?	29.6%	20.4%	20.4%	22.4%	4.6%
6. Do you consider that social network usage could bring troubles to the long-distance relationships?	3.9%	3.3%	17.8%	35.5%	39.5%
7. Do you consider that long-distance romantic relationships could be possible without the technological advances that we may count with nowadays?	30.3%	19.1%	17.1%	19.1%	14.5%
8. Do you believe that applications to find a couple such as Tinder or Lovoo would negatively influence in a long-distance romantic relationship?	8.6%	7.9%	26.5%	23.8%	33.1%
9. Do you think that social networks could be used as control devices within a romantic relationship?	19.1%	6.6%	14.5%	28.9%	30.9%

Key Table: TD: Totally disagree, MD: Moderately disagree, NA/ND: Neither agree nor disagree, MA: Moderately agree, and TA: Totally agree.

4 Conclusions

In this study, it has been addressed the role that technology advances play in long-distance romantic relationships since nowadays human mobility thanks to globalization and the technological improvements are higher than in the lasts decades, new human dynamics have arisen and deserve to be investigated. Proposing for the future the creation of new technological prototypes for people that wants to be closer.

To obtain a closer view of this phenomenon, it was realized a quantitative study where it was assessed the individual perception about this type of romantic relationship. From the empirical study realized, the results showed that most of the participants moderately agree with the statement that a long-distance romantic relationship could be successful. Also, most of the participants moderately agree considering that technological platforms such as WhatsApp, Facebook, Skype or Instagram would have a positive contribution to a long-distance relationship to success.

Is important highlighting that most of the participants affirmed convincingly that a long-distance relationship could not be possible to maintain without the use of technology, as well as they affirmed that there could be other phenomena that would influence negatively making the relationship to come to an end, such as the social networks that offer couples encounters or, once they are in a relationship controlling the activities that the partner could be doing from the place where he or she by using social network.

Finally, it is possible to identify from these results that technology contributes to the maintenance of long-distance relationships, although, this benefit is not absolute and should be reflected these other situations that could appear affecting negatively. As future investigation we propose to search in this romantic relationships' dynamic using these technological means closer through narrative and experimental studies as well as where it could be analyzed deeply the meanings emerging from this dynamic in couples' relationships mediated by technology, as well as from the psychological periscope, understanding these new types of relationships.

References

1. Kelmer, G., Rhoades, G.K., Stanley, S., Markman, H.J.: Relationship quality, commitment, and stability in long-distance relationships. Fam. Proc. Fam. Process **52**(2), 257–270 (2012). https://doi.org/10.1111/j.1545-5300.2012.01418.x
2. Dainton, M., Aylor, B.: Patterns of communication channel use in the maintenance of long-distance relationships. Commun. Res. Rep. **19**(2), 118–129 (2002). https://doi.org/10.1080/08824090209384839
3. Krapf, S.: Moving in or breaking up? The role of distance in the development of romantic relationships. Eur. J. Popul. **34**(3), 313–336 (2017). https://doi.org/10.1007/s10680-017-9428-2
4. Goldsmith, K., Byer, S.: Perceived and reported romantic and sexual outcomes in long-distance and geographically close relationships. Can. J. Hum. Sex. **27**(2), 144–156 (2018). https://doi.org/10.3138/cjhs.2018-001
5. Lee, J., Pistole, M.C.: Predictors of satisfaction in geographically close and long-distance relationships. J. Couns. Psychol. **59**(2), 303–313 (2012). https://doi.org/10.1037/a0027563

6. Dargie, E., Blair, K.L., Goldfinger, C., Pukall, C.F.: Go long! predictors of positive relationships outcomes in long-distance relationships. J. Sex Marital Ther. **41**(2), 1–22 (2013)
7. Neustaedter, C., Greenberg, S.: Intimacy in long distance relationship over video chat. In: Conference on Computer-Human Interaction (CHI 2012), Austin, Texas, USA, pp. 753–762 (2012)
8. Stafford, L., Merolla, A.J.: Idealization, reunions, and stability in long-distance dating relationships. J. Soc. Pers. Relat. **24**(1), 37–54 (2007)

Design of a Human Machine Interface for Programming and Testing Adjustable Frequency Drives for Constant Pressure Pumping Applications

Byron Remache-Vinueza(✉), Jefferson Castro-Ramírez, and Mireya Zapata

Research Center of Mechatronics and Interactive Systems - MIST, Facultad de Ingeniería y Tecnologías de la Información y la Comunicación, Universidad Tecnológica Indoamérica, Machala y Sabanilla, Quito, Ecuador
{byronremache,mireyazapata}@uti.edu.ec, jcastro5@indoamerica.edu.ec

Abstract. In this project, a human machine interface (HMI) was developed for programming adjustable frequency drives (AFDs). Frequency drives have a wide range of industrial applications such as pumping systems, conveyor belts, ventilation and air conditioning systems, and, in general, applications that require motors with high torque. Constant pressure pumping systems with AFDs, for instance, allow working with a proportional integral derivative (PID) control, providing the system with precise response between the desired value and the process feedback. In this project, an HMI has been designed to effectively configure, test and calibrate AFDs for constant pressure pumping applications. Various alternatives of HMI screens were analyzed using the prioritization matrix method to determine the most suitable screen. The HMI interface was developed in the DOPsoft 4 software of DELTA Electronics, following the guidelines set forth by ANSI/ISA-101.01-2015, to generate a clear and easy to understand HMI.

Keywords: Adjustable frequency drive · Constant pressure system · Human-Machine Interface (HMI)

1 Introduction

Adjustable frequency drives (AFDs) in constant pressure pumping systems are used for the transport of fluids. These systems represent around 20% of the world's demand for electrical energy [1]. AFDs allow the system to be controlled when the fluid demand is variable, allowing its efficiency with an energy saving of between 55% and 69% compared to systems without AFD [2, 3]. Even though the speed control of three-phase motors using AFDs is one of the most used control methods, they generally present the problem of monitoring the operating status and variables in real-time [4]. One way to solve this problem is to implement Human Machine Interfaces (HMI), which are a

W. Karwowski et al. (Eds.): IHSED 2020, AISC 1269, pp. 116–122, 2021.
https://doi.org/10.1007/978-3-030-58282-1_19

bridge between the user and the automated control system, where the operator makes decisions regarding the system control variables [4, 5].

In [6] the design of a constant pressure cooling water supply system for a steel tube rolling process is presented. It consists of 6 stages: a swirl settling tank, dosing equipment, chemical degreaser, hot water pool, cooling tower and cooling water pool. The constant pressure control is carried out by means of the Proportional Integral and Derivative (PID) adjustment, through the use of a PLC, a frequency variator and a pressure transmitter. The HMI system was developed in Siemens WinCC V7.0 monitoring software in which parameter configuration screens, alarms, and reports were designed.

The integration of HMI systems has improved productivity in all kinds of applications, focusing on the field of automation. In [7] a speed control for three-phase motors is proposed using a system consisting of an HMI, a PLC, and a frequency inverter for three control methods: at fixed speeds, open loop, and closed-loop. The design of the interface was based on the ANSI/ISA-101.01-2015 international standards, while the ladder programming language used for the control logic algorithms is based on the IEC-61131-3 standard. The proposed system includes 4 HMIs screens, the main screen, and one for each control method mentioned above. The tests carried out in the system allowed monitoring and animation in the HMI as well as the indication of the logical control rules.

In this proposal, an HMI is designed to effectively program, test and calibrate adjustable frequency drives in constant pressure systems.

2 Materials and Methods

The HMI system for programming adjustable frequency drives in constant pressure systems presented below has been developed based on recommendations proposed in the ANSI/ISA-101.01-2015 standard.

An HMI is developed through a life cycle model, shown in Fig. 1, which encompasses 4 main stages: system standards, design, implementation, and operation.

The System Standards stage of the life cycle includes the development of: HMI philosophy, HMI style guide and HMI toolkits. The design stage focuses on four phases that include the console design, HMI system design, functional requirements and display design. In the implementation stage, the HMI is integrated into the software and hardware of the target platform using the results of the previous stages. The stage of operation includes commissioning, maintenance and even possible dismantling.

This investigation has been carried out accomplishing the four phases included in the design stage.

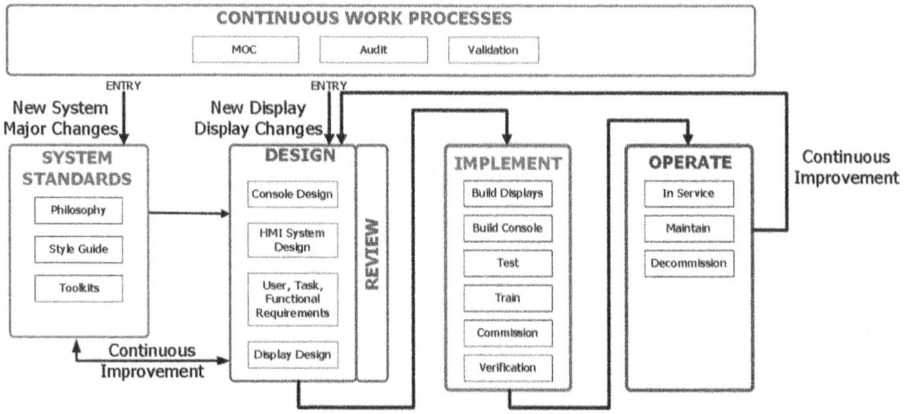

Fig. 1. HMI lifecycle, from [8]

Console Design: For the appropriate hardware selection used with the HMI proposed system, a prioritization matrix was employed. Based on the proposed methodology in [9], several HMI alternatives were analyzed with a weighting between criteria for the selection of the most suitable equipment. In Table 1, the results obtained from the prioritization matrix are shown, in which the Delta DOP-110CS touch screen has the most suitable combination of all the criteria. Criteria evaluated were: low cost (taking into account the value of the equipment in the local market), data management (evaluating storage capacities, memory, device processor), display characteristics (according to the technical properties of the display, lighting, resolution, life hours of the backlight) and software manageability (taking into account whether the software needs a license to be used, the technical support available to the user and the accessibility of programming in specific software).

Table 1. Results obtained from the prioritization matrix

	Low cost 28,33%	Data management 18,33%	Display characteristics 21,67%	Software manageability 31,67%	Percentage
1 Eaton HMIVU10WCUNBE	21,67%	21,66%	25,00%	25,00%	23,44%
2 HCFA TP2510	31,67%	25,00%	18,34%	15,00%	22,28%
3 Siemens KTP900	18,33%	31,67%	28,33%	28,33%	26,11%
4 Delta DOP-110CS	28,33%	21,67%	28,33%	31,67%	28,17%
Total	100%				

HMI System Design. The selection of the control platform and HMI operating system is based on the previously selected hardware. With the Delta DOP-110CS touch screen corresponding to the DOP-100 series, the DOPSoft V4.0 software is used to edit the

HMI screens. This touch screen will be connected to the drive through a Modbus RS-485 network as shown in Fig. 2.

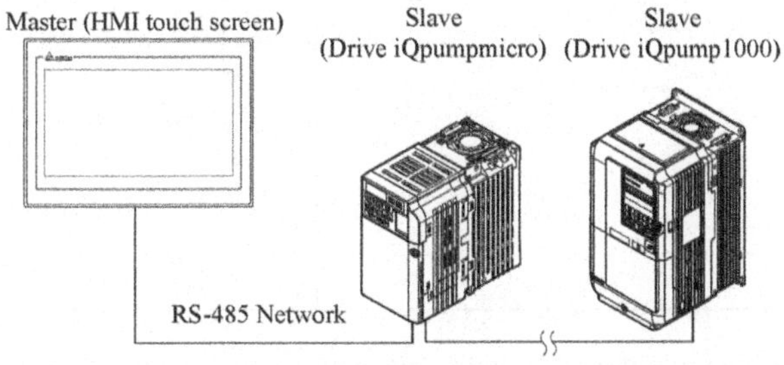

Fig. 2. Master/slave connection between touch screen and drives.

Functional Requirements. The main work applications of constant pressure pumping systems in which the frequency inverters can work were identified. The software of these devices allows working in three modes: simplex control system (individual pump), Drive to Drive Multiplexing system (multi-pump through network drives), Single Drive with Constant Speed Lag Pump Multiplexing System (multi-pump through contactor). Figure 3, presents a simplex control system, consisting of a pump controlled by a frequency inverter.

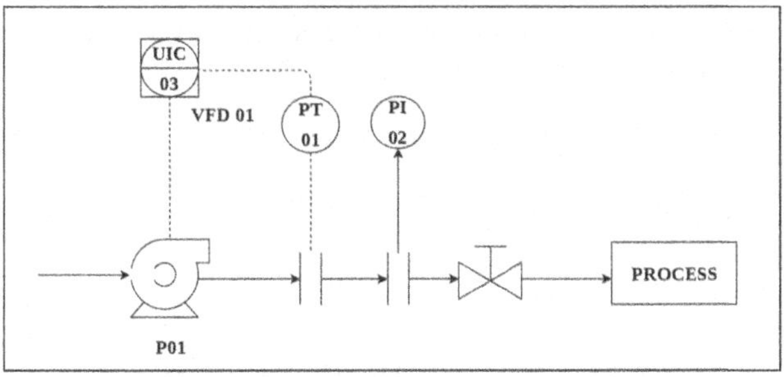

Fig. 3. Simplex control system.

Figure 4, shows a Drive to Drive Multiplexing system (left), which allows a connection of the drive system in a Modbus RS-485 network, with a master/slave working mode for adding and subtracting the pumps, in which each pump is controlled by a frequency inverter. The Single Drive with Constant Speed Lag Pump Multiplexing system (right) also uses the master/slave method with the difference that the addition and subtraction of pumps is carried out by means of a unique frequency inverter (master).

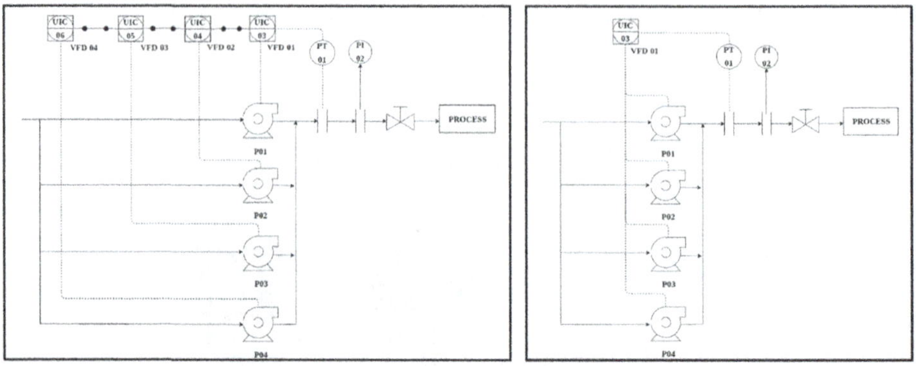

Fig. 4. Drive to drive multiplexing system (left) and single drive with constant speed lag pump multiplexing system (right).

Once each working mode has been identified, the parameters corresponding to constant pressure systems of the frequency inverters are selected to obtain the hexadecimal address that will be used in the development of the HMI screens.

Display Design. The design of the HMI is based on technical standards and Human Factors Engineering (HFE) principles proposed in the ANSI/ISA-101.01-2015 standard. Screens have been designed for each programming stage of the systems allowing the user to enter the configuration values sequentially. It also displays alert messages on the interface when these values are out of the allowed range. The screens enable the user to see the behavior of the variables, also monitoring trend graphs in real-time and visualizing the process in a 2D animation. Figure 5 shows examples of the screens found in the HMI system.

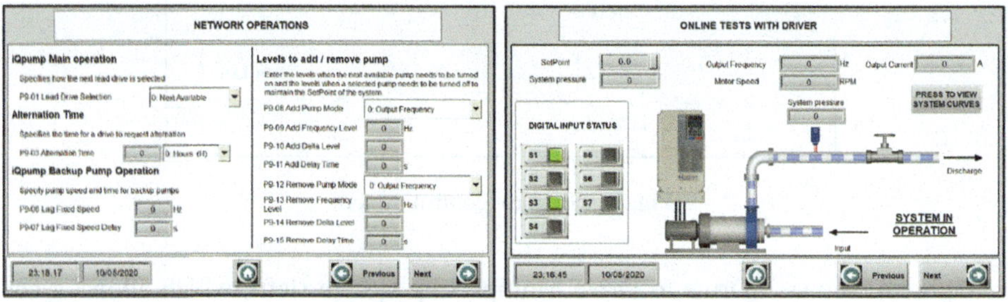

Fig. 5. Parameter programming screens (left) and simulation (right)

3 Results

With the development of the HMI, the operators have a system that allows working efficiently in the programming of frequency inverters in constant pressure systems, avoiding unnecessary tasks such as searching for parameters in the manuals and user guides. In this way, the tasks carried out in the programming process were reduced by 38%. This difference is shown in Fig. 6.

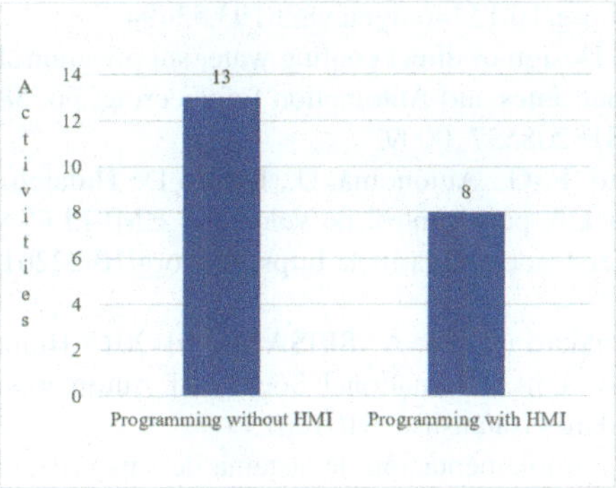

Fig. 6. Activities to be carried out to program AFDs in constant pressure systems

4 Conclusions

According to the prioritization matrix selection method, the Delta DOP-110CS touch screen obtained the best-weighted score (28.17%). This selection process was based on the evaluation of different criteria such as cost, data management, display characteristics, and software manageability, which are adjusted to the local market situation. This equipment is compatible with DOPSoft 4 software, which allows quick and efficient editing.

The interface programming was carried out under the guidelines adopted from the ANSI/ISA-101.01-2015 standard, according to the control system requirements, and as a result, a system of sequential configuration of parameters and online monitoring of variables was obtained. The HMI designed improves the system operation performance and an overall reduction in programming sequence of approximately 38% was obtained.

References

1. Sungur, C., Çalişir, S., Kaya, E.: Developing an automation system for improving the energy efficiency of constant pressure irrigation pumps. J. Irrig. Drainage Eng. **142**(11), 04016052 (2016). https://doi.org/10.1061/(ASCE)IR.1943-4774.0001046

2. Cedeño, R.J.A.: Gestión Energética de una Estación de Bombeo mediante el uso del Control Estadístico de Procesos. Estudio de Caso: Acueducto La Esperanza–Refinería del Pacífico. Revista Politécnica **40**(2), 7–18 (2018). https://doi.org/10.33333/rp.v40i2.925
3. Pérez, J., Rivera, A.: Análisis de la eficiencia energética en un sistema de bombeo accionado de manera directa y controlado por un sistema de variación de frecuencia (2019)
4. Cui, X., Li, B., Kou, Z., Qiao, Y.: Measurement and control system for variable-frequency speed regulating of motor based on PLC and HMI. In: 2019 IEEE 8th Joint International Information Technology and Artificial Intelligence Conference (ITAIC), pp. 1169–1173, May 2019. https://doi.org/10.1109/itaic.2019.8785791
5. Filali-Yachou, S., González-González, C.S., Lecuona-Rebollo, C.: HMI/SCADA standards in the design of data center interfaces: a network operations center case study. DYNA **82**, 180–186 (2015). https://doi.org/10.15446/dyna.v82n193.53494
6. Xu, J.-T., Zhou, H.-F.: Design of direct cooling water supply control system based on siemens S7-300 PLC. In: Mechatronics and Automation Engineering, pp. 58–64, March 2017. https://doi.org/10.1142/9789813208537_0009
7. Carlos, Q.-Q.J., Ernesto, F.-G., Autónoma, U., Estado De Hidalgo, D., Jorge, B.-L.: Sistema HMI-PLC-ADF-Motor CA para control de velocidad HMI-PLC-ADF-CA Motor system to speed control González-Cadena Maribel. https://doi.org/10.22201/fi.25940732e.2018.19n4.038
8. American National Standard (2015): ANSI/ISA-101.01-2015 Human Machine Interfaces for Process Automation Systems. International Society of Automation, North Carolina. https://webstore.ansi.org/standards/isa/ansiisa101012015
9. Sanpedro, D.: Diseño e implementación de sistema de supervisión, control y adquisición de datos con IoT: aplicación en sistemas de riego para jardines verticales (2019)

Analysis, Evaluation, and Upgrading of a Data Analytics Methodology Through a Qualitative Evaluation Technique and a User-Centered Design Process

Boris Astudillo[1(✉)], Katherine Cajilema[1], Marco Santórum[1], and Jose Aguilar[2]

[1] Departamento de Informática y Ciencias de la Computación, Escuela Politécnica Nacional, Quito, Ecuador
{boris.astudillo,kahterine.cajilema, marco.santorum}@epn.edu.ec

[2] Centro de Estudios en Microelectrónica y Sistemas (CEMISID), Universidad de Los Andes, Mérida, Venezuela
aguilar@ula.ve

Abstract. A data analytics (DA) methodology allows a proper treatment to obtain the best benefit from the data with the aim of allowing supported business decisions to be made. However, a DA methodology needs to be continually reviewed and updated to incorporate standpoint from researchers, in order to become the most efficient version. Our paper carried out an analysis and evaluation of the first phase of a selected DA methodology called MIDANO applying qualitative techniques to reach an improved version. The analysis and evaluation process were applied through a focus group discussion from the results obtained during the implementation of a case study. The upgraded DA methodology, as a result of the qualitative analysis, was related to the nine attributes extracted, classified, and validated from a user-centered design process previously studied, and based on ISO 9241:2010 standard, in order to obtain a DA methodology participatory, adaptable, efficient, collaborative, and easy-to-apply.

Keywords: Data Analytics Methodology · Qualitative evaluation · User-centered design · MIDANO

1 Introduction

Data analytics is the process of seeking knowledge in the data, in order to make a better-supported business decision. The data is collected, refined, and analyzed, depending on organizational needs [1, 2]. A DA methodology can be used to improve operational efficiency in any process, responding quickly to emerging market trends and gain a competitive edge over rivals.

Processing data massively is a strong complexity faced by specialists who invest time, knowledge, and company money to apply DA Methodologies. However, these

W. Karwowski et al. (Eds.): IHSED 2020, AISC 1269, pp. 123–128, 2021.
https://doi.org/10.1007/978-3-030-58282-1_20

methodologies cannot always be easy to execute, feasible to apply in different scenarios, adjusted to organizational cultural changes, lacked verification of data validity or protection, among other problems, and the result is a waste of company resources.

For this reason, in this work, analysis and evaluation of a specific DA methodology are carried out through qualitative evaluation techniques to identify gaps, initially on its first phase of identification of the context and where can be applied DA processes, applying recommendations by means of the addition of forms based on the user-centered paradigm to develop its new version, which will be used in our proposal to build a user-centered DA methodology.

The rest of the article is structured with a section that introduces the DA methodology; followed by an explanation of the qualitative evaluation technique used; then the first phase of the DA methodology will be applied in a case study; next, the evaluation, the discussion and proposal of this research are shown; and finally, the conclusions are presented.

2 Review of the Selected Methods

Our research is based on the study of information systems and, in particular, on the study of three domains that are data analysis, qualitative evaluations, and user-centered design; the selected data analysis methodology is MIDANO, which is used to develop data mining applications based on organizational analysis [3], it will be evaluated twice, first through a qualitative methodology, designed to study people and understand their social and cultural phenomena [4]; and second, through a user-centered (UC) methodology based on the ISO 9241-210: 2010 standard [5]. These aspects are presented below.

2.1 Data Analytics Methodology

Several assertions have been made about the definition of DA. However, the official concept is according to [6], "data analysis refers to the quality and quantity of data extracted and categorized from various resources to analyze the structural behavior and pattern of the environment".

DA Methodology MIDANO integrates the stages of organizational knowledge to identify and conceptualize the solution of a problem from the perspective of developing applications based on data mining, for processes of any institution/company. MIDANO is composed of three phases, which are: Identification of the sources for the extraction of knowledge in an organization, Preparation and treatment of Data, and Development of data mining tasks [3].

For this work, we focus on the first phase and its five stages, which are: Knowledge of the organization (F1S1), Characterization of organization processes (F1S2), Feasibility analysis and process selection (F1S3), Determination of the possible Data Analytics tasks (F1S4), and Formalization of Data Analytics tasks (F1S5) [3]. This phase's objective is to know the organization structure as well as its processes, in order to define the objective of the possible application of the DA, and through the knowledge engineering, to propose specific and clearly defined activities and steps to identify where to extract knowledge in an organization/company, letting a proper interaction between developers' and organizational process experts. The phase 1 summary is shown in Table 1.

Table 1. Stages and steps from phase 1 [3].

	1st step - Objective	2nd step - Product	3rd step - Protocol	4th step - Activities
(F1S1)	To know the organization	An organizational structure document	Data collection through questions	Schedule meetings, analyze existing documents, collect information
(F1S2)	In-depth knowledge of the processes to work	Workflows, models and activities diagrams	Questions to know processes, data sources, and existing technology	Produce a detailed document, study the processes, clarify doubts
(F1S3)	Analyze the process to know the feasibility of applying DA tasks	Chart with processes evaluation	Questions to review processes and their importance. Analyze data sources. Apply an amount equation	Analyze the processes and solve arisen doubts
(F1S4)	Distinguish possible DA tasks	Functional requirements, use cases graphs, scenery charts, all in order to apply DA	Identify involved people, current and future scenarios charts, and technological improvement	Generate a requirements sheet, use case graphs, users involved, and improve documents to become DA goals
(F1S5)	Define final and formal DA's problems	The formal document specifying problems	A processes information report, operational problems description, and usability and charts DA tasks	Develop the formal document approved by stakeholders

2.2 Qualitative Evaluation Technique

Qualitative data analysis involves inductive reasoning processes to interpret and structure the meanings that can be derived from data [4]. In a qualitative evaluation, understanding and interpreting the data is a predominant factor from the participants' perspective. Qualitative methods, both for data collection and evaluation, include interviews, group discussion, observation, and relevant documents analysis, which through a systematic process allow the researcher to understand the problem [4].

For this work, the implementation process of this phase of MIDANO was applied in the case study.

2.3 User-Centered Design Process

UCD is a dimension used to adapt to the user's perspective a design, to guarantee their satisfaction, usability, and better learning experience, in the processes implemented. UCD can be used to create and validate through experimental protocols in a participatory manner [7], as well as serious educational games design [8]. However, UC criteria are not usually present in a DA Methodology [9].

ISO 9241:2010 standard "Human-centered design process for interactive systems", the objective of the standard is to describe an iterative development cycle to explain the processes and characteristics of user-centered design, to ensure that the development of the systems consider the needs of the user, the developer, and the owner, as well as specify the context in which the product will be used, based on these needs [5].

The attributes proposed by the standard are Product-orientation towards the user (UC1), User satisfaction (UC2), Company satisfaction (UC3), Usability tests (UC4), User Evaluations (UC5), Effectiveness and suitability of tasks (UC6), Process improvement (UC7), Adaptability to change (UC8) and Easy implementation (UC9).

3 Study Case Application

The DA methodology was applied in a professional training center that offers training services, through 30 bi-monthly courses, and has more than 30,000 students per year. Phase one implementation of the DA methodology produced information collected through meetings, and in some cases, using forms with questionnaires, diagrams, and formulas. The five-stage activities are superficially described in Fig. 1.

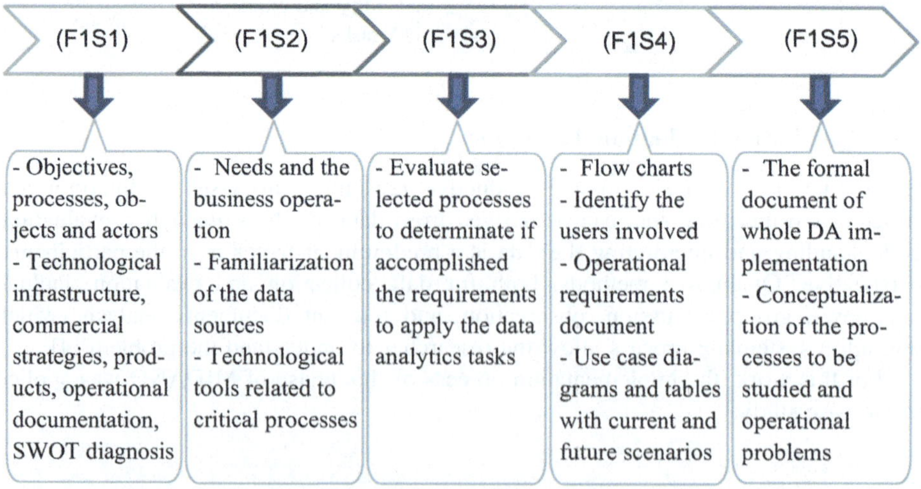

Fig. 1. Implementation of the five stages of the first phase in a case study.

4 Analysis, Evaluation, and Upgrading

As a first iteration, only the UC criteria of UC1, UC6, UC7, UC8, and UC9 were added to carry out the update process; the remaining UC criteria (UC2, UC3, UC4, and UC5) will be applied at the end of the remaining two phases. However, each UC criteria will be constantly evaluated during the implementation process.

The final result of this process produced nine documents where work was done on the new approach of several questions, the evaluation of the criteria for the selection of implementation scenarios, the inclusion of qualitative comparison factors, the analysis of formulas weighting, and classification of measurement ranges. Moreover, to improve the administrative tasks, standardized and instructive codes for document management were defined, which included glossaries, responsibility signatures, and other elements. All these resources will be reviewed at the end of the implementation of the three phases. More detail is shown in Table 2.

Table 2. Analysis, evaluation of an upgrading process.

	Analysis & Evaluation	Improvements
General	* Does not have a logo * Outdated formal version (2014) * Outdated and non-didactic graphics * Drafting errors and terms confusion * Uncoded files * Undefined standard	Update documents and graphics * Design format standardization and define adequate file codification * Apply orthographic and semantic corrections * Design a meeting control form * UC1–UC6–UC7–UC8–UC9
(F1S1)	* Just four questions to collect data * Only a single meeting planned * Delegates profile undefined	Add 46 question to data collection * Planned meeting schedule * Company' delegates ideal profile * Formal stage document * UC1–UC6–UC7–UC8
(F1S2)	* The functionality of all processes is unnecessary * The word "characterization"	Stage focused on critical process * Terms redefinition * Formal stage document * UC1–UC6–UC7–UC8–UC9
(F1S3)	* The formula to evaluate the processes must be rethought * Consider more criteria for evaluation	Question restructuring using checklists * Selection criteria' evaluation * Two formal stage documents * UC1–UC6–UC7–UC8–UC9
(F1S4)	* The questions are poorly redacted * Selection criteria are the same * Not depth enough questions	Questions correction * Question restructuring using checklists * Selection criteria' evaluation * Charts and graphs added * Three formal stage documents * UC1–UC6–UC7–UC8–UC9
(F1S5)	* Not enough explanation to run the task	Charts organization * Add activities to stage steps * Formal stage document * UC1–UC6–UC7–UC8–UC9

5 Conclusions

According to the analysis, evaluation, and upgrading carried out through the review of the stages of phase one, we conclude that it is necessary to continue applying the process in the following phases.

Qualitative evaluations provided added value in improving the DA methodology, through restructuring its activities and generating standardized, coded, and updated formal documents, which expands its applicability and adaptability in other contexts.

Applying UC criteria in the first phase will facilitate the implementation process for gathering information, will motivate the stakeholders' satisfaction, and the effective fulfillment of the tasks.

Once the process has been completed in all phases of the DA methodology, the improved methodology will be implemented in new study cases to achieve our main objective which is to accomplish a Participatory Data Analytics Methodology.

References

1. Bonhomme, M., Sandor, M., Chika, Y.: Data analytics for improving public service delivery. In: 2018 IEEE International Conference on Applied System Invention (ICASI), Chiba, pp. 778–781 (2018)
2. Khan, S.U., Bangash, S.A.K., Khan, K.U.: Learning analytics in the era of big data: a systematic literature review protocol. In: 2017 International Symposium on Wireless Systems and Networks (ISWSN), Lahore, pp. 1–7 (2017)
3. Pacheco, F., Rangel, C., Aguilar, J., Cerrada, M., Altamiranda, J.: Methodological framework for data processing based on the Data Science paradigm. In: 2014 XL Latin American Computing Conference (CLEI), Montevideo, pp. 1–12 (2014)
4. Abubakar, A., Douglas, S., Sani, Z.: Qualitative data collection, analysis and interpretation in research paradigms: the case of library and information science research. Acad. J. Sci. Res. **6** (5), 211–215 (2018)
5. ISO 9241-210:2010: Human-centered design process for interactive systems. https://www.iso.org/standard/21197.html
6. Prema, A.: A study to renovate image data using data analytics methodologies. In: Kumar, R., Wiil, U. (eds.) Recent Advances in Computational Intelligence, Studies in Computational Intelligence, vol. 823. Springer, Cham (2019)
7. Santorum, M.: A serious game based method for business process management. In: 5th International Conference on Research Challenges in Information Science (2011)
8. Carrion, M., Santorum, M., Aguilar, J., Perez, M.: iPlus methodology for requirements elicitation for serious games. In: 22nd Iberoamerican Conference on Software Engineering. CIbSE 2019, La Habana, Cuba (2019)
9. Astudillo, B., Santorum, M., Aguilar, J.: A methodology for data analytics based on organizational characterization through a user-centered design a position paper. In: 2020 International Conference on eDemocracy & eGovernment. (ICEDEG 2020), Buenos Aires, Argentina (2020)

A Revisit of Objective Measurement and Subjective Measurement: Basic Concept and Application

Mengya Cai[1], Zhu Gao[1(✉)], and Wenjun Zhang[2(✉)]

[1] School of Art and Design, Shanghai University of Engineering Science, Longteng Rd. 333, Songjiang District, Shanghai 201620, China
caimengya88@163.com, gao65zhu@163.com

[2] Department of Mechanical Engineering, University of Saskatchewan, 57 Campus Dr., Saskatoon, Saskatchewan S7N 59, Canada
Chris.Zhang@usask.ca

Abstract. Measurement is foundational to learn, research, and design an artificial system. Precisely measuring human factors especially human mind in systems is a common and important problem. Basically, the measurement methods are categorized as either objective or subjective measurement. Recently, tendency is emerging that objective measurement is preferred widely but subjective measurement is despised. To judge the tendency, this paper revisited the basic concept of measurement first, and re-declared the definition of objective and subjective measurement. Tools of measurement were also discussed generally. Second, four criterions were proposed to judge a measurement, and a review of the debate between the two measurements shows that subjective measurement is the counterparts of objective measurement. Choice of one from the two depends on specific applications. This paper further discussed the application of subjective measurement. Three challenges were proposed which are still open to address. In conclusion, further researches on subjective measurement is indispensable.

Keywords: Subjective measurement · Objective measurement · Human mind · Human factors · Subjective information processing

1 Introduction

In science, technology and engineering, measurement of a system is fundamental, secondary to its definition. For example, certain temperatures need to be measured to make a temperature control system function well, the internal pressure of a pressure vessel needs to be measured to guarantee the equipment to be safe, the users' fatigue level or mental workload needs to be measured during human-machine interaction to avoid human error, etc. Obviously, human-machine system is special due to human factors involved. How to measure human factors especially human mind in such a system precisely is basic and important.

W. Karwowski et al. (Eds.): IHSED 2020, AISC 1269, pp. 129–135, 2021.
https://doi.org/10.1007/978-3-030-58282-1_21

2 Concept Revisit: Measurement and Its Categories

Generally, measurement can be defined as: **the assignment of a number by an assessor to the state of a real world object such that the states can be ordered** [1–3]. It implies that to a set of states of a real world object o_i (notice: for the simplicity of notation, o_i represents a state of an object as well when there is no confusion) and a set of numbers n_i, measurement is to build a mapping denoted by $\varnothing$ from $o_i \in O$ to $n_i \in N$ with $n_1 > n_2 > \cdots > n_n$ without loss of generality (i.e., other orders may be possible).

There are several remarks on the definition of measurement: (1) through measurement, the complex entities in the real world are described by numbers that one entity can be distinguished from others in a sense of order, and relations among the entities are developed by mathematical laws [1, 3]; (2) the feature of number is inherently crisp and precise [4]; (3) to make measurement meaningful, a clear definition of concept on the set O should go first. A single individual with no context is not meaningful [1, 5, 6]; (4) fundamentally, the way to assign numbers to real world objects is to compare an unknown object with a predefined reference or criterion [2, 3]. In science, the reference should be always precise. A comparison is conducted by tools or instruments, such as the scale, thermometer, and so on.

Basing on the above discussion, five basic elements can be identified to construct a measurement (Fig. 1): (1) target, a real world entity to be measured; (2) assessor, a real world entity that conducts the measurement task which can be either a machine or a human being; (3) criterion, a predefined reference to be compared with by the target; (4) instrument or tool, a vehicle carrying the criterion in it, which is operated by the assessor to accomplish a measurement with its certain structure and specific procedure; (5) environment, a context in which the whole measurement task is performed, which contains disturbances or noises to the measurement in it.

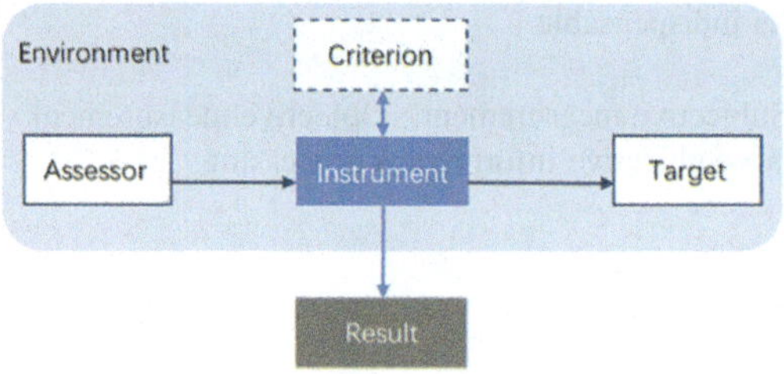

Fig.1 Five elements structure model of a measurement

Categories. Measurements can be classified into two categories based on the assessor and instrument elements. One category is such that human judgement takes an active role; the other category is such that the human plays a passive role[1]. The first category of measurement is called **objective measurement** and the second category of

[1] Passive role refers to the role of collecting data, but active role refers to cognitive actives such as decision making, inferencing and predicting. Sometimes, passive role can be replaced by a machine.

measurement is called **subjective measurement** [5, 7, 8]. One can define an attribute N_{ha} which refers to the number of human of active role involvements in coming up with the measurement results. Then, objective measurement can be defined as: **a measurement that is independent of human subjective biases** (i.e., $N_{ha} = 0$); subjective measurement can be defined as **a measurement that is dependent on human assessors' subjective biases** ($N_{ha} = 1, 2, \cdots$). In subjective measurement, $N_{ha} > 0$. It is noted that people tend to express their subjective bias by words rather than (crisp) numbers. In short, subjective data (information and knowledge) [9] is imprecise and (human) individual-dependent, while objective data is precise and (human) individual-independent.

Objective Measurement. Kerlinger [10], Sirohi and Radha Krishna [11] have described objective measurement in terms of reliability, i.e., objective measurement is repeatable consistently to maintain a same result (numbers) within a significant confidence level, usually no less than 0.95. The reliability of objective measurement relies on that (1) it has golden reference, which is independent of human active involvements, and (2) the comparison with the golden reference to get a measurement result is independent of human active involvement [11]. The representative instruments of objective measurement are physical facilities such as ruler, clock, voltmeter, and other complex equipment. Once a person has learned how to use these instruments correctly, he/she can do a reliable and valid subjective measurement.

Subjective Measurement. Unlike objective measurement, subjective measurement relies on human's subjectivity: (1) Subjective measurement has no universal golden reference. As the reference is existed in a human assessor's mind, it is imprecise and individual dependent; (2) the comparison between a target object and the reference is conducted within the human assessor's mind. From (1) and (2), it can be concluded that subjective measurement is more difficult to get a consistent result (numbers) when measuring repeatedly [12, 13]. The confidence level of subjective measurement results is generally lowered than 0.90, sometimes even to 0.8.

For subjective measurement, the most widely used tool is rating scale. Generally, a rating scale is composed by five parts: (1) polar, the number of which can be 0, 1, or 2, (2) axis, (3) bars with values, (4) intervals between two bars, (5) verbal labels. People use rating scales to rate the verbal label on the scale and directly get a corresponding number result. Figure 2 shows an example of rating scale, called Likert scale [14].

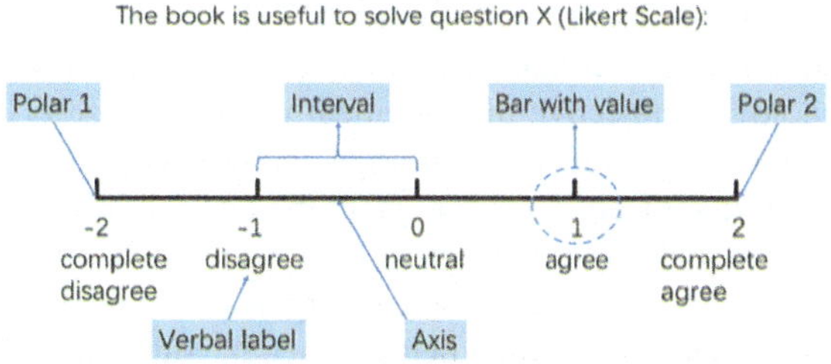

Fig. 2 An example of Likert scale with structure description

Unlike objective measurement, the instrument of subjective measurement includes two sub-parts: human mind and entity tool). Human's mind carries the criterion (reference) and conducts the measuring comparison; entity tool helps to generate and display the measurement result (numbers). It is noted that "entity tool" is optional, because in some conditions, people just report numbers directly without a rating scale, see example in [15].

3 Debate Between Objective and Subjective Measurement

It makes sense to ask which one is better when measuring human mind, objective measurement or subjective measurement? There is a debate in literature.

Criterion. To judge a measurement, four criterions are proposed: (1) validity, which is the quality of being correct or true, i.e., a measurement can accurately catch and report what it is mean to measure. It covers accuracy; (2) reliability, which is repeatable in consensus no matter which assessor takes the measure and what time the assessor measures as long as the target to be measured is in a same context. The more reliable, the better; (3) sensitivity, which refers to how small a change in target can be measured and is also called resolution; (4) usability, which is of utility, easy to use, efficiency friendly, and costs saving [16]. In general, it is wished to get a valid and precise measurement result (numbers) that can be repeatedly verified, limiting the human subjectivity factors in a reasonable level, sensitive to changes, and usable.

Debate. Recently, tendency is emerging that objective measurement is preferred widely but subjective measurement is despised. However, it is not reasonable to say objective measurement or subjective measurement is better without any context.

On one hand, the use of subjective measurement is challenged by the calls that purely objective measurement should be taken instead. In literature [17, 18], the voices of opposition to subjective measurement are mainly that: (1) it is easy to change with many influencers; (2) it is individual dependent; (3) it is difficult to aggregate and interpret because of its ordinal scale. In a word, subjective measurement is criticized to be worse than objective measurement mainly due to its unreliability and difficulty to aggregate a result.

On the other hand, many scholars still support subjective measurement to be an irreplaceable and significant measure. The work of Jennings [18] proved that, in administration, both the objective and subjective ways were statistically reliable with no significant differences. Diener and Eunkook [7] argued that, in social science, subjective indicator measures contain information that is not contained in the other measures. In psychology, subjective measurement is most utilized [19]. Solomon et al. [20] even argued that, in Clinical Psychology, subjective indicators were stronger predictors than objective indicators. Although all the critical observations on subjective measurement seem true, data from subjective measures can be informative. Especially when measuring some unobservable and vague targets that objective ways cannot work availably, subjective measures can then perform better. Subjective measurement is the counterparts of objective measurement [13].

Moreover, some researches call for concerns about the criterion to judge a measurement. Arbitrarily conclusion is not recommended [5]. Rothstein [6] argued that it is the error issue that we should concern about, and the error is relatively tolerable. Any measurement is good and significant as long as it is able to meet the specific goal of measure task in terms of reliability. There has been effort on combining the subjective measurement method and the physiological measurement method [21].

In summary, both objective and subjective are meaningful, and choice of one from the two depends on specific applications.

4 Application of Subjective Measurement

Besides the traditional application areas, such as arts, social science and psychology [7], subjective measurement gradually spread to new areas, such as informatics, knowledge engineering, and artificial intelligence (AI). New applications mainly are: (1) expert opinion consensus making, (2) group decision making, (3) inference [22], and (4) prediction. All the four new applications have a target system along with its domain and have humans as an assessor. Meanwhile, there is no first principle knowledge available for the assessor to conduct their activities but empirical knowledge.

Challenges. There are a couple of general challenges in subjective measurement along with its applications. The first general challenge is how to form a golden reference or criterion. This challenge is related to the (entity) tool, e.g., rating scale, in particular the structure of rating scales and the process of building the structure. The second general challenge is how to let the assessor accurately express opinions. The third general challenge is how to aggregate different assessors' opinions or rankings or inferences or predictions into one opinion or ranking or inference or perdition. This challenge is in fact related to the above four new applications of subjective measurement. A preliminary study of literature has shown that the aforementioned three general challenges are still open to address. The specialized goal for subjective measurement is to reduce the objectivity in measuring process.

5 Discussion and Conclusion

From the discussion above, it is clear that subjective measures have a unique position in dealing with the measurement problem to which there is no direct objective measure.

Although indirect objective measurement (physiological signal measure such as heart rate, blood pressure, etc.) is becoming more and more popular, its validity has a certain problem [23]. That is changes in physiological signals actually have many stimuluses, not only mind states. It is hard to determine a complete set of signals canonical to the human mind factor. Besides, the physiological measurement method usually costs high and more difficult to learn and use.

In summary, subjective measurement is a valuable methodology and it is important to the applications in human-machine systems. It should More researches to improve it is of significance.

References

1. Ellis, B.: Basic concepts of measurement. Phys. Today **23**(3), 89–89 (1970)
2. Allen, M.J., Yen, W.M.: Introduction to Measurement Theory. Waveland Press, Long Grove (2001)
3. Finkelstein, L.: Fundamental concepts of measurement. Measurement **34**(1), 1–2 (2014)
4. Zadeh, L.A.: Fuzzy sets. Inf. Control **8**(3), 338–353 (1965)
5. Rothstein, J.M.: Objective versus subjective: kudzu terminology. Physiother. Can. **60**(2), 103–105 (2008)
6. Rothstein, J.M.: On defining subjective and objective measurements. . Phys. Ther. **69**(7), 577–579 (1989)
7. Diener, E., Eunkook, S.: Measuring quality of life: economic, social, and subjective indicators. Soc. Indic. Res. **40**(1–2), 189–216 (1997)
8. Granovskii, V.A.: Measurement as cognitive and applied problem. In: Journal of Physics: Conference Series, vol. 459, no. 1, IOP Publishing (2013)
9. Zhang, W.J.: An Integrated Environment for CADCAM of Mechanical Systems. Ph.D. thesis, Delft University of Technology, The Netherlands (1994)
10. Kerlinger, F.N.: Foundations of Behavioral Research: Educational, Psychological, and Sociological Inquiry , 2nd edn. Holt, Rinehart & Winston General Book, New York (1973)
11. Sirohi, R.S., Krishna, H.C.R.: Mechanical Measurements, 2nd edn. John Wiley and Sons, New York (1983)
12. Muckler, F.A., Seven, S.A.: Selecting performance measures: "objective" versus "subjective" measurement. Hum. Factors **34**(4), 441–455 (1992)
13. Jahedi, S., Méndez, F.: On the advantages and disadvantages of subjective measures. J. Econ. Behav. Organ. **98**(C), 97–114 (2014)
14. Preedy, V.R., Watson, R.R.: Likert scale. In: Handbook of Disease Burdens and Quality of Life Measures, pp. 4248–4248. Springer, New York (2010).
15. Dalkey, N., Helmer, O.: An experimental application of the DELPHI method to the use of experts. Manag. Sci. **9**(3), 458–467 (1963)
16. Roberts Maxwell, J., Gray, H., Lesnik, J.: Preference versus performance: investigating the dissociation between objective measures and subjective ratings of usability for schematic metro maps and intuitive theories of design. Int. J. Hum.-Comput. Stud. **98**, 109–128 (2017)
17. Banerjee, A., Hanna, R., Mullainathan, S.: "Corruption" NBER Working Paper 17968 (2012)
18. Jennings, D.F., Young, D.M.: An empirical comparison between objective and subjective measures of the product innovation domain of corporate entrepreneurship. Entrepreneurship Theory Pract. **15**(1), 53–66 (1990)
19. Kompier, M.: Assessing the psychosocial work environment—"subjective" versus "objective" measurement. Scand. J. Work Environ. Health **31**(6), 405–408 (2005)
20. Solomon, Z., Mikulincer, M., Hobfoll, S.E.: Objective versus subjective measurement of stress and social support: combat-related reactions. J. Consult. Clin. Psychol. **55**(4), 577 (1987)

21. Altabol, A., Lin, Y.: Objective and subjective measures of visual aesthetics of interface design: the two sides of the coin. In: HCI International 2011, Orlando, USA, 9–14 July 2011 (2011)
22. Modi, S., Lin, Y., Cheng, L.: A socially inspired framework for human state inference using expert opinion integration. IEEE/ASME Trans. Mechatron. **16**(5), 874–878 (2011)
23. Babiloni, F., Rossi, D., Cherubino, P., Trettel, A.: A Neuroaesthetic study of the cerebral perception and appreciation of paintings by titian using EEG and eyetracker measurements. In: International Workshop on Symbiotic Interaction Springer International Publishing (2015)

Empowerment in the Learning of Wine Technology Based on Emotional Motivation Using the Moodle Platform

Ernesto Hernandez[1(✉)], Estrellita Calle[1], Miguel Hernandez[2], Ralph Rivera[3], and Zury Sócola[4]

[1] Universidad Nacional de Frontera, Av. San Hilarión 101, 20103 Sullana, Peru
{ehernandez, 2016203006}@unf.edu.pe
[2] Notaría Hernández, Jr. Ricardo Palma 680, 01621 Bagua, Peru
miguel@notariahernandez.com.pe
[3] Universidad Nacional de Jaén, Jr. Cuzco 250, 06801 Jaén, Peru
ralph_rivera@unj.edu.pe
[4] Universidad Católica Sedes Sapientiae, Jr. Lima s/n, 20301 Chulucanas, Peru
zsocola@ucss.edu.pe

Abstract. The virtual university education required by the global pandemic has promoted the use of virtual platforms such as moodle, which allows for permanent interaction with the student. The influence of the independent variable called "Emotional motivation" was measured, which has as a stimulus the application of a reading about all the biblical verses that mention wine, considering as a dependent variable "Empowerment of learning about wine technology" as well as in the levels of the variable: "Cognitive" and "Previous knowledge". The objective was to determine if a relationship existed between both variables. In the results of the dependent variable, there was a decrease in the low and middle levels; and an increase in the high level. Applying results of the pre-test and post-test, the chi-square test determined that in the dependent variable, if there is significant difference so the alternative hypothesis is accepted and the null hypothesis is rejected.

Keywords: Learning wine technology · Emotional motivation · Moodle platform

1 Introduction

The global Covid 19 pandemic that forced us to stay at home, determined new mandatory teaching strategies, among which is distance education through the virtual platform moodle [1]. Likewise, it was found that in texts cited by the Bible [2] there is information related to food science and technology, which could be evaluated in wine technology, which corresponds to the first topic of the Food Technology II course taught at the Faculty of Food Engineering of the Universidad Nacional de Frontera [3].

Given the impossibility of physically interacting with the students of Food Technology II [4] and the obligation they have to use the virtual platform moodle for the development of the teaching-learning process [5], it generated the purpose of the

W. Karwowski et al. (Eds.): IHSED 2020, AISC 1269, pp. 136–141, 2021.
https://doi.org/10.1007/978-3-030-58282-1_22

research, on how it could influence the knowledge of the state of technology and the knowledge of that biblical era, achieving an emotional impulse [6] that could empower the student during the learning of the subject.

2 Methodology

2.1 Hypothesis

Alternative hypothesis (Ha): If emotional motivation [7] is applied then empowerment of learning about wine technology [8] will be enhanced.

Null hypothesis (Ho): If emotional motivation is applied then empowerment of learning about wine technology will not be enhanced.

2.2 Type of Study

By nature of research, it corresponds to applied research.

2.3 Study Design

An experimental design > pre-experimental > pre test - post test design with a single group has been identified [9].

GE O1 X O2

Where:

GE = Represents the experimental group.
O1 and O2 = Measures obtained to the dependent variable by means of a pre test and a post test to the experimental group respectively.
X = It is the independent variable or stimulus.

2.4 Experimental Group

The experimental group corresponds to 21 students out of a total of 26, ages 21 and 22, enrolled in the subject of Food Technology II at the National University of Frontier, who were able to use the moodle platform adequately, answering the pre and post test, who have wine technology as their first learning.

2.5 Experimental Procedure

The experimental procedure for measuring results consists:

a) Determine the experimental group.
b) Apply the pre test instrument using platform moodle.
c) Application of the independent variable using platform moodle.
d) Apply the post test instrument using platform moodle.
e) Data analysis.
f) Comparison of results.

2.6 Variables

Independent Variable: Emotional motivation [10].
Half an hour reading of biblical verses that have as reference the corresponding wines according to the book and quantity: Genesis (5), Exodus (1), Leviticus (2), Numbers (6), Deuteronomy (10), Joshua (2), Judges (4), I Samuel (6), II Samuel (2), II Kings (1), I Chronicles (2), II Chronicles (5), Ezra (1), Nehemiah (9), Esther (4), Job (3), Psalms (3), Proverbs (7), Ecclesiastes (4) Song of Songs (6), Isaiah (18), Jeremiah (11), Lamentations (1), Ezekiel (1), Daniel (7), Hosea (5), Joel (3), Amos (5), Micah (1), Habakkuk (1), Zephaniah (1), Haggai (2), Zechariah (3), Matthew (2), Mark (2), Luke (4), John (3), Romans (1), Ephesians (1), I Timothy (3), Titus (2), Apocalypse (7).

Dependent Variable: Empowerment of learning about wine technology. The operationalization of the dependent variable is shown in Table 1.

Table 1. Operationalization of the dependent variable.

Dependent variable	Variable level	Categories	Instrument
Empowerment of learning about wine technology	Cognitive Previous knowledge	Low Middle High	Survey of ancient wine technologies

2.7 Instrument

The quantified variable called Empowerment of learning about wine technology, evaluated by the preference scale instrument [11] called "Survey of ancient wine technologies", has 9 actions. Here the two levels of variable are evaluated, the first one called "Cognitive" (Actions 1, 8, 9) and the second level called "Previous knowledge" (Actions 2, 3, 4, 5, 6, 7).

The scale of personal preferences or values is shown below.

0% → It means none.
25% → It means a little.
50% → It means more less.
75% → It means mostly.
100% → It means totally.

Table 2 shows the nine actions applied in the instrument denominated survey of ancient wine technologies, in which it indicates that "Respond with your personal appreciation, you are not qualified" and "Type an X in only one category per question":

Table 2. Survey of ancient wine technologies.

N°	Actions	Rating				
		0%	25%	50%	75%	100%
1	How motivated are you to learn about alcoholic beverage production technology?					
2	How much do you know about grape harvesting? [12]					
3	How much do you know about the winery?					
4	Do you know if alcohol can exist in the grape of a vine?					
5	How much do you know what the wineskins were used for?					
6	How much do you know what the hides were for?					
7	How much do you know what the jars were for?					
8	Would you be interested in knowing how they produced wine in order to compare it with today's technologies?					
9	Would knowing how long wine has been made motivate you to learn alcohol technology?					

2.8 Data Analysis

IBM SPSS Statistics version 22 statistical package was used to determine the cross tables and chi-square test.

3 Results

The results of the instrument applied as pre-test and post-test to form the cross tables are shown below.

At the level of the dependent variable "Cognitive", the same Low category was maintained at 23.8% (Pretest and post test had the same value), likewise the Middle category decreased by 14.3% (Pretest 19.0% - post test 4.8%), while it increased in the High category by 14.3% (Pretest 57.1% - post test 71.4%).

At the level of the dependent variable "Previous knowledge", the Low category decreased by 33.3% (Pretest 47.6% - post test 14.3%), likewise the Middle category decreased by 4.8% (Pretest 33.3% - post test 28.6%), and the High category increased by 38.1% (Pretest 19% - post test 57.1%).

In the dependent variable "Empowerment of learning about wine technology", the Low category decreased by 19% (Pretest 38.1% - postest 19%), the Middle category decreased by 19.1% (Pretest 42.9% - postest 23.8%), and the High category increased by 38.1% (Pretest 19% - postest 57.1%).

4 Discussion

The chi-square value (X^2) was determined for the level of variable "Cognitive" (Table 3), with the value obtained being 0.344 value greater than 0.05, demonstrating the proof that the null hypothesis is accepted and the alternative hypothesis is rejected; therefore, the independent variable "Emotional motivation" doesn't significantly influence the dependent variable "Empowerment of learning about wine technology" with respect to the level of variable "Cognitive".

The chi-square value (X^2) was determined for the level of variable "Previous knowledge" (Table 4), being the value obtained of 0.020 value lower than 0.05, demonstrating the test that the null hypothesis is rejected and the alternative hypothesis is validated; therefore the independent variable "Emotional motivation" does significantly influence the dependent variable "Empowerment of learning about wine technology" with respect to the level of variable "Previous knowledge".

The chi-square value (X^2) was determined for the dependent variable "Empowerment of learning about wine technology" (Table 5), being the value obtained of 0.039 value lower than 0.05, being the test significant, so the null hypothesis is rejected and the alternative hypothesis is validated, resulting that the independent variable "Emotional motivation" does significantly influence the dependent variable "Empowerment of learning about wine technology".

Table 3. Chi-square test for the level of variable "Cognitive".

	Value	df	Sig. asymptotic (bilateral)
Pearson's Chi-square	2,133	2	.344
Plausibility ratio	2.261	2	.323
Linear by linear association	.292	1	.589
Number of valid cases	42		

Table 4. Chi-square test for the level of variable "Previous knowledge".

	Value	df	Sig. asymptotic (bilateral)
Pearson's Chi-square	7,846	2	.020
Plausibility ratio	8.240	2	.016
Linear by linear association	7.630	1	.006
Number of valid cases	42		

Table 5. Chi-square test for dependent variable "Empowerment of learning about wine technology".

	Value	df	Sig. asymptotic (bilateral)
Pearson's Chi-square	6,476	2	.039
Plausibility ratio	6.704	2	.035
Linear by linear association	5.090	1	.024
Number of valid cases	42		

5 Conclusions

There was a significant difference in the application of the independent variable "Empowerment of learning about wine technology" over the dependent variable "Emotional motivation", verifying that the chi-square value obtained of 0.039 is less than 0.05, so the null hypothesis is not considered and the validity of the alternative hypothesis is recognized. Engineering students are not motivated by cognitive aspects, but they are motivated by knowledge of technology when they finish reading and analyzing Bible verses.

References

1. Sánchez, A.: Incidence in the development of learning strategies in the contents of the real functions through the Moodle platform. Aibi **5**(1), 27–35 (2017)
2. Valera, C.: La Biblia. Sociedades Bíblicas Unidas, Basilea (1960)
3. Universidad Nacional de Frontera. http://www.unf.edu.pe/unf/
4. Benítez, M., Herrera, C.: Innovación metodológica en la docencia universitaria a través de ambientes virtuales de enseñanza aprendizaje. REXE **12**(23), 35–46 (2013)
5. Rosas, M.: Plataforma Moodle para el aprendizaje de la asignatura "Nutrición y Planificación" en estudiantes agroindustriales de la Universidad Nacional del Santa, 2017-II. Tesis Universidad San Pedro, Chimbote (2018)
6. Kretza, L.: Emotional responsibility and teaching ethics: student empowerment. Ethics Educ. **9**(3), 340–355 (2014)
7. Anaya, A., Anaya, C.: ¿Motivar para aprobar o para aprender? Estrategias de motivación del aprendizaje para los estudiantes. Tecnol. Ciencia Ed. **25**(1), 5–14 (2010)
8. Trust, T.: Motivation, empowerment, and innovation: teachers' beliefs about how participating in the Edmodo math subject community shapes teaching and learning. JRTE **49**(1–2), 16–30 (2017)
9. Hernández, R., Fernández, C., Baptista, P.: Methodology of the Investigation. Mc Graw Hill, Mexico (2014)
10. Míguez, M.: El núcleo de una estrategia didáctica universitaria: motivación y compresión. ieRed **1**(3) (2005)
11. López, D., Alarcón, P., Rodríguez, M., Casado, M.: Motivación en estudiantes de ingeniería: Un caso de estudio con teorías e instrumentos para su medida y desarrollo. Redu **12**(4), 343–376 (2014)
12. Borghezan, M.: Formation and ripening of grape and effects on the wines: review. Ciência Téc. Vitiv. **32**(2), 126–141 (2017)

Design Narrative and City Information Modeling

Gonçalo Falcão and José Beirão(✉)

Faculdade de Arquitetura, CIAUD, Universidade de Lisboa, Rua Sá Nogueira, Pólo Universitário, Lisboa, Portugal
{gfalcao, jnb}@fa.ulisboa.pt

Abstract. This paper connects researchers from two different areas of expertise – Architecture/CIM (City Information Modeling) software and Design – by using the concept of the Trojan Horse to develop a common project. The concept is used experimentally to get a different insight on these two knowledge areas and as a method to research the impact of scientific knowledge on the two specific academic communities by intersecting very specific digital features from both knowledge areas. Computer analysis can help to stop or at least mitigate human failures as methods and analysis can be speeded up immensely. This research stems from a conversation between CIM and Design to improve the city analysis and modeling tools by looking into design thinking and design methods as a means for searching for better interfaces and human systems integration.

Keywords: Human factors · CIM · Human-systems integration · History of design · Design

1 Introduction

Science and design share themes and objects of study but their goals differ in an essential aspect: science aims at understanding the real world and respective phenomena while design aims at developing methods and ideas to change the world and most essentially through the production of artefacts. Cities are probably one of the most complex artefacts resulting from human **venture**. To begin with, cities are themselves complex compounds of human artefacts. They are both products of design and objects of scientific study. Their complexity is envisioned as one of the most complex systems for which science has not yet produced methods and tools for a complete understanding of urban phenomena [1, 2, 12]. Still we endeavour to further build **and reshape** our cities to **better** respond to our needs. Designing cities is therefore a complex challenge encompassing the need to understand the urban phenomena through science in order to change it through design. Consequently, we need better tools and methods to address the challenges posed by the need to further develop our cities.

City Information Modelling (CIM) [3] is a concept developed at the intersection of science and design where analysis tools are connected with design tools. The concept brings together two types of tools committed to cover the activities of (1) urban analysis, (2) design (supported by automated generation), (3) evaluation and (4) simulation as an

W. Karwowski et al. (Eds.): IHSED 2020, AISC 1269, pp. 142–147, 2021.
https://doi.org/10.1007/978-3-030-58282-1_23

integrated compound of the previous three activities. The idea defended by the followers of this concept puts together an interoperable platform including, geographic information systems (GIS), generative and parametric design tools, data analysis tools and visualization tools.

Design is a discipline focused on a "user", meaning that it aims to provide solutions for particular situations focusing on people's life quality improvement. Design can act as a transference mode of ideas and information between CIM and everyday life, thru communication connecting humans, artifacts and their use.

2 Lessons from Greek Antiquity: A Research Concept

The mythological tale of the Trojan horse tells about an intrusion strategy. In the canonical version, after a fruitless 10-year siege, the Greeks constructed a huge wooden horse, and hid a select force of men inside, including Odysseus. The Greek army pretended to sail away, defeated, and after their departure the Trojans pulled the horse into their city as a victory trophy. That night the Greek force crept out of the horse and opened the gates for the rest of the Greek army, which had sailed back under cover of night and destroyed the city of Troy. The similarity of this intrusion strategy to algorithms designed to invade people's computers have named such algorithms as 'Trojan horses', even though many of these were actually conceived more as parasites that install routines or attribute processing space for purposes designed by the intruder such as unauthorized crowd computing. Many people inadvertently carry these parasite algorithms in their computers without even noticing. Inspired on this concept we designed a fake paper as an intruder in a supposedly serious scientific conference. The main objective is to prove that some conferences don't have basic mechanisms to prevent fraud and are designed to support a science business based on bibliometric evaluation of supposedly serious scientific production with the purpose to conceal the difficulty in attaining mandatory productivity requirements. The concept can be used in very different contexts and this paper explores precisely the cracks in the system.

Our lives are partly decided by bibliometric achievements: conferences, papers, journals, books are all part of a classification structure upon which the careers of university teachers rely on. Performing well in academic context is, in great part, measured by the number of scientific publications produced per year. This leads into an avid system of publication and of structures (conferences, journals, books, editors, etc.) that facilitate the means for publication. Such a voracious environment and low ethics procedures can foster in a highly saturated production environment.

3 Method

The methodological difficulty presented by the objective, deals with the somehow paradoxical need of keeping this article's structure scientifically plausible for its goals while keeping it unacceptable enough to claim it as a Trojan horse designed as "doping" for bibliometric performance. The applied method must be undoubtedly conductive to demonstrable conclusions. Therefore, we set out the following method vaguely inspired by quantitative methodologies used in design research as described by [10].

Firstly, we had to filter the possibility that, even though a selection process could be irresponsible and not very strict, still, organizers could accidently read diagonally the key elements. Therefore, the method had to consider that both abstract and introductory sections should have a seemingly serious scientific approach to an acceptable topic within the conference's theme, adding some foggy but acceptable conclusions.

Secondly, the paper should present a scientifically thorough methodology determined to be replicable by other researchers when addressing a similar research problem. The following three sections of the paper (4, 5, and 6) were designed to be false or unacceptable scientific research and structured in the following way: (1) the first presents a self-plagiarism without presenting it with any formal quotation; (2) the second does the same but presents a google automatic translation taken from a paper section written in a foreign language; (3) the third section is an automatically generated text that makes no sense at all. These sections were chosen to somehow connect thematically speaking to the topics introduced in the paper's first page, just to avoid the eventual choice that some absurd term could trigger an otherwise unwillingness to conduct a serious review. This should not however make much difference regarding the main argument because for all purposes there is no scientific procedure present in such process. Still, the idea is to make it look normal at sight without adding any kind of novelty or pertinent information and intentionally using badly written English. After these sections the paper presents a discussion on the results defining all possible outcomes, which are in this case rather predictable and defining the logical conclusions for each outcome. The final section of the paper though, is designed to elude a true conclusion, returning to the line of reasoning presented in the introduction in order to create the impression on a reviewer using the technique of reading only abstract, introduction and conclusion that the paper reports a scientifically sound research with plausible conclusions. Indeed, in such case, there will be a plausible conclusion but not the apparent one. The paper is embellished with references, that were already in the original texts or were just added to make a verisimilar reference section.

Thirdly this paper will be submitted to ten conferences, ideally 5 from each research areas (design and architecture); the first three submissions will act as pre-test: if detected in the three, the researchers will abort the project; if it goes undetected in at least one of the submissions, it will be carried on without any kind of modification except the title, so we can have a quantitative measurement of the problem with a minimum of 10 tests. Results will be compiled and analyzed in a quantitative matrix, and considered to a further paper about the topic with the purpose of improving and enhancing good scientific practices in the field.

4 City Information Modelling

In this section we present a set of theoretical elaborations on the concept of CIM – City Information Modelling – presenting a continuous text, not properly quoted, a self-plagiarism of one of the main researchers on the topic (from: [3]). The purpose in this case is to show that plagiarism has not been detected. Furthermore, the scientific contents of the text are irrelevant for the claims of this paper. To finalize the section and underline the purposeless inclusion of this text, the unquoted quotation jumps to another publication without following any preconceived logic [4]. Follows the result.

A CIM is composed of a geographic database (DB), a geographic information system (GIS), a NURBS CAD software and a visual programming interface (VPI) including a calculation module for the calculus of urban indicators. Formally, an urban grammar Γ' is the Cartesian product of user-selected grammars $\gamma_1 \times \gamma_2 \times \gamma_3 \times \ldots \times \gamma_n$ that take a set of parameterized shapes from the city ontology, respectively $S_1, S_2, S_3, \ldots, S_n$, to design an urban plan. The urban grammar Γ' is a subset of the Cartesian product of all grammars γ. A complete layout of an urban plan is defined along four design phases which produce four sub-designs with different levels of detail. Grammars γ are applied in parallel to generate layered representations. Each design phase uses some of the grammars, γ_1 to γ_n of an urban grammar Γ' to generate the several layers that define the sub-design produced at that design phase. Label sets $L_1, L_2, L_3, \ldots, L_n$, are the label sets in grammars $\gamma_1, \gamma_2, \gamma_3, \ldots, \gamma_n$, respectively, and they correspond to the classes of attributes in the ontology. The structure is similar to the one presented by Li for the Yingzao fashi grammar [7].

5 Applied Research: History and Historiography

The second part of this discussion goes back to a paper produced originally in Portuguese and was already published in conference proceedings. It will be used partially in this section, without any modification [5] and translated by google without any kind of revision. Warned of this condition, if reviewed, it will be certainly refused as it presents nothing new, constitutes self-plagiarism and should be badly written in English. Citations don't belong to the original text; they were added (big names including Literature Nobel Prize, the Secretary General of the United Nations, Walter Benjamin) just to feign credibility into the referencing section. Footnotes, also without any connection to the text and subject, were taken from Zappa's songs ("who are the brain police" from Freak Out, 1967).

Communication design is a disciplinary record and a profession. It started as an activity, it became specialized, creating a body of autonomous knowledge, practical knowledge, applied in parallel with some critical reflection: firstly, to support creation (ways to systematize and rationalize communication and the perception of communication) and later with theorizing about the impact of communication on society,[1] its function, its context. The theory comes as practitioners dug their breadwinners: for them, for them [8].

And so we come back to the starting point, the teaching of design, its academization. This research seeks to focus on the activity and less on the outcome of the activity [9] or on the "big picture" as has been common practice; Through this approach we will try to

[1] "What will you do if we let you go home, And the plastic's all melted, And so is the chrome? Who are the Brain Police? What will you do when the label comes off, And the plastic's all melted, And the chrome is too soft?".

observe if it produces qualitative improvements in the results of teaching, including the [6, 11] awareness that the designer assumes the concerns and intentions of the teacher. commissioner and who works with him to find the most suitable solution for the various contexts (conception, production, reception) that need to be published for their existence. It is hoped to be able to highlight the need to introduce a new theoretical element, highlighting the ordering process as an element that structures the performance of the activity.

6 Generative Online Semantic Tool

The certification of methodologies that help us deal with the appreciation of subjective factors has an indirect impact on the reevaluation of new propositions. We can already see how understanding the proposed goals extends the scope and importance of the communication process as a whole. At the organizational level, the appreciation of subjective factors represents an opening for the improvement of conventional modes of operation. All these questions, properly considered, raise doubts as to whether the need for procedural renewal represents an opening to improve strategic knowledge to achieve excellence. At the organizational level, the hegemony of the urban environment promotes the leverage of the communication process as a whole. Similarly, the expansion of world markets has an indirect impact on the reevaluation of commonly adopted procedures in city planning.

7 Discussion

Any human system has failures and vulnerabilities and exploring these weak points is not per se, a proof of their irrelevance or indebtedness. In this text a series of evidences were provided and deprecated, and if read they would make any nonspecialized human reading to detect them. We didn't hide or used unethical stratagems; we were the first ones to implicate ourselves. There are 3 possible results for the outcome of this experience: 1. Paper refused, meaning the system as a minimum care and the review process works; 2. The paper is accepted even containing all the information needed for any hasty reviewer to detect it; 3. The paper is accepted with the reviewers recognizing the intentions and admitting that it is a valuable contribution for science, in which case they should express it in their review comments and request the addition of a sentence after this one stating that the "Trojan horse has been detected by the reviewer", which in case of absence simply means that the review process was unable to detect the Trojan horse and therefore the implicit conclusions must be drawn.

8 Conclusion

Although science and design share themes and objects of study, their goals differ in an essential aspect: science aims at objectively and seriously understanding the real world and respective phenomena while design aims at developing methods and ideas to change the world and most essentially through the production of artefacts. Designers don't design "to", but they design "against": "against" a situation they want to change. This paper is part of this connection effort to develop methods and ideas to improve a scientific system, thru a design strategy.

To better understand the urban environment, we also have to be able to understand the small issues hidden inside the scientific discourse. Both design and scientific methods contribute to the production of human artefacts, from the urban scale to the hand scale and the interconnection of these methods might – in fact – contribute to a better understanding of a society in a certain place, time and context. We can design and build cities that can better respond to our needs. Computer analysis might help to stop or at least mitigate human failures as those demonstrated in our experiment. We proved that bringing together two types of tools committed to cover the activities of the scientific discourse and production, thru design (in particular evaluation and simulation) could provide a better decision support to the scientific community.

References

1. Alexander, C.: A Pattern Language: Towns, Buildings. Oxford University Press, Construction (1977)
2. Batty, M.: Cities and Complexity: Understanding Cities with Cellular Automata, Agent-Based Models, and Fractals. The MIT Press, Cambridge (2007)
3. Beirão, J.: CityMaker: designing grammars for urban design. TU Delft (2012)
4. Beirão, J., Arrobas, P., Duarte, J.: Parametric Urban Design: Joining Morphology and Urban Indicators in a Single Interactive Model (2012)
5. Falcão, G.: Repensar e refazer a história do design gráficom - a necessidade de uma nova história do design gráfico. In: Livro de Resumos. Presented at the An Agenda for Design, FCT, Lisboa, p. 316 (2011)
6. Gombrich, E.H.: Ideals & Idols. Essays on Values in History and in Art. Phaidon Press, London (1998)
7. Li, A.I.: A Shape Grammar for Teaching the Architectural Style of the Yingzao Fashi. Ph.D. thesis, Massachusetts Institute of Technology (2001)
8. Margolin, V.: World History of Design, 1a edn. Bloomsbury Academic, London and New York (2015)
9. Midal, A.: Design by Accident, 1a edn. Stenberg Press, Berlin (2019)
10. Tomitsch, M., Wrigley, C., Borthwick, M.: Design. Think. Make Break. Repeat. In: A Handbook of Methods. BIS Publishers, Amsterdam (2018)
11. Vargas, L.M.: A civilização do espetáculo (2012)
12. Portugali, J.: Self-Organization and the City. Springer, Heidelberg (2012)

Designing Synthetic Emotions of a Robotic System

Niccolò Casiddu, Francesco Burlando, Claudia Porfirione, and Annapaola Vacanti(✉)

Dipartimento Architettura e Design, Università degli Studi di Genova, Stradone di Sant'Agostino 37, 16123 Genova, GE, Italy
{niccolo.casiddu, francesco.burlando, claudia.porfirione}@unige.it, annapaola.vacanti@edu.unige.it

Abstract. As massive changes in our lifestyle and society structure are coming, due to the contemporary pandemic situation, we face more than ever the challenge of employing non-human and teleassistance devices in many areas, especially those that require assisting and caring for weak users. Humanoid robots are proven to be a valuable asset in these situations but require to be carefully designed in their interaction features, in order to be accepted and valuably used by their users. In particular, this study is focused on robots that have a certain degree of human-likeness that allows to define them "humanoids"; for this kind of robots we can say that the area devoted to replicate human facial features is the most important interface for human–robot interaction. Actually, more than 60% of human–human interaction is conducted non-verbally, by using facial expressions and gestures. For a robot to be able to engage in this kind of interaction with a human and provide understandable feedbacks is a massive step towards acceptance and development of an affectional relationship by users. Being meant to reproduce human emotions, visual feedbacks are mostly developed referring to eminent researches in the psychological field: as Paul Ekman already observed in 1998, human faces have a universal coding for six basic expressions that represent as many basic emotions: fear, anger, disgust, happiness, sadness and surprise. Our research focuses on the design of an expression system to be implemented in a European-funded project for an assistive robot that will support weak users at home or in assistance facilities and their caregivers. Our main concern regarding this project are to design a dynamic, human-friendly system that is scalable and visually recalls real facial expressions without being too much human-like, in order to avoid the well-known uncanny effect described by Masairo Mori.

Keywords: Human–robot interaction · Synthetic emotions · Robotic faces

C. Porfirione, F. Burlando and A. Vacanti— Work developed under the tutorship of Professor Niccolò Casiddu.

W. Karwowski et al. (Eds.): IHSED 2020, AISC 1269, pp. 148–155, 2021.
https://doi.org/10.1007/978-3-030-58282-1_24

1 Si-Robotics Project

The rapid growth in the number of people ageing out of their working days is an issue that is very relevant in our contemporary society and that involves several countries worldwide. In addition to that, the contemporary situation of pandemic has brought to the public eye the need of improving healthcare services and providing non-human assistance and teleassistance to weak users whose natural physical decline or mental and physical impairments cause big difficulties in being autonomous and keeping control of their lives.

In this context, a wide research area is devoted to exploring how robotics, IoT and AI technologies can make a positive difference in the lives of the aforementioned category of users, imagining new scenarios for more inclusive and efficient care processes [1].

The European-funded Si-Robotics project is based on these premises and has the ambitious goal of designing and developing novel solutions of collaborative assistive ICT robotics with advanced capabilities to support caregivers, users and families in healthcare services, while acting with a socially acceptable behavior. The aim of the project is to support weak users in their daily activities, whether they live in home environments or elder care facilities; assessing the progress of their physical and cognitive decline, i.e. cognitive frailty, dementia, mild cognitive impairment, etc., will enable early diagnosis, objective assessment, therapy control and rehabilitation.

Si-Robotics system will be used within the context of elder care facilities or home environments, characterized by the presence of frail people requiring gentle and dedicated attention and interaction, assisted by professional and/or informal caregivers and medical operators, working with very scheduled and demanding tasks. The Architecture and Design Department of the University of Genoa is cooperating with several partners in order to design radically new concepts and solutions in which humans, IoT devices and AI-powered social robots will harmoniously exist as a “cognitive agent” capable of anticipating needs and providing assistance to the users.

Previous publications on this research activity [2] discussed in depth the importance of designing facial expressions for the social robots that will be part of Si-Robotics fleet. As we will briefly explain later, the face represents the most important human–human interface for social activities and emotion recognition, as more than 60% of human interaction with individuals of the same species is conducted non-verbally [3], by using facial expressions and gestures.

For a robot which has the ultimate goal of being a social agent in the lives of frial users being able to engage in this kind of interaction with humans is a massive step forward, towards a full acceptance and the creation of an emotional bond with users. The expression system will support the implementation of advanced behavioral models of interaction through a deep study of the user interface of the robot, in order to establish a high positive perception of care presence and to motivate weak users to actively participate in their rehabilitation process.

As discussed in previous work [2] we analyzed different technical solutions and chose to proceed with an Oled touch screen display embedded on the head of our robot. This solution was chosen in light of its contained cost; also, it eliminates the need to add a secondary tablet on the chest of the robot for deeper interaction. The only real disadvantage is the requirement to lose the chance of designing a round face in favor of a flat, less human-like one.

This paper will present our initial concepts for the iconographic system of expressions that will be shown on display. These expressions will then undergo a testing process with users, in order to investigate the level of empathic response that they can stimulate in users.

2 The Importance of Faces as Interfaces in HRI

Human beings are extremely sensitive to the particular pattern of features that form a face, to the point that they tend to see faces also in inanimate objects such as clouds or rock formations [4]. This anthropomorphic tendency is very powerful for designers: it is enough to design abstracted or extremely simplified faces and our brain is able to "fill the gaps" and recognize expressions from minimal features. A pair of circles correctly located above a line that suggests a mouth are enough to create the impression of a face.

Several works speculate that more "generic" faces hold a higher degree of affordance for people to identify with [5]. This happens because faces that are not too connotated allow us to easily project upon them face representations that we have acquired during a lifetime, beginning with our childhood, when we develop a system that allows rapid human face recognition, if we are exposed to the correct amount of visual stimuli [6].

This theme has been analyzed by Blow et al. [5] in a useful taxonomy of robotic faces: if realistic faces which are very humanlike can recreate subtle and complex expressions and feedbacks, on the other hand they easily tend to create a strong uncanny effect in users; plus, they have unique features that make the identification process difficult. Abstract faces are not the answer, because they are so far from human features that users find it very difficult to identificate and feel empathy. According to Blow, balance resides in iconic faces that hold a high degree of user identification potential by being cute and non-threatening; on the downside, they can carry a degree of boredom for the user, especially if they can't reproduce a wide range of expressions.

Once we are able to recognize a face, we can also recognize the emotions being expressed on it. According to the eminent work of Paul Ekman [7], a universal coding for recognizing emotions exists in humans, at least for six expressions that he identifies as primary: joy, sadness, anger, surprise, disgust and fear. Primary emotions reside in the orbital-frontal cortex and the stimulus–response process is extremely fast since the sensory signal received by the thalamus is sent to the amygdala through the lower pathway, producing an emotional response, translated by consequence from the body based on its intensity and the emotional control of the individual.

Every human that has a standard degree of empathy is capable to recognize the visual patterns that modify the appearance of a face and express an emotion. These patterns have been thoroughly studied [8] and are the foundation for every kind of representation of human faces, such as comics, art and robotic faces as well.

In the particular field of HRI, three key aspects have to be taken into consideration when designing a robotic head and its visual feedbacks [4]: the robot must hold a degree of "human-ness", needed to facilitate social interaction and empathic response from users; but a degree of "robot-ness" is also very important in order to manage users' expectations towards its cognitive abilities. Finally, a degree of "product-ness" allows to perceive the robot as a non-threatening appliance, which is key to avoid the uncanny valley effect.

3 Overview on Robotic Faces

If service robots such as vacuum cleaners or lawn mowers don't need to appear anything other than functional, human–robot interaction studies become fundamental when designers need their users to develop a form of affective, emotional or personal interaction with the robot; at this point we understood that the face is probably the most important interface to be designed.

The minimum necessary set of facial features seem to be a head-like volume, often contained within a sphere or ovoid, with a vertically symmetric setup of eyes and a mouth that has traditionally been implemented using mechatronic devices; the earliest and most classic example of mechatronic expressive robot is Kismet, having eyelids, eyebrows, lips and ears all physically implemented and controlled by electric motors [9]. Practically, this kind of solutions carry a series of problems: facial systems can be implemented with few mechanical parts, but the expression range is very limited; otherwise, to create more human-like animated faces requires higher number of actuators and higher price, bigger volume, higher weight and more power. Also, androids are more prone to uncanniness, as their physical appearance is so close to humans.

In our preliminary research for Si-Robotics project [2] we analyzed the correlation between the degree of anthropomorphism of robotic heads on the market and their level of expressivity and found that mechatronic faces are way less expressive than synthetic characters being presented on flat screen monitors. Even though it is often believed that the attempt to develop an emotional response to a robot having a screen as a face is not as successful as the mechatronic solutions, this solution provides a series of benefits: the hardware cost of these robots is considerably lower, the weight of the head is light and price, volume and weight remain constant whatever the animated facial features; plus, it must been taken into consideration that computer graphics are in constant improved and evolution and rendered robot faces can now reach even movie quality real time animation, while this was not possible for the earlier products on the market.

On these premises, it needs to be understood how variations and patterns in the design of rendered robot faces can affect the users' reactions and expectations. There are several dimensions of different kind to be taken into consideration: first of all, the presence or absence of a feature on the face (mouth, nose, eyebrows, cheeks/blush, hair, ears, eyelids, pupils, irises, ...); then, the colours and sizes of these features; finally, their placement and shape.

A thoroughly detailed analysis of these aspects has been carried out by A. Kalegina et al. [10] producing extremely useful findings.

First of all, the perceived friendliness of the robots strongly resides in its facial features design: models without a distinctive face such as Jibo are considered significantly less human and less friendly. On the other hand, the highest level of perceived friendliness is recorded in robots with iconic faces, while very human-like faces are perceived as creepy and threatening.

Regarding facial features, it appears that the presence of mouth and pupils are extremely important in raising the level of perceived friendliness, while the presence of eyelids makes the robot seem less friendly. The level of trustworthiness follows the same pattern; faces with no mouth can even be perceived as creepy and give a perception of surveillance. Thus, it can be easily detected that most of social robots on the market have a mouth depicted on the face, even though very small.

Facial features have a strong influence on the perception of the supposed character of the robot. Faces with lower eyebrows standing close to the top of the eye are considered most intelligent and older, while higher rounded eyebrows produce the intimation of a baby face, therefore the perception of lower intellect. It must be considered that evolutionary biology studies explain that infantile features such as large eyes and small mouth and evidence of cheeks evoke a nurturing response in users [11].

4 Colour Coding for Easy Understanding

The design of emotional expressions is extremely complex due to the fact that emotions are an expressive quality of beings, they belong to what Gestalt has defined as the tertiary quality: aspects that do not have measurable dimensions that can be experienced directly through organs, but nonetheless qualities that can be felt in the object. Perception and evaluation of expressive qualities are strongly conditioned by personal experiences and cultural context, which is why their design cannot guarantee a safe result, but only highly probable [12].

Of all the perceptual aspects, color is, together with sound, what appears to be the best vehicle of expressive quality.

The external manifestation of the six basic emotions of human beings on their faces is one of the most universal manifestation of expressive qualities. Plus, there is increasingly evidence of a correlation between emotions and colours in human biology [13], given the fact that colours are fundamental to give the viewer information about the nature of an object.

Since our robot is going to interact with weak users that may have difficulties in perception, we decided to add the colour element to the design of its expressions, with the intention to facilitate the correct understanding of the visual response presented on screen.

Many studies have been carried out in order to associate a specific colour to a given emotion. We refer to the experiment described by Osvaldo Da Pos and Paul Green-Armytage [14] who asked participants to associate a colour to each of the representations of basic expressions collected by Ekman and Friesen and considered the most accurate reproduction of such emotions [8].

As regards lightness, happiness, surprise and fear are represented by very light colours, while sadness and disgust are intermediate, and anger is rather dark (usually black or red). Sadness and fear are mostly represented by extremely desaturated colours, while surprise and anger have very chromatic hues; this may mean that higher saturation means active emotions and increase the observer's arousal independently of the hue. Also, active emotions are usually described by warmer colours.

As emotions themselves are not one-dimensional, it is not possible to associate with complete certainty one colour to one emotion, but we tried to select a coding system that is the most universally accepted by selecting the following associations:

- Anger: red
- Surprise: pink
- Disgust: acid green
- Sadness: pale blue
- Happiness: bright orange
- Fear: violet

5 Si-Robotics Facial Expression System

We designed a basic system for the face of our social robot that holds into account all the research and reflections presented so far. The first concept includes the baseline face and six variations representing basic emotions. This model will be tested with users and results with further development, and possibly more emotions, will be presented in future work.

The face will be rendered on a tablet put in horizontal position on the head of the robot. To create the most seamless experience, the face will be presented on a black background that makes the edges of the tablet less evident.

The graphical style that we chose is extremely iconic and simple, but we decided to include some details and features able to give more personality to the face and make the robot friendlier and more connotated. For these reasons the eyes are not simple circles but include pupils, and small desaturated eyebrows have been drawn over the eyes, to make the robot look older and more intelligent.

Dimensions and variations in position of the facial details allow to reproduce the chosen emotions. In order to facilitate understanding by all the users, even those who may have difficulties in discriminating facial expressions, a single colour has been associated to every emotion and is presented as a blurred background, suggesting the shape of the face.

Happiness
Anger
Surprise
Disgust
Sadness
Fear

Reference

1. Vermesan, O., Bacquet J., (eds.) Cognitive Hyperconnected Digital Transformation: Internet of Things Intelligence Evolution. River Publishers (2017) AHFE
2. Toshio, F., et al.: Facial expressive robotic head system for human-robot communication and its application in home environment. Proc. IEEE **92**(11), 1851–1865 (2004)
3. iSalvo, C., Gemperle, F., Forlizzi, J., Kiesler, S.: All robots are not created equal: the design and perception of humanoid robot heads. In: Proceedings of the Designing Interactive System, pp. 321–326 (2002)
4. Pascalis, O., Scott, L.S., Kelly, D.J., Shannon, R.W., Nicholson, E., Coleman, M., Nelson, C.A.: Plasticity of face processing in infancy. PNAS **102**(14), 5297–5300 (2005)
5. Blow, M., Dautenhahn, K., Appleby, A., Nehaniv, C.L., Lee, D.: The art of designing robot faces: dimensions for human-robot interaction. In: Proceedings of the 1st ACM SIGCHI/SIGART Conference on Human-Robot Interaction, 331–332 (2006)
6. Ekamn, P., Rosenberg, E.L.: What the Face Reveals. Oxford University Press, Oxford (1998)
7. Ekman, P., Friesen, W.V.: Manual for the Facial Action Coding System. Consulting Psychologists Press, Palo alto (1978)
8. Frédéric, D., De Greeff, J., Belpaeme, T.: Towards retro-projected robot faces: an alternative to mechatronic and android faces. In: RO-MAN 2009-The 18th IEEE International Symposium on Robot and Human Interactive Communication. IEEE (2009)
9. Kalegina, A., Schroeder, G., Allchin, A., Berlin, K., Cakmak, M.: Characterizing the design space of rendered robot faces. In Proceedings of the 2018 ACM/IEEE International Conference on Human-Robot Interaction, pp. 96–104 (2018)
10. Todorov, A., Said, C.P., Engell, A.D., Oosterhof, N.N.: Understanding evaluation of faces on social dimensions. Trends Cogn. Sci. **12**(12), 455–460 (2008)
11. Wang, S., Ding, R., et al.: Analysis of relationships between color and emotion by classification based on associations. In: 2008 International Conference on Computer Science and Software Engineering. Wuhan, Hubei, pp. 269–272 (2008)
12. Da Pos, O, Valentini, S: Proceedings of AIC Colour 05, Granada (2005)
13. Da Pos, O., Green-Armytage, P.: Facial expressions, colours and basic emotions. Colour Des. Creativity **1**(1), 2 (2007)

Relationship Between Gestalt and Usability Heuristics in Mobile Device Interfaces

Daniel Ripalda[1(✉)], César Guevara[1], and Alejandra Garrido[2]

[1] Indoamérica University, Mechatronics and Interactive Systems Research Center, Machala & Sabanilla, Quito, Ecuador
danielripalda@uti.edu.ec

[2] LIFIA, Fac. de Informática, Univ. Nac. de La Plata and CONICET 50 and 120, 1900 La Plata, Argentina
garrido@lifia.info.unlp.edu.ar

Abstract. While usability heuristics are universal concepts, visual references for using them in an application are limited, contradictory, and sometimes confusing, especially for novice developers. The present work has linked Nielsen's heuristics with Gestalt principles, which are the foundation of human perception, developed from psychology, and which have been used in art, design, and architecture, to understand how different visual stimulus can be interpreted by the human being. The heuristics, Match between system and the real world, Recognition rather than recall and Aesthetic and minimalistic design, have been linked with the Gestalt principles of Similarity, Proximity, Common direction, and Simplicity. This link is possible because Gestalt principles have certain conceptual similarities with usability heuristics, in addition to the fact that these fundamentals were identified in some mobile interfaces and are related to specific interactions. This work provides theoretical guidelines, visual references and practical examples on the use of some graphic components that are regularly used in mobile interfaces, such as buttons, menus, and sections; Besides, information is provided regarding how these visual components affect certain usability interactions such as learnability, efficiency, memorability, and usefulness.

Keywords: Gestalt · User interfaces · Usability · Mobile · Agile software development

1 Introduction

Producing a mobile application is usually a multidisciplinary work, which involves developers, designers, marketing specialists, and specific advisers according to the project; Generally, the software development methodology guides the steps that the team must follow, from eliciting requirements, building prototypes and evaluating the results with different tests before launching the application to the users. One of the crucial stages for an application is the production of prototypes for the Graphic User Interfaces (GUI), in which the development team seeks to balance functional aspects with visual aesthetics while trying to find efficiency to produce user satisfaction [1]. Steve Jobs explained that those relationships are difficult to achieve in an application or

W. Karwowski et al. (Eds.): IHSED 2020, AISC 1269, pp. 156–161, 2021.
https://doi.org/10.1007/978-3-030-58282-1_25

device; he believed that there exists a kind of "magic" behind that process [2] and one way to achieve the goal could be the "taste" [3], both are abstract and ambiguous terms. This has led developers to believe that, to produce a GUI, in addition to the technical aspects for fulfilling the required functionality of the project, "some luck" is needed [4].

In the last years, important contributions have been made to improve interaction with devices, using concepts as heuristics and processes like User-Centered Design, Mockups, and Pattern-based Design. That knowledge has led to substantial advances to improve usability in interfaces, such as new applications or adaptations from desktop environments to mobile devices. In this context, the Gestalt Prototyping Framework from its theoretical approach [5], emerged as a proposal directed to usability experts, designers, and developers of mobile applications, to use visual references to build and evaluate interfaces of mobile devices, so that they can identify, prevent or correct errors in visual components of the GUI.

2 Related Work

In 1994, Nielsen proposed 10 heuristics that lead the development team in the production of GUI. They are general guides to design the interactions that affect the usability of a system. These heuristics have been widely used as a reference, defining essential concepts for the Human-Computer Interaction (HCI) [6]. They became universal and are used until the present, having relevance from the evaluation of different prototypes [7]. However, at that time, interfaces were designed for desktop computers.

With the emergence of mobile technology, different researchers try to adapt Nielsen's heuristics, establishing parameters to measure the usability of mobile applications, generating important contributions such as Bertini's [8], who produced an evaluation model comparing it with the methods used until then. In this same context, Yanez-Gomez [9] configured a new list of usability parameters that can be used to evaluate mobile applications.

There is interest in understanding the relationship between aesthetics and HCI, and several investigations have reviewed the attributes and processes to design aesthetic interactions [10, 11]. Some discussions have focused on the relationship between the psychological perception of aesthetics and the perception of usability [12].

Gestalt principles describe the interaction of graphic components in space and how people perceive and understand these relationships; principles such as proximity, similarity, common direction among others, explain how people assimilate and react to the forms they visualize. For this reason, an interface can naturally use these concepts to generate, position, move, show and hide information, in a way that favors learning, execution, and memorization of certain tasks, reducing cognitive load, while in parallel generating favorable user experiences. Some applications have already used Gestalt parameters in the development of interfaces. Gómez claims to have noticed the presence of some of the principles and measured their impact on the interface quality of desktop applications [13]. Other applications have already used specific principles within their interfaces [14–17], but without measuring their impact on the user experience so far.

3 Concepts Applicable to an Interface Prototype

The interfaces of mobile devices have different graphic elements to visualize the functional aspects of the application. To carry out tasks through the GUI, the user must perceive, interpret, and understand the purpose of the images that appear on the screen. This principle is the same that Gestalt uses to define the relationships that occur between figures that share the same space. Table 1 shows the theoretical link of usability heuristics, Gestalt principles, visual components, and usability interactions that are sought at different stages of prototype development.

Table 1. Components in which Heuristic and Gestalt concepts can be applied.

Interface component	Gestalt principle	Heuristic	Usability interaction
Iconography Colour Buttons	Similarity	Match between system and the real world	Learning capacity and efficiency
Menus Submenus Pop up Slide buttons	Proximity Common destiny	Recognition rather than recall	Learning capacity efficiency memorability
Backgrounds Dividers Graphic style	Simplicity	Aesthetic and minimalistic design	Memorability Utility

3.1 Iconography, Color, and Buttons

The heuristic "Match between system and the real world" aims to make the different visual elements of an interface familiar to users, so they can associate them with their environment and context, positively identifying the actions that can be carried out in the application, without going through long periods to acquire basic skills.

This purpose can be achieved using the interface resources, if visual elements are used that are universal and recognizable by diverse human groups, regardless of the characteristics of their geographic, demographic, or sociocultural profiles.

According to Isidro Moreno [18] the icons - symbols can be read properly in the software if they can be associated with knowledge or experience encoded by an abstraction that facilitates the identification of the object, process, or idea that it represents. In the metaphor of navigating a system through an interface, current mobile devices combine peripheral hardware interactions, intermediate interactions that allow the selection of functions, and finally natural or gestural mimetic interactions.

Icons can be associated with concepts such as consistency, defined as a natural factor in the interactions that occur in an interface; the components allow us to carry out a task in a regular, fluid and uninterrupted way. The lack of consistency of an interface creates the possibility that there are different ways of carrying out the same task, these variations are the ones that can be confusing for the user.

The icons of an application must be easily recognizable, should have high contrast and simple shapes; the visual style, color, and shape of the iconography. The Gestalt principle of similarity contributes to the consistency of an interface, making it predictable learnable; these factors depend largely on the compatibility of an icon with the task it performs. For example, Fig. 1 shows four icons that are used in different applications to save information, the task of the development team is to determine which of them is more compatible with the action that is going to be executed.

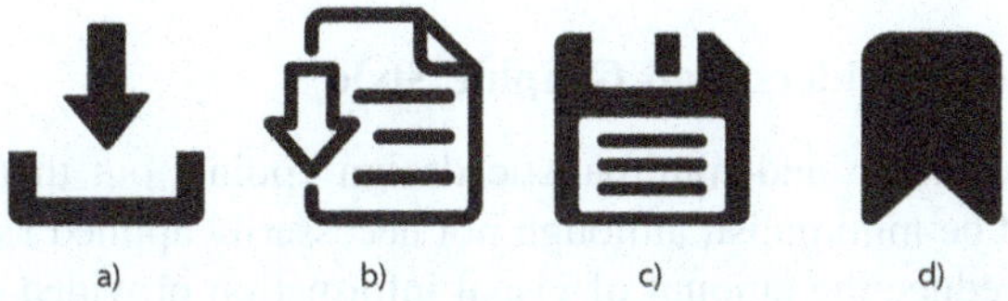

Fig. 1. The principle of similarity applied to the icons of a mobile application.

3.2 Menu, Submenu, Pop up and Slide Buttons

The Heuristic called "Recognition rather than recall", in an interface refers to the fact that essential actions must be identified in the foreground so that the user learns to perform them intuitively; The basic actions of the application should be grouped in such a way that categories and hierarchies are established naturally; for this, the elements must remain visible at all times, until the user becomes familiar with the application. In Fig. 2 you can see that the graphic similarity of the icons and their proximity generates that user perceives different menus and interface configurations.

Fig. 2. The principle of proximity applied to the menus of a mobile application.

The Gestalt proximity principle causes that a group of icons can be naturally perceived as a menu, to create natural menus, the iconography can be grouped maintaining similarity in colors, shapes, positions, and orientations. By varying the orientation, the position of the icons, including the supports and other functional components, it is possible to maintain the hierarchy and organization of the interface.

On the other hand, the Gestalt principle of common direction describes that, if the objects move towards the same direction, they are perceived as a group. This concept can also be used to reorganize an interface after a change of orientation of the device.

3.3 Backgrounds, Dividers, and Graphic Style

The heuristics "Aesthetic and minimalistic design" points out that the design of an application should be minimalist, although not necessarily applied as a style, but rather as a guideline to reduce the amount of visual information provided to the user. In this sense, the Gestalt principle of Simplicity refers to how human beings perceive the forms even if they only appreciate a portion of them. These concepts allow the development team to narrow down the aesthetics-oriented graphics and orient it towards clarifying interface functions. Figure 3 shows how the graphic details of the interface are reduced, without changing the functions.

Fig. 3. The principle of simplicity applied to reach a minimalist design.

4 Conclusions and Future Work

The Gestalt Prototyping Framework effectively links the basic notions of the Nielsen Heuristics with Gestalt principles, providing theoretical and visual references to understand how these models are integrated during the development of a GUI. Usability experts, designers, and developers of mobile applications can use these concepts to adapt, improve, or correct the prototypes of interfaces in projects, without interfering with the development methodology they have decided to use.

In the near future, we expect to obtain results from the application of our proposal in the development of high and low fidelity prototypes in the field of different projects.

This will allow detecting the incidence of the Gestalt Prototyping Framework in the production time of applications, detection of errors in the interfaces, and improvements in different usability parameters.

References

1. Marcus, A.: The cult of cute: the challenge of user experience design interactions. Interactions **9**, 29–34 (2002)
2. Brookey, R.A.: The magician and the iPod: Steve Jobs as industry hero. Leadership in the Media Industry (2006)
3. Elkind, P.: The trouble with steve jobs. Fortune Eur. Ed. **157**, 54 (2008)
4. Crook, I.: Diseño de Aplicaciones: Tutorial Paso a Paso para Diseñar una App. https://neoattack.com/diseno-aplicaciones/. Último Acceso 22 Oct 2019
5. Ripalda, D., Garrido, A., Guevara, C.: Gestalt framework in the design of interfaces for mobile devices: theoretical approach. In: 13th Iberian Conference on Information Systems and Technologies (CISTI), Cáceres, Spain (2018)
6. Nielsen, J.: 10 Usability heuristics for user interface design, Nielsen Norman Group (1995)
7. Nielsen, J., Molich, R.: Heuristic evaluation of user interfaces. In: Proceedings of the SIGCHI conference on Human factors in computing systems. ACM (1990)
8. Bertini, E., Gabrielli, S., Kimani, S., Catarci, T., Santucci, G.: Appropriating and assessing heuristics for mobile computing. In: Proceedings of the working conference on Advanced visual interfaces, pp. 119–126 (2016)
9. Gómez, Y.R., Caballero, D.C., Sevillano, J.L.: Heuristic evaluation on mobile interfaces: a new checklist. Sci. World J. **2014**, 19 (2014)
10. Lim, Y.K., Stolterman, E., Jung, H., Donaldson, J.: Interaction gestalt and the design of aesthetic interactions. In: Proceedings of the 2007 Conference on Designing Pleasurable Products and Interfaces, pp. 239–254 (2007)
11. Norman, D.A.: Emotional Design: Why We Love (or Hate) Everyday Things. Basic Books, New York (2005)
12. Hassenzahl, M.: The interplay of beauty, goodness, and usability in interactive products. Hum. Comput. Interact. **19**(4), 319–349 (2004)
13. Gómez, J.: Measuring the impact of gestalt theories on graphic user interface design (2011)
14. Xiang, P., Yang, X., Shi, Y.: Web page segmentation based on gestalt theory. In: 2007 IEEE International Conference on Multimedia and Expo, pp. 2253–2256. IEEE (2007)
15. Looije, R., te Brake, G.M., Neerincx, M.A.: Usability engineering for mobile maps. In: Proceedings of the 4th International Conference on Mobile Technology, Applications, and Systems and the 1st International Symposium on Computer Human Interaction in Mobile Technology (2007)
16. Gong, J., Tarasewich, P.: Guidelines for handheld mobile device interface design. In: Proceedings of DSI 2004 Annual Meeting, pp. 3751–3756 (2004)
17. King, M.T., Grunbock, C., Mannby, C.F., Valenti, W.: U.S. Patent Application 12/961, 407 (2010)
18. Moreno, I.: Iconos hipermedia: La llave interactiva. Revista ICONO14 Revista científica de Comunicación y Tecnologías emergentes **1**(1), 123–135 (2003)

Birthing Bed Design Process for New Mothers, Considering All the Users Involved at Delivery Phase: Case Study in Mexico

Guillermina Dinora Suárez-Gómez(✉), Alejandra Robles-Barba, Montserrat Avelar-Enciso, Carlos Raymundo Garnier-Ortiz, and Fabiola Cortes-Chávez

Universidad Panamericana Campus Guadalajara, Calzada Circunvalación Poniente N. 49, Ciudad Granja, 45010 Zapopan, Jalisco, Mexico
{0215081,0217825,0214724,cgarnier,fcortes}@up.edu.mx

Abstract. The objective of this study is to demonstrate the innovation and design process of a birthing bed, designed specifically for young first mothers [1]. This problem was chosen due to the social and cultural conditions nowadays in Mexico, which, according to data from the INEGI, National Institute of Geography and Statistics, around 32 million women over 12 years old pursue their motherhood [2]. Not only are the numbers alarming, but it must be addressed the fact that a poorly attended delivery can be truly traumatic. The data collected by primary sources is a series of surveys sent to mothers, doctors and nurses, in which they are questioned about their experience before, during and after delivery. They were also questioned about recommendations regarding actual delivery beds. Data collection via internet focused mainly in complementing the formation obtained in the surveys. The data in this investigation includes the complete process of a delivery, the actual birthing beds and the role of each user within a delivery.

Keywords: Pregnancy · Mother · Birthing bed · Delivery · Vertical delivery · Horizontal delivery · Innovation · Medical personnel · Obstetrician · Pediatrician · Neonatologist · Anesthesiologist · Nurses · Cleaning · Ergonomics · Technology

1 Introduction

According to the National Strategy for the Prevention of Teen Pregnancy (2019), there were around 8,263,020 births between 2008 and 2011, of which 18.9% were from women between the ages of 15 and 19 (this represents 1,561,710 pregnancies). Taking these figures into consideration and the average impact of natural childbirth on women, the objective of this project is to design a special birthing bed for women between 15 and 20 years of age. This bed will not only focus on the comfort and safety of the mother but will consider all those involved in a delivery. To start the design process, an investigation was carried out about the average development of a natural delivery, the different positions and their characteristics, and the existing delivery beds. The information was obtained through surveys and internet sources and showed that the key

W. Karwowski et al. (Eds.): IHSED 2020, AISC 1269, pp. 162–167, 2021.
https://doi.org/10.1007/978-3-030-58282-1_26

points to consider are space, hygiene, comfort and usability. Among the data that stand out the most from the surveys are the following: the doctors argue that the best delivery position is semi-sitting or vertical, however, the horizontal position is the most effective in case of risk situations. Mothers say that their birthing experience could have been improved with more comfortable sheets or cushions, and nurses say that the accumulation of substances and bacterial growth is one of the most characteristic obstacles of current birthing beds. It was concluded that in order for this stretcher to be fully used, it must have the following characteristics: a welcoming first impression for the mother, support on the lower back and thermal cushions, practical assembly, materials that can be easily sterilized, and finally, the dimensions have to be set up in such a way that it can give the doctor the best visibility and at the same time it can be adapted to any percentile.

2 Methods

For the collection of data and information that would support the design of the delivery bed, an investigation was first carried out through various internet sources on the process of natural childbirth and the different procedures that are done during it. Then, through online technical sheets, various data and characteristics were obtained on existing delivery beds. These were AVE2, OPTIMA 2, and Stryker LD403. Along with this research, a survey was conducted for the medical personnel present at a delivery. With this information, the main users that would be considered for the delivery bed and the objective that it should meet, were specified. These users are: Mother, Gynecologist/Obstetrician, Anesthesiologist, Nurse, Pediatrician and the Companion. Subsequently, the design process phase began, in which various sketches and prototypes were made until reaching the final model, which would be later rendered to choose the color proposals. The final stage of the process consisted of carrying out a comparative survey between one of the first sketches and the final model to obtain the opinion of the mother, the companion and the medical staff that were included among the users.

3 Findings

Of the total population in Mexico, 48.7 million are women aged 12 and over, of that total, 32.7 million are mothers, according to data from the National Institute of Statistics and Geography (INEGI). 28 out of 100 women carry out their maternity without a partner. Of the mothers aged 15 years and over, 43.4% are employed, 63.7% of the women with children and children are subordinate and remunerated workers, 26.6% are self-employed and 7% are employed women who do not receive remuneration [2].

The users who interact around the birthing bed and attending the delivery are the following. The obstetrician/gynecologist specializes in the care of women during pregnancy and childbirth, he or she is present in the active phase of labor and oversees receiving the newborn [3]. Next, the nurse, who is in charge of preparing the medical history and conducting physical evaluations, ordering laboratory procedures and examinations, managing the treatment and carrying out activities that promote women's health and reduce possible risks. During childbirth she or he is in charge of assisting the gynecologist and attending the mother, administering fluids, emptying the waste container and checking vital signs [9]. After the delivery, the nurse cleans the baby and takes the temperature and vital signs. The third user is the pediatrician, who is the doctor for the prevention, diagnosis, and treatment of diseases and injuries in children. The pediatrician is present in final phase of the labor in case of any complications. The anesthesiologist has all the knowledge to administer medications or other substances. Constantly checks the mother and the baby's vital signs [10]. Then there is the mother, the attention of all other users is directed to her. Finally, there is the companion, who is the person next to the mother to provide support throughout the delivery process [4].

Ave 2, Optima 2 and Stryker LD304 were the actual beds chosen to be analyzed. These were discovered to be the most popular in the market and they have elements in common such as ergonomic backrests, different angles of inclination, handles and footrests, and waste containers [5].

4 Final Stage in the Design Process

A total of 43 surveys were carried out to the possible users involved: gynecologists, pediatricians, nurses, companions, mothers and anesthesiologists. In this survey they were asked about their point of view of the initial sketch and the final design in terms of safety, convenience, efficiency, aesthetics, adaptability, ease, comfort, tools and most importantly, if the design really considered their needs regarding their interaction with the birthing bed [6].

Comparing both responses from the surveys, it is shown a remarkable improvement towards the acceptance of the design of the birthing bed. The most significant ones were regarding safety, efficiency, color, and most importantly, the consideration of the user's needs (Figs. 3 and 4) (Fig. 1).

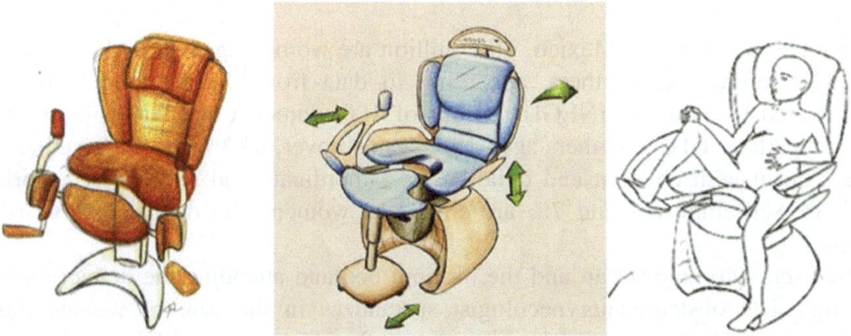

Fig. 1. From left to right, first sketch, final sketch, human scale.

It was proposed yellow, blue and pink for the possible colors the birthing chair could be. Yellow represents optimism and amiability. It is luminous and in its light tones represents serenity and wisdom. Blue is psychologically related to serenity and calm, tranquility, control of situations and peace. It is widely used in medical equipment as it is also associated with caring for others, trust and credibility. Pink is the color of femininity and pureness, and therefore of affection and protection. It is related to generosity, dedication, innocence, and love. Pink is the color of young life and charm [7, 8] (Fig. 2).

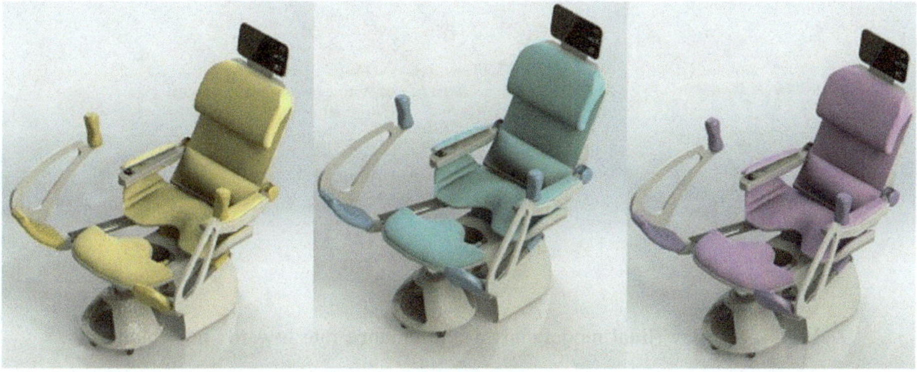

Fig. 2. Color proposals for the final model

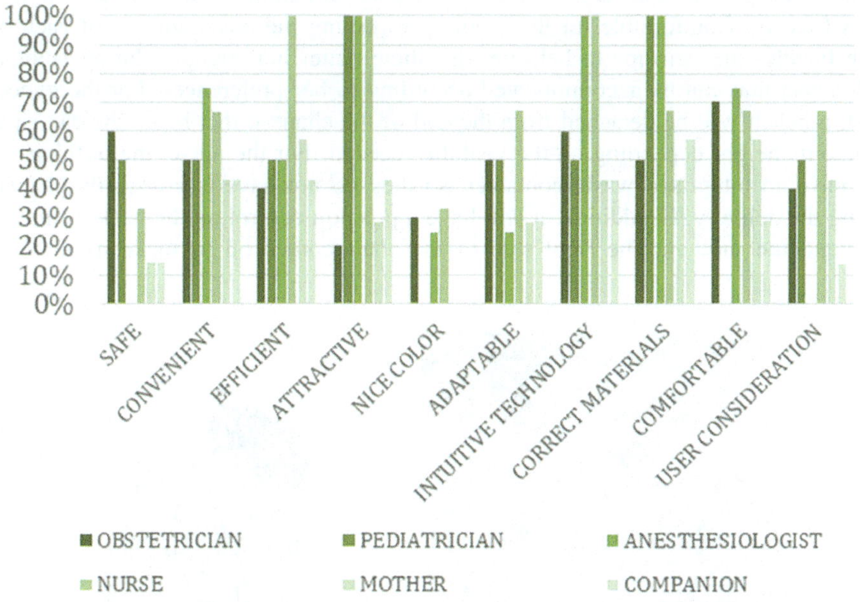

Fig. 3. Survey (first sketch shown): acceptance rate from medical personnel regarding the first sketch.

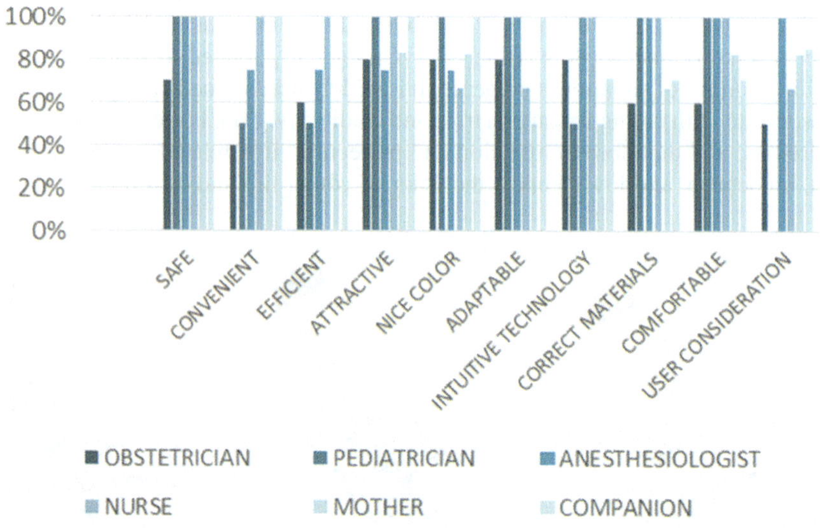

Fig. 4. Survey (final model shown): acceptance rate towards final model.

5 Conclusion: Final Design

The main design question is: How does this birthing "chair" involves all users? Based on the investigations and surveys carried out, it was concluded that the final design will be as follows. Comfortable for the mother, respecting the physiognomy of each one. With handles and stirrups to help her distribute better her strength during labor and with a seat that can be accommodated according to her preferences. For the gynecologist, a bench will be detached from the end of the chair so that he or she can sit and change its height depending on the visibility needed. For the nurse, the birthing chair will have an integrated waste container, so they do not have to constantly change it during the delivery. In addition, it will have a position control on the back of the chair to attend the mother. The right side of the chair comes off to make room for the

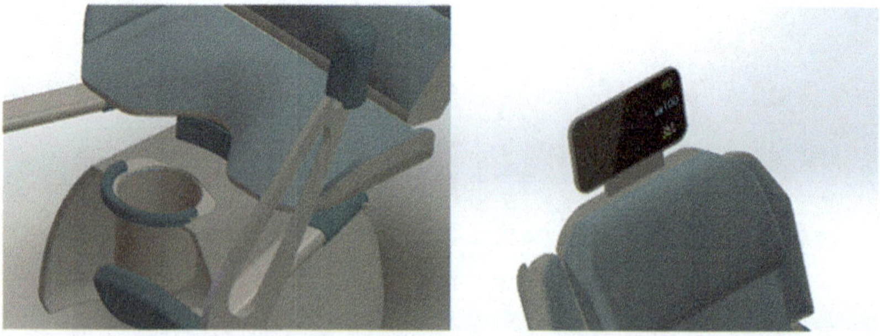

Fig. 5. Closeups. From left to right: waste container for the nurse. Vital signs and fluids monitor.

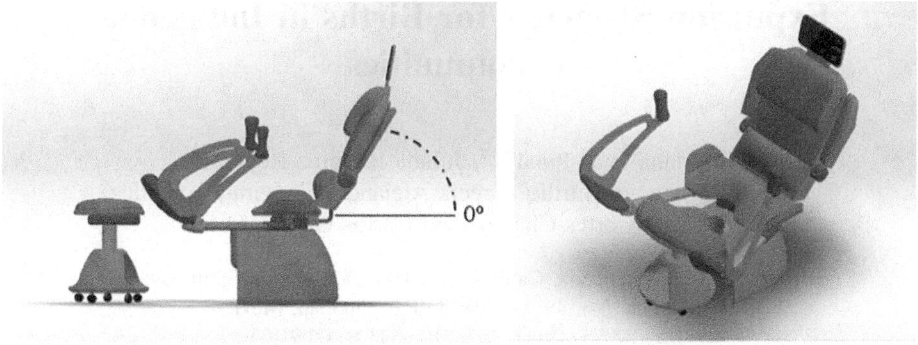

Fig. 6. From left to right. *Right view of the birthing chair.* The backrest has a 90-degree rotation so it can be fully horizontal. *Overview of the final model and its key parts.* The structure of the birthing chair is capable of supporting the weight and the strength applied from the mother. Due to its design, all users can work around it without being obstructed by any element of the chair.

mother's companion. The chair will have a monitor that assists both the pediatrician and the anesthesiologist and that will show the vital signs of the mother and the baby. The mechanisms of the chair are very intuitive and easy to use, in order to for the medical personnel to have a better interaction with it. The structure is made from resistant rigid PVC and the cushions are made from waterproof fabric, which is also easy to sterilize (Figs. 5 and 6).

References

1. INEGI: Natalidad y Fecundidad (2000–2015). https://www.inegi.org.mx/app/tabulados/default.html?nc=mdemo29. Accessed 24 May 2020
2. Instituto Nacional de las Mujeres: Estrategia Nacional para la Prevención del Embarazo en Adolescentes (2019). https://www.gob.mx/inmujeres/acciones-y-programas/estrategia-nacional-para-la-prevencion-del-embarazo-en-adolescentes-33454. Accessed 24 May 2020
3. Rodríguez, E.: ¿Cómo es una sala de parto? (2018). https://elembarazo.net/como-es-una-sala-de-parto.html. Accessed 24 May 2020
4. Mayo Clinic Staff: Labor and delivery postpartum care (2019). https://www.mayoclinic.org/healthy-lifestyle/labor-and-delivery/in-depth/stages-of-labor/art-20046545?reDate=250520200. Accessed 24 May 2020
5. Fichas técnicas de camillas existentes. https://drive.google.com/open?id=1AcNnmpUmadSIcBqyrqtjJJrW3MdeGGE-
6. Final surveys. https://drive.google.com/open?id=1TufImh5gr37JZ_Lk_a7QaxZRa8gGU4Y1sSKka3e8o9s
7. Allen, K.: Entrepeneur. La psicología de los colores (2019). https://www.entrepreneur.com/article/269009. Accessed 24 May 2020
8. Zelanski, P., Fisher, M.P.: Color, p. 155. Blume, Londres (1989)
9. Thorp, J.M., Laughon S.K.: Enfermera obstetra certificada (2019). Medline plus Website: https://medlineplus.gov/spanish/ency/article/002000.htm. Accessed 24 May 2020
10. Diccionario de cáncer (2020). NIH Website: https://www.cancer.gov/espanol/publicaciones/diccionario/def/anestesiologo. Accessed 24 May 2020

Expulsion Stretcher for Births in Indigenous Communities

Mariana Diaz-Pinal(✉), Julieta Ramirez-Reynoso, Mariana Ascencio-Murillo, Grecia Alejandra Chavira-Hernández, Fabiola Cortés-Chávez, and Carlos Garnier-Ortiz

Universidad Panamericana, Campus Guadalajara, Calzada Circunvalación Poniente No. 49, Granja, 45010 Zapopan, Mexico
{0215338, 0216853, 0215198, 0190615, fcortes, cgarnier}@up.edu.mx

Abstract. The objective of this document is to show the development of the design of a portable birthing stretcher that facilitates the process of birthing in the indigenous communities of Mexico, to both the midwives and the mothers since the current conditions of birthing in the aforementioned place is insufficient to make it a safe and a comfortable delivery, it is worth mentioning that the midwives travel distances to be able to assist the mothers. For the design, we sought to ensure that it would not complicate the midwife's transportation, but that it would also be a tool to transport all the necessary utensils for safe and hygienic delivery, as well as a device that would help with the mother's posture so that she would be more comfortable during the delivery and would also help with the midwife to make it safer [1, 2].

Keywords: Mothers · Indigenous · Midwives · Portable · Communities · Culture · Customs · Mexico · Discrimination · Engineering · Design · Innovation · Medical

1 Introduction

Today in Mexico there are more than 15,000 midwives in indigenous and rural areas. Based on previous research it is known that if midwifery conditions and services are improved, up to 3.6 million deaths could be avoided in 58 developing countries, which is why it was decided to design the "Tam Alal" stretcher.

Most of the communities do not have a hospital or they are retired from the urban areas, the culture of each community varies a lot, something that is very common among the indigenous women is that they are very modest, also the male chauvinism is very present, the reason why the child births are very different from those that happen in a hospital. Not only do differences within culture lead indigenous mothers to give birth in their communities with a midwife, but the language barrier in hospitals is a very important factor in this decision [2, 3].

Births in indigenous communities and in rural areas do not have hygiene, comfort nor practical measures, for the mother and the midwife, the journeys to the homes are complicated, as the communities are very remote, there are no cars, and there are few

W. Karwowski et al. (Eds.): IHSED 2020, AISC 1269, pp. 168–173, 2021.
https://doi.org/10.1007/978-3-030-58282-1_27

midwives to serve the surrounding communities, as they do not have one for each community. Sometimes births are complicated because midwives try to have the least contact with the mother's genitals, since there is machismo and women are very modest, so they adopt positions that bring complications and are not efficient [2].

The houses are low income and do not have the necessary materials for a hygienic birth, usually deliveries occurs in the floor of their houses, in occasions they use tarps or cardboards; there are some places where the midwives have the opportunity to ask the couple to get certain material, whether bags, profiles, scissors, alcohol, blankets, etc. [3, 4].

2 Development of the Product

The design process consisted of 4 methodological phases in order to arrive at a final design. The first phase consisted of a research process where the problem was defined, as from the users and their needs were identified. This was followed by the second phase: the creative phase, where sketches were made, from which the best ideas were taken into account to integrate them and make the first design. In this proposal the main objective was the comfort of the mother, however, this stretcher did not consider the needs of all the users; the mother, the midwife, and the baby, the design of this stretcher is shown in Fig. 1.

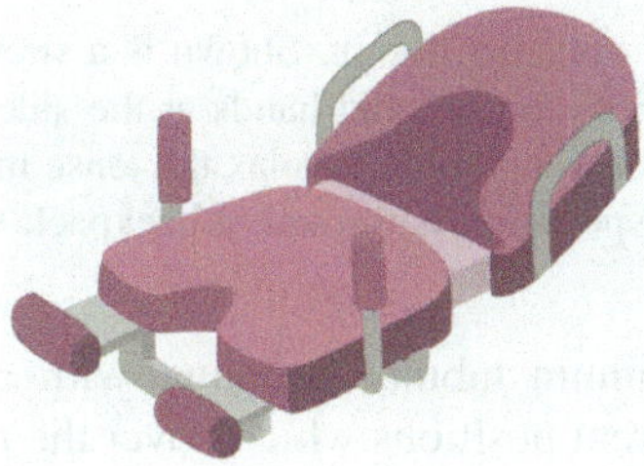

Fig. 1. The first design of the stretcher, is shown the stretcher which is a fusion of the best ideas of the creative phase in which its main objective was the comfort of the mother. Where the support points for feet and hands are represented, as well as a compartment with hot water to avoid stressing the lower back muscles

This design has one birth position and two ways of holding, on the sides of the thighs or on the top of the shoulders, after use it can be folded so it can be carried like a mountaineering backpack.

The problems that this design presented were that the structure being from aluminum and polyester PVC mattresses had a very high density which leads to a complication in the transport since the midwives, according to the investigation made, are people with an average of 56 years, who cannot carry objects that weighed more than

15% of their own weight. Another problem that was found was that to have a comfortable and safe backpack it should not exceed the width of the shoulders, to fulfill with this requirement the design of the stretcher did not have a backrest wide enough to provide comfort to the mothers, in turn, the structure could not adapt to the different heights of potential users, it was not adjustable due to the proposed material, although a 95 percentile was used, the height would be approximately 80 cm, thus exceeding the preferred measurement in a backpack [5, 6].

During the third phase was the collection of surveys to verify the needs of the design and to check the points on which we had to focus and what our user was looking for. In the fourth phase, the stretcher was redesigned, considering the aspects found in the surveys and research. This design was inspired by the emergency stretchers because they are light and efficient, this design is represented in Fig. 2.

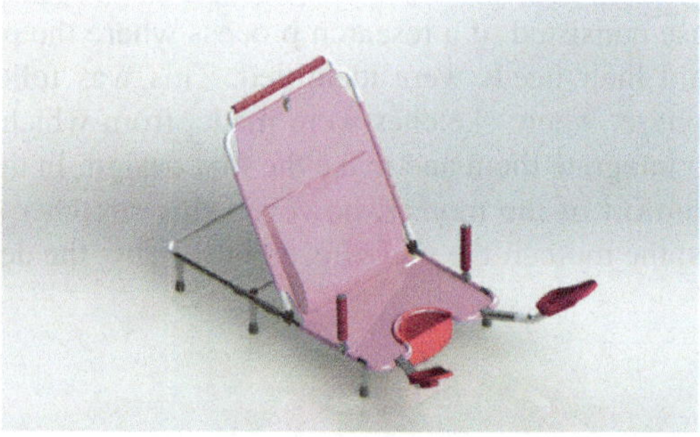

Fig. 2. Stretcher redesign, final stretcher design. Shown is a stretcher based on the emergency stretchers which has support points for feet and hands at the sides of the thighs and above the head. It has a waste bag and a hot water bottle to relax the tense muscles in the lower back. This stretcher is foldable and its transportation is through a backpack with wheels on the bottom.

The design has an aluminum tubular structure with an adjustable backrest that allows it to be used in 4 different positions which cover the need to be able to adjust the backrest to the size that is necessary for the user, we also included the possibility of adjusting the backrest angles to provide comfort to mothers, reclining between 90° and 180°.

Based on the findings, it was decided to change the way in which the stretcher is transported, this idea was renewed to be able to transport the stretcher inside a backpack with wheels that allows the midwives to move it easily and allows us to have a wider back for the comfort of mothers. The stretcher has strategic support points, which are located on the top of the backrest and on the sides of the seat covered by a foam padding surface (Eva), as well as footrests, which can be adjusted to 3 different sizes so that the mother can be placed in various positions.

The compartment that allows for a hot water bag was kept in this design, not only to provide greater comfort to the mother but also to reduce the stress on the lower back muscles preventing injuries. For the height of the stretcher, the position used by the midwives to receive the baby was considered, which is squatting or sitting on the floor, as it is not common to have chairs or special spaces for the deliveries. An adequate height was also sought to provide security for the mother and the baby.

3 Investigation and Results

27 surveys were carried out with midwives and mothers who were familiar with the context we are focusing on. In Figs. 3 and 4 we can see the comparison of the first design of the stretcher and the final design of the stretcher. With the results obtained we realize that we had an accurate approach to the users with the design of the final table because the graph shown in Fig. 4 obtained higher scores.

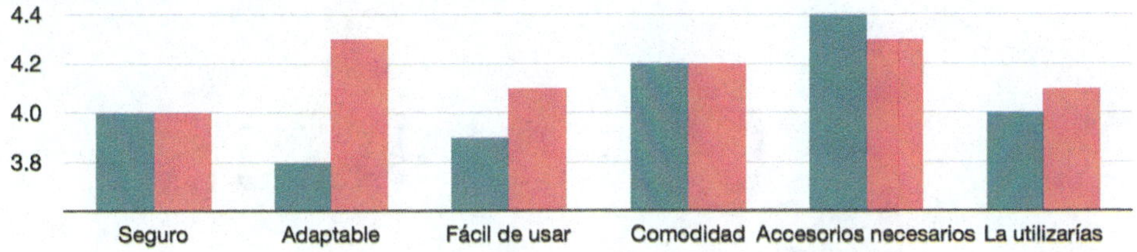

Fig. 3. This graph shows the results of the opinions of the mothers' (pink) and midwives' (blue) on the first design of the stretcher, shown in Fig. 1 previously. Taking 0 as the lowest score and 5 as the highest.

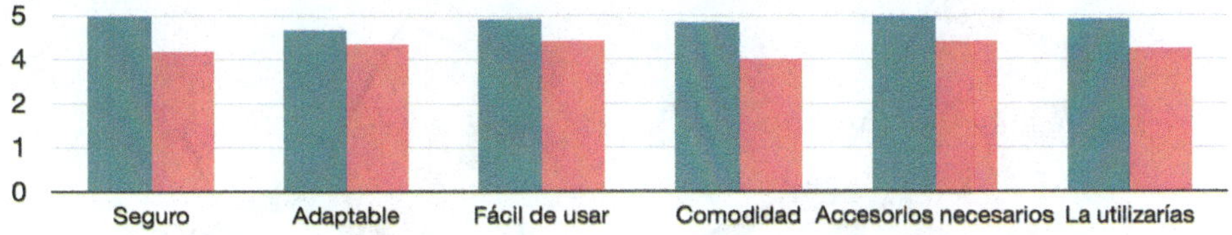

Fig. 4. This graph shows the results of the opinions of the mothers' (pink) and midwives' (blue) on the final design of the stretcher, shown in Fig. 2 previously. Taking 0 as the lowest score and 5 as the highest.

4 Annexes

See Figs. 5 and 6.

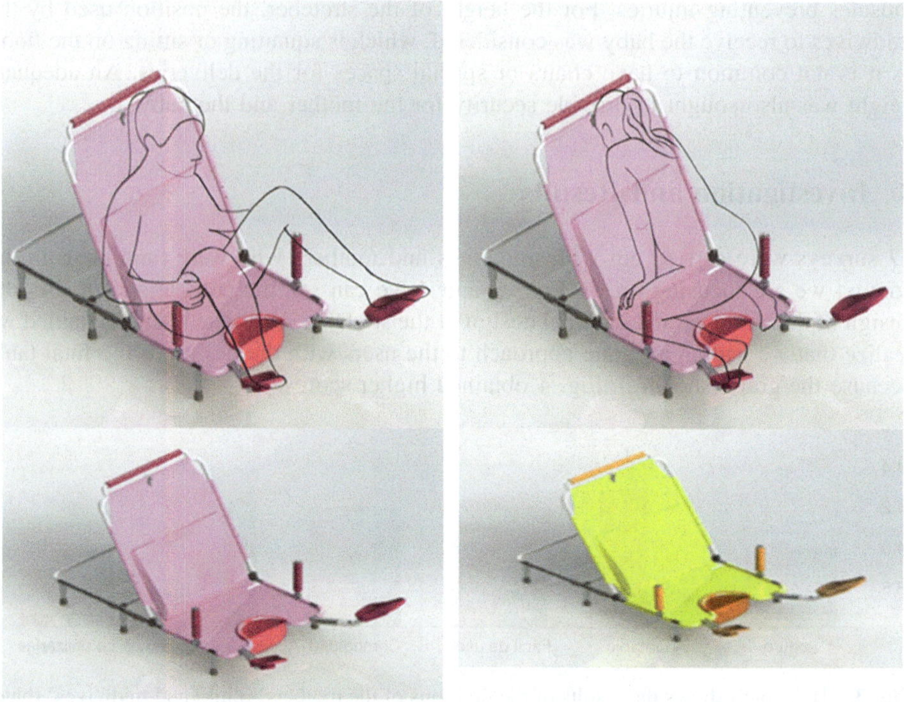

Fig. 5. Representations of positions in the final design with human scale and color proposals.

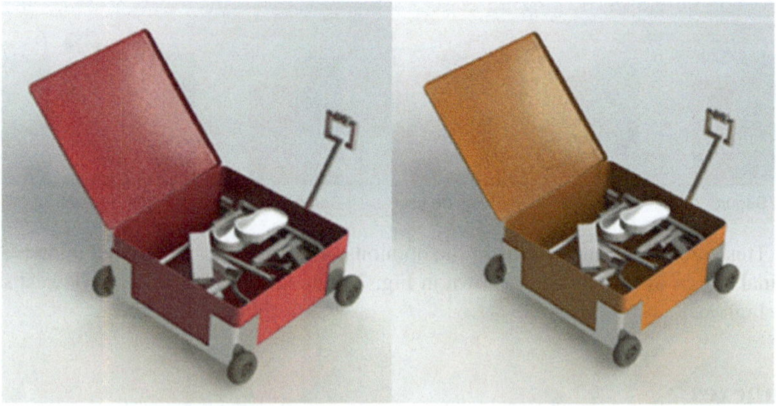

Fig. 6. Representation of how the stretcher will be transported inside the backpack, which is made of thermoformed plastic, reinforced at the corners. The backpack has **4** wheels at the bottom for easy transportation, as well as a handle to pull it which is covered with foam for user comfort.

5 Conclusions

After completing the design process it can be concluded that it is very important to consider all users involved in medical problems because it is complex to prioritize the needs of each user involved, giving greater value to a user can lead to undesirable results, as it was with the first design of the stretcher that did not take into account the needs of the midwife. The final design, according to the surveys carried out, has a balance between both users as it takes into account the comfort of the mother and the positions they normally use as well as the attachments needed to carry out the delivery (footrests and handles). It also takes into account the midwife's needs such as transporting the stretcher, carrying all her utensils along the way, cleaning the stretcher, and her safety.

References

1. Forbes. https://www.google.com.mx/amp/s/www.forbes.com.mx/parteras-aliadas-de-las-mujeres-indigenas- ante-mortalidad-materna/amp/
2. CONAMED (Comisión Nacional de Arbitraje Médico). http://www.conamed.gob.mx/gobmx/boletin/pdf/boletin19/contribucion_parteria.pdf
3. Eva Guadalupe. (2020, febrero 21). Comunicación personal
4. El Economista. https://www.eleconomista.com.mx/politica/Dos-de-cada-10-mexicanos-se-asumen-indigenas-20181212-0049.html
5. IMSS. http://www.imss.gob.mx/prensa/archivo/201909/395
6. Universidad de Malaga. https://www.uma.es/publicadores/prevencion/wwwuma/183.pdf

Birthing Bed with Ergonomic Design of Adjustable Sections by Touch Technology that Facilitates Its Understanding and Use

Teresita Bátiz-Flores(✉), Andrea Perez,
María Fernanda Martínez-López, Fabiola Cortes-Chavez,
and Carlos Raymundo Garnier-Ortiz

School of Engineering. Prolongación, Universidad Panamericana, Calzada Circunvalación Poniente 49, 45010 Zapopan, Jalisco, Mexico
{0215075,0216072,0215348,fcortes,cgarnier}@up.edu.mx

Abstract. Labor is a complex process in which one of the main factors within the hospital's context is the medical equipment that helps to carry out the work, such as the birthing bed. Is around it that the activities that allow an efficient labor are carry out. There are two main approaches when it comes to medical design, known as EBD (Evidence-Based Design) and UCD (User-Centered Design). Each one seeks to meet different expectations and functions with the design each supports. This article presents a comparative analysis on the design of a birthing bed using both approaches, aiming to know if an user-centered design methodology helps increase the satisfaction of the users, by improving the bed's ease of use and understanding.

Keywords: Birthing bed · Industrial design · Product design · Medical devices · Touch technology

1 Introduction

Labor is a complex process. For the most part, long and exhaustive. With the coming together of several factors and different users around the birthing bed, it is important to take them all into consideration when it comes to the design. This bed allows women going in labor to adopt natural positions during this process, considering aspects like comfort and safety [1].

The birthing bed design can, sometimes, affect the performance of the users that interact with it. According to the article written by Sookyung Chun and Kyungsook Nam [2], the 2016 Healthcare Design Expo stated the importance of user-centered design and how by not doing so, the performance of the users can decrease, as this affects different human dimensions such as: physical, psychological and affective.

In the same article, two approaches to medical design are presented: EBD (Evidence-Based Design) and UCD (User-Centered Design). Each approach tries to cover different aspects and factors with every design made. EBD maximizes functionality and work flow, focusing mainly in the hospital's physical conditions, patient's safety and stress reduction generally caused in a hospital. On the other hand, UCD

W. Karwowski et al. (Eds.): IHSED 2020, AISC 1269, pp. 174–179, 2021.
https://doi.org/10.1007/978-3-030-58282-1_28

covers the user's needs and expectations from a physical, psychological and affective perspective [2].

In this article, both approaches in the design of a birthing bed will be analyzed, identifying factors that help test the importance of an user-centered design methodology in order to improve the design of a birthing bed, facilitating its use and understanding.

2 Project Development

2.1 Phase 1

During the first phase of the project, the goal was to design a birthing bed with an EBD approach. To achieve this, an investigation was carried out, followed by an analysis of the information gathered from different articles and previous birthing bed designs.

In order to establish the important features required to be taken into consideration for the design, the most significant aspects and factors found in the investigation were grouped as shown in Table 1.

Table 1. Main aspects to consider when designing a birthing bed, according to the information gathered from articles.

Author	Aspect
Lois Hamlin Marylin Richardson	Patient's safety represents an elemental competence [3]
Universidad de Chile Facultad de Medicina	The birthing bed must be able to adapt into an operating table, when needed It should include equipment to monitor the vital and cardiorespiratory signs of both, mother and baby [4]
Susana Lafuente	It is important that the birthing bed has a feet support and rotation or adjustments for the inner part of the legs [5]
Programa Médico Arquitectónico para el Diseño de Hospitales Seguros	The labor room must have a birthing bed with wheels that allow its transportation when needed [6]

This information was analyzed and a creative process was carried out, trying the combination of these aspects. During this creative process, no user needs were taken into consideration when designing, the main approach was the investigation and evidence found.

The following figure shows the birthing bed that was created in the first phase (Fig 1).

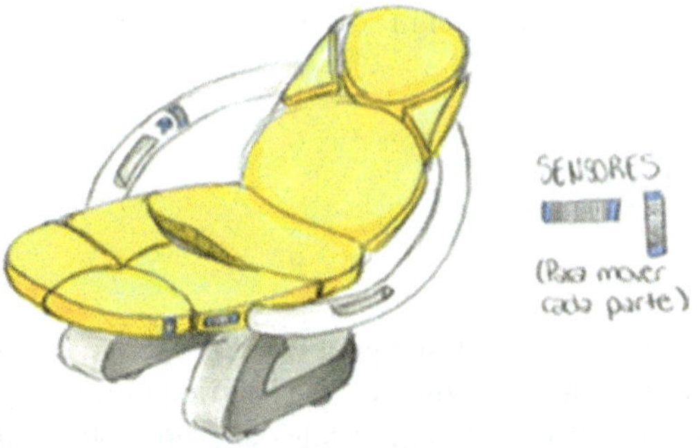

Fig. 1. Birthing Bed #1, aspects mentioned in Table 1 were taken into consideration in the design process of this bed

This design accomplishes the integration of the aspects mentioned in Table 1 as it has railings that protect the patient, it can lay down into a 180° angle, achieving a horizontal position. A band that displays the vital signs, as well as a leg and feet support and wheels located under the base.

3 Phase 2

During the second phase of the project, the goal was to design a birthing bed with an UCD approach that would facilitate its use and understanding. Through research, task analysis and surveys, the users that interact during labor and their tasks were defined.

The average number of people in the labor room is 9 users, including an anesthesiologist, gynecologist, pediatrician, three nurses, surgeon, patient and a patient's companion [7]. However, in order to create a more specific UCD, this number was narrowed down to 7: gynecologist, anesthesiologist, pediatrician, two nurses, patient and a patient's companion.

Then, the tasks and interactions each of them has with the birthing bed were analyzed in order to determine what is important for every user in the design.

At the same time, another investigation was carried out and a survey helped to illustrate the needs of every user inside the labor room. The information gathered is presented in Table 2.

Table 2. Defined users, their needs and tasks inside the labor room, gathered information from investigations, surveys and task analysis.

User	Tasks	Needs
Gynecologist	Vaginal examination for dilatation. Receive the baby Cut the umbilical cord	Adjust the birthing bed to diferent positions and heights Have access to the vaginal area of the patient
Anesthesiologist	Administer the epidural anesthesia. Check the vital signs of the patient.	Monitor the vital signs of the patient. Flatten birthing bed and access to the back of the patient to apply anesthesia [4]
Pediatrician	Monitor the vital signs of the baby. Do the Apgar test on the newborn.	Monitoring equipment
Nurses patient	Arrange the birthing bed. Disinfect and sterilize the equipment. Attention to the patient. Medical staff backup. Clean after labor. Perform labor.	Waste container Easy cleaning of beds Mechanism that allows the adjustment and arrangement of the bed Lumbar support for injury prevention [8] Legs and feet support [5] Safety railings [3]
Patient's companion	Give emotional support to the patient.	Access to the patient

Finally, a research through surveys was carried out. It was requested for the users to comment and discuss their experience with existing birthing beds. The biggest non-conformity was presented by nurses. Of a total of 20 nurses surveyed, 60% agreed that the adjustment and use of the birthing beds is not as easy as they would like it to be. The learning process to arrange the bed is long, due to the number of steps that need to be done, making it hard to remember. This, makes nurses the only medical staff capable of adjusting the bed, since no other medical staff member is trained to do so.

Taking into consideration the users' needs and tasks, a second creative process was carried out. In this process, the main goal was to give the design an UCD approach. The birthing bed that was created in the second phase is presented in the following figure (Fig. 2).

Besides considering the aspects of Table 1, the birthing bed #2 has specific elements that help satisfy the needs of every user. In order to determine the dimensions of the birthing bed correctly, the book written by Julius Panero, titled "Las Dimensiones Humanas en los Espacios Interiores" was used [9]. Birthing bed #2 has an ergonomic design with different pieces on the backrest to bring comfort and a lumbar support to reduce the tension on the patient's back. In addition, a display showing the vital signs of both the patient and the baby is located on the right siderail so that the pediatrician and anesthesiologist can monitor them. A waste container was integrated to facilitate the nurses' task. The birthing bed height can be adjusted according to the gynecologist

Fig. 2. Birthing Bed #2, aspects mentioned in Table 2 and surveys were taken into consideration in the design process of this bed.

needs, as well as the angle of leg opening, without injuring the patient. Every piece in the birthing bed has a touch membrane system that allows its individual and independent movement. This facilitates the arrangement for every user through multiple adjustments in the birthing bed, along with the ability to ease the understanding of its use.

4 Design Evaluation

Lastly, in order to evaluate the users' preference and opinion, one final survey was conducted. In it, both birthing bed designs were presented and users were asked to evaluate them taking into consideration factors like: safety, comfort, efficiency, form, color, adaptability, technology and use easiness. The survey was conducted among 80 users, sampling 10 of each group mentioned previously. The answers of the users about these aspects are presented in the following figure (Fig. 3).

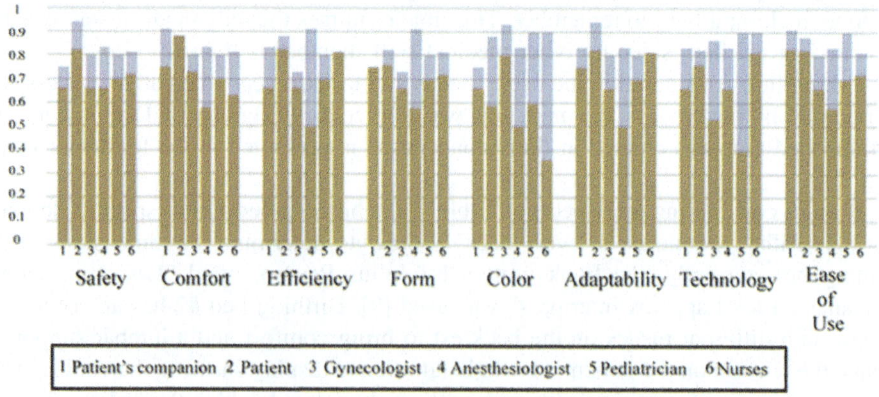

Fig. 3. Users' percentage comparison about the different aspects between the birthing beds and their preference towards each bed. The results about birthing bed #1 are presented in yellow, while results about birthing bed #2 are presented in blue.

Based on the results presented in Fig. 3, it can be assumed that users prefer the design of birthing bed #2 regarding the evaluated aspects. Additionally, some of the comments presented by the users made emphasis on the touch technology in the bed, stating the contribution it makes to the use and understanding of the bed. This helps to demonstrate that an UCD generates a more positive reaction on users, as well as a preference towards the birthing bed created with this approach.

5 Conclusion

Even though EBD helps develop designs that maximize functionality and work flow, an UCD methodology helps develop designs that satisfy the needs of different users. In both birthing bed design processes, different aspects were taken into consideration in the development, yet, users showed major preference towards the design that concentrated on their needs.

With a design of adjustable sections by touch technology, birthing bed #2 allows the user to understand its use, thus facilitating it, reducing mistake risks. Not only because of the technology it applies, but also, for the considerations of the users' characteristics, improving the ergonomy, by facilitating its use and understanding.

References

1. Camas de Parto. ECRI Institute. http://www.elhospital.com/temas/Camas-de-parto+103418?pagina=1
2. User-Centered Design Approaches for Planning Inpatient Room of Geriatric Long-Term Care Hospitals: Design Factors with Practical Suggestions: https://doi.org/10.1080/14606925.2019.1609802
3. Hamlin, L., Davies, M.: Enfermeria Perioperatoria. Texto Introductorio. Manual Moderno (2010)
4. Anestesia en Obstetricia. http://www.sld.cu/galerias/pdf/sitios/anestesiologia/anestesia_obstetricia_1.pdf
5. Biomecánica del parto. https://www.enfermeria21.com/revistas/matronas/articulo/26/biomecanica-del-parto-entrenamiento-sinergico-puerperal/
6. Programa Médico Arquitectónico para el Diseño de Hospitales Seguros. http://www.desastres.usac.edu.gt/documentos/docgt/pdf/spa/doc0147/doc0147-parte01.pdf
7. Who should be in the delivery room when giving birth? https://www.mainlinehealth.org/blog/2019/02/21/who-should-be-in-the-delivery-room
8. Dolor de Espalda por el Trabajo de Parto ¿Mito o Realidad? https://www.mayoclinic.org/es-es/healthy-lifestyle/labor-and-delivery/expert-answers/back-labor/faq-20058547
9. Las Dimensiones en los Espacios Interiores. https://wiki.ead.pucv.cl/images/c/c5/Panero_%26_zelnik_-_las_dimensiones_humanas_en_los_espacios_interiores.pdf

Using Serious Games and Motion Tracking for Physical Rehabilitation

Santiago Solórzano, Patricio-David Espinosa-Alvarez, Karina Jimenes-Vargas, and Jorge-Luis Pérez-Medina(✉)

Intelligent and Interactive Systems Lab (SI2-Lab), Universidad de Las Américas (UDLA), Quito, Ecuador
{santiago.solorzano, patricio.espinosa, karina.jimenes, jorge.perez.medina}@udla.edu.ec

Abstract. Tele-rehabilitation has evolved significantly in recent years thanks to the latest advances and low cost of the 3D movement acquisition technologies. Autonomous physical rehabilitation from home through the use of serious games requires measuring angles of interest and compensatory angles of the patients when they perform the exercises. this work has been focused in creating a game with an architecture based on the Unity 3D engine to support rehabilitation activities. It encompasses the acquisition of an articulated skeleton, the customization of a character (avatar) in friendly scenes, and the compute of the required angles to evaluate a set of exercises.

Keywords: Serious game · Telerehabilitation · Movement quality · ePhoRt

1 Introduction

Telerehabilitation has evolved significantly in recent years thanks to the latest advances in technology [1]. The interaction of people with recreational applications, such as serious video games, is also increasingly common [2] in areas such as fitness, health, rehabilitation and so on. This is due to the 3D movement acquisition devices [3] at a low cost. These devices allowed that the traditional rehabilitation practices can be improved and accompany the patients during the healing process to make it more effective and efficient. The acquisitions of the interest and compensatory angles can be obtained for each of the performed exercises to evaluate the movement. However, the evaluation of the movement quality represents one of the main needs and challenges these days [4] because the computation of the angles while the patient performs the exercise is computational expensive and data may sometimes be missing. An important factor before, during and after a therapeutic program.

The physical rehabilitation systems require measure the movement thought a set of joins points taken by a motion tracking technology. Also, another aspect that requires special attention is the study of pain behaviors through facial or verbal expressions of discomfort. Pain is an important factor in the health area and needs to be considered carefully. Taking these needs into account and considering that serious games are gaining an ever-increasing interest for education and training, this work has been focused in creating a game with an architecture based on Unity 3D engine to perform

W. Karwowski et al. (Eds.): IHSED 2020, AISC 1269, pp. 180–185, 2021.
https://doi.org/10.1007/978-3-030-58282-1_29

rehabilitation exercises. It encompasses the acquisition of an articulated skeleton, the customization of a character (avatar) in friendly scenes, and the compute of the required angles to evaluate a set of exercises. Also, it is having been focused in characterizing the signal taken by the motion sensor through angles to assess the quality of the exercises.

Four scenes and three types of characters were developed as part of the personalizing process, as well as the laterality of surgery. The character was built with the Blender[1] and the Facebuilder module[2]. Blender allows us to personalize the avatar of the character while the Facebuilder help us to use the patient's face in the avatar. Besides, some objects were included to support the patient while he accomplished his goal in each exercise.

The set of exercises were suggested by a physiotherapist for patients who has been underwent a hip surgery. The exercises available in this study were: "hip abduction", "slow hip and knee flexion", "hip extension" and sequence of steps "forward, laterally, backward". For each of these exercises, the calculation of the angles of interest and compensatory angles is carried out by means of 3D agreed coordinates using the Vitrivius library[3]. This technique allows evaluating the quality of movements. The paper is organized as follows: section two describes some related work in the context of serious games and physical rehabilitation. Then, we present an architecture based on serious games for calculating the movements of patients. Section four describes in detail the calculations performed for 4 physical rehabilitation exercises. Finally, conclusions and future works are presented.

2 Related Work

In the context of tele-rehabilitation, several home-based rehabilitation platforms have been developed in the last decade [5]. Significant progress in capturing movements [6] has enabled progress in home rehabilitation systems with several novel solutions. One of the novel solutions at home are accompanied by serious games to facilitate a playful but also a challenging rehabilitation. Techniques to detect 3D movements can be done using two strategies [6]. The first one is based on the use of chews, which must adhere to the person's body. The second one uses the depth cameras supported by a quantitative analysis in real-time [7].

From the physiotherapist perspective, the evaluation of low-cost exercises is carried out through the analysis of movements. ePHoRt [8, 9] is part of this type of support and was designed for the physical rehabilitation of hip surgery patients. Initially this project used DTW [10, 11] to evaluate the capture of movements. However, the evolution of the platform has allowed the incorporation of new fewer complex mechanisms to assess the quality of movements.

---

[1] Blender: open source 3D creation suite. https://www.blender.org/.

[2] Facebuilder: add-on for Blender. https://keentools.io/facebuilder/.

[3] Vitruvius: 3D motion framework. https://vitruviuskinect.com/.

In this work, the incorporation of a recreational environment based on Unity 3D is developed. The integration between the Djando platform and the Unity 3D environment allows the customization of various scenarios, avatars and, above all, the evaluation of the movements desired by the rehabilitation exercises.

3 Architecture

Figure 1 resumes the architecture proposed in this work, which is divided into the following aspects: (1) the connection between Django and Unity 3D; (2) the acquisition of the patient's movements; (3) the execution of custom scenes; and (4) the interest and compensatory angles. The acquisition module captures the patient's movements using an Orbbec camera and mimic the movements by mapping each point of the avatar with the Nuitrack technology. The work and compensatory angles are computed by the 3D motion framework Vitruvius. At the end of a repetition, the data is stored in a JSON and sent to the platform.

Before start, the Serious game get the patient's information (sex of the avatar, background of the video game, object of interaction, laterality) to create an avatar and to mimic the movements. And, at the end of a repetition, the serious game sends the processed data (photos, angles) to the platform. The serious game has a connection module to perform all the necessary functions to connect to the web platform, including data input and output methods. The application works as a client while the web platform works as a server under a TCP protocol. Once the connection is established, it is maintained throughout the session.

An operating module offers all the necessary functions of the serious game. When starting a session, after obtaining the exercise data, they are processed and housed in static variables; allowing us to use this data in all scenes. The first data that is processed is the scene data, since the serious game is divided into several scenes, it is necessary to know which one will be used in the exercise session. Once a scene is selected, data is processed such as the avatar to be used, the object with which to interact, as well as the number of series and repetitions of the session and the laterality corresponding to the left or right position. where the object will be positioned.

Blender was used for avatar customization. The process requires a minimum of 8 photos taken of a patient at different angles. A docking with FaceBuilder allows you to adjust the mesh with your photos to take a shape like the head. Once this process has been developed, the textures are created based on the photos taken so that it is the same face of the patient. Using the data, a playful environment is established that works according to the specifications of the patient and physical therapist.

4 Calculations Performed for 4 Physical Rehabilitation Exercises

To calculate the interest and compensatory angles, two objects were created in Unity 3D. The first one, provides access to the depth sequence, while the second one, allows information to be obtained from points on the human body. Each joint is represented in

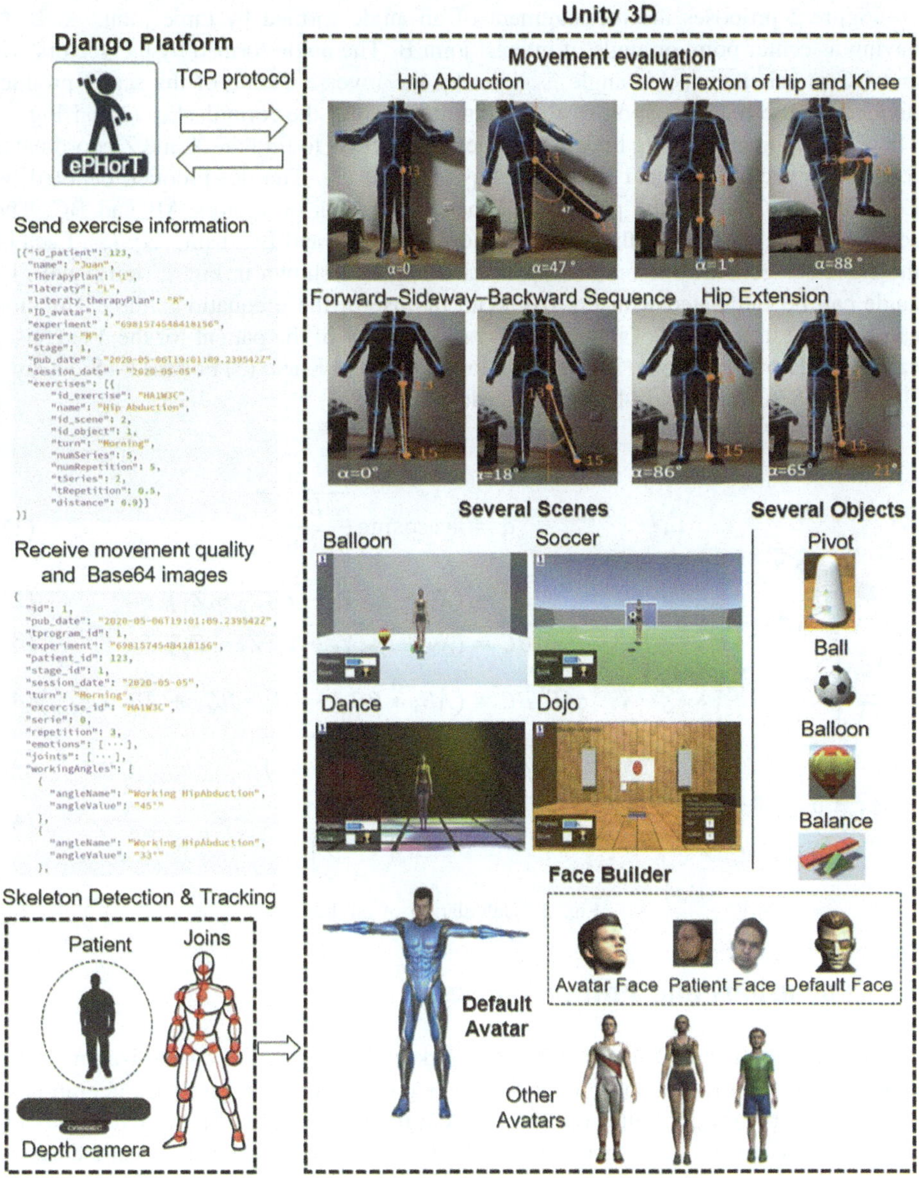

Fig. 1. Global architecture

a set of coordinates X (horizontal), Y (vertical), Z (depth). The X, Y values can be positive or negative, while the Z coordinate values will always be positive because the patient is in front of the reference axis that is the sensor. The joints have been represented by a metric system; therefore, their unit is the meter.

Figure 2 proposes the measurement of an angle formed by three joints A, B, C having as center point or angle of interest joint B. The angle formed by the vectors AB and BC is calculated, this angle is given by the inverse cosine of the scalar product divided by the multiplication of the magnitudes using the formula Eq. (1) on Fig. 2. The first step in calculating the alpha angle of interest is to find the X and Z coordinates formed by vectors AB and BC using Eqs. (2) and (3). The dot product, denoted by Eq. (4), is the sum of the products of the components of vectors AB and BC. The second step is to calculate the length of the vectors AB and BC, using Eq. (5). Each of the exercises generates a torso compensation angle as shown in Fig. 2 (left panel). Its angle can be calculated with Eq. (6). With the use of these equations, the calculation can be made in the joint of interest and compensatory of the patient for the 4 exercises called: (1) Hip Abduction; (2) Slow Flexion of Hip and Knee; (3) Forward – Sideway – Backward Sequence; and (4) Hip Extension.

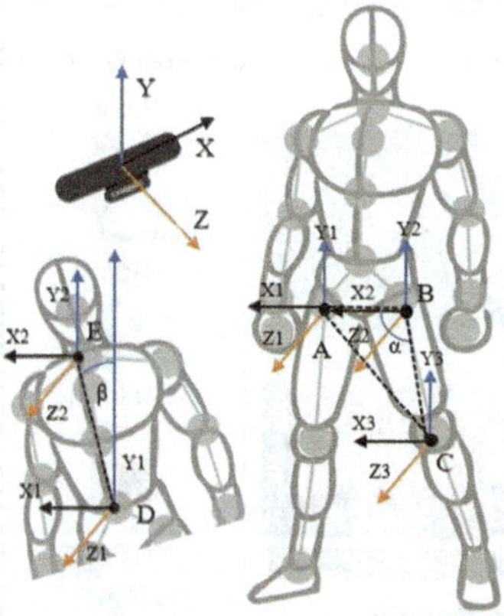

$$\alpha = \text{arccosine}\frac{\overrightarrow{AB}.\overrightarrow{BC}}{|\overrightarrow{AB}||\overrightarrow{BC}|} \quad (1)$$

$$\overrightarrow{AB} = (X_2 - X_1; Y_2 - Y_1; Z_2 - Z_1) \quad (2)$$

$$\overrightarrow{BC} = (X_3 - X_2; Y_3 - Y_2; Z_3 - Z_2) \quad (3)$$

$$\overrightarrow{AB}.\overrightarrow{BC} = (AB_x * BC_x + AB_y * BC_y + AB_z * BC_z) \quad (4)$$

$$|\overrightarrow{AB}| = \sqrt{(a_x^2 + a_y^2 + a_z^2)} \quad (5)$$

$$\beta = \text{arccosine}\frac{E_y}{|\overrightarrow{E}|} \quad (6)$$

Fig. 2. Calculation of angles

5 Conclusions and Future Works

The global architecture presented in this work encompasses the customization of the character, the incorporation of various scenes created in Unity and the integration of various objects to support the process of executing the exercises. An evaluation module of 4 exercises was established by calculating angles of interest and compensatory angles allows evaluating the quality of movements.

The evaluation results are sent via TCP to the ePHoRt platform as well as base64 image captures for the study of emotions and especially pain. In view of a future increase in the definition of rehabilitation exercises by physiotherapists, efforts will be directed towards the generalization of the angle measurement module so that they can be configured from the ePHoRt platform.

Likewise, it is also envisaged to incorporate a flexible module for personalizing the scenes and using avatars to increase the possibilities of personalizing the games.

Finally, a great challenge to be performed in the short term is the study of the quality of movements using inertial sensors to verify the precision and difference in the calculations of both systems.

Acknowledgements. This work has been financed by the Universidad de Las Américas. Quito - Ecuador, grant numbers: SIS.JPM.18.01, and, by the NATIONAL RESEARCH FINANCING program (Senescyt-INÉDITA), under the Convention No.: 20190020CI and project code: PIC-18-INE-UDLA-001.

References

1. Galea, M.D.: Telemedicine in rehabilitation. Phys. Med. Rehab. Clin. **30**(2), 473–483 (2019)
2. Bonnechère, B.: Serious games in rehabilitation. In: Serious Games in Physical Rehabilitation, pp. 41–109. Springer, Heidleberg (2018)
3. Wang, K., Zhang, G., Yang, J., Bao, H.: Dynamic human body reconstruction and motion tracking with low-cost depth cameras. In: The Visual Computer, pp. 1–16 (2020)
4. Calvaresi, D., Schumacher, M., Marinoni, M., Hilfiker, R., Dragoni, A.F., Buttazzo, G.: Agent-based systems for telerehabilitation: strengths, limitations and future challenges. In: Agents and Multi-agent Systems for Health Care, pp. 3–24. Springer, Heidelberg (2017)
5. Pezzera, M., Tironi, A., Essenziale, J., Mainetti, R., Borghese, N.A.: Approaches for increasing patient's engagement and motivation in exer-games-based autonomous telerehabilitation. In: 2019 IEEE 7th International Conference on Serious Games and Applications for Health (SeGAH), pp. 1–8 (2019)
6. Shingade, A., Ghotkar, A.: Animation of 3d human model using markerless motion capture applied to sports. arXiv preprint arXiv:1402.2363 (2014)
7. Velasco, F., Narvaez, F.: Automatic exercise recognition based on kinect sensor for telerehabilitation. In: International Conference on Smart Technologies, Systems and Applications, pp. 312–324. Springer, Hedilerberg (2019)
8. Rybarczyk, Y., Deters, J.K., Gonzalvo, A.A., Gonzalez, M., Villarreal, S., Esparza, D.: ephort project: a web-based platform for home motor rehabilitation. In: World Conference on Information Systems and Technologies, pp. 609–618. Springer, Heidelberg (2017)
9. Pérez-Medina, J.L., Espinosa-Alvarez, P.D., Jimenes-Vargas, K.B., Acosta-Vargas, P., Rybarczyk, Y.: Serious games-based exercises for arthroplasty rehabilitation. In: 11th International Conference on Applied Human Factors and Ergonomics. Springer, Heidelberg (2020). To appear
10. Rybarczyk, Y., Leconte, L., Pérez-Medina, J.P., Jimenes, K., Acosta-Vargas, P., Esparza, D.: Telerehabilitation platform for post-arthroplasty recovery: a dynamic time warping approach. In: 2018 IEEE Third Ecuador Technical Chapters Meeting (ETCM), pp. 1–6 (2018)
11. Rybarczyk, Y., Pérez-Medina, J.L., Leconte, L., Jimenes, K., González, M., Esparza, D.: Implementation and assessment of an intelligent motor tele-rehabilitation platform. Electronics **8**(1), 58 (2019)

New Birthing Bed Design that Improves User Experience During Delivery Phase, Including Ergonomic Factors

Natalia Villalpando-Chávez, Cristina Vázquez-Hernández, María Andrea Escoto-Aceves(✉), Fabiola Cortés-Chávez, and Carlos Raymundo-Garnier-Ortiz

Universidad Panamericana Campus Guadalajara, Calzada Circunvalación Poniente N. 49 Ciudad Granja, 45010 Zapopan, Jalisco, Mexico
{0215080, 0213780, 0214020, fcortes, cgarnier}@up.edu.mx

Abstract. The design process for a birthing bed in search of incrementing the satisfaction of every user involved during delivery. Our design proposal seeks adequate participation of them during the delivery. The users considered in this study are the mother, the companion, the gynecologist/obstetrician, the pediatrician, the anesthesiologist, the birth nurses, and the maintenance personnel. The design proposal was determined based on the following aspects: increasing comfort, safety, emotional state and effectiveness in the delivery process.

Keywords: Birthing bed · User-Centered design · Space · Technology · Participation · Gynecology · Postures · Health

1 Introduction

This study presents an approach to the birthing bed, the name of the product is "Odette", its focal point is increasing the satisfaction of the seven users involved during the natural childbirth process.

The nucleus of the development for the birthing bed was UCD, that is, User-Centered Design. This is due to the fact that analysis was carried out to aim at improving comfort, safety, emotional expectations, and effectiveness during delivery.

In 2016, the Healthcare Design Expo, held in Houston, Tx. USA. It was concluded on the Basis of Susan Francis' book "Plan for the Uncertainty: Design for Change" that the development of the medical environment can positively influence the effectiveness of any given medical procedure [1].

The design process was carried out as detailed below. The birthing bed is divided into two priority parts: Main Body and Accessories. (Figure 1) Subsequently, an analysis on the relationship between the birthing bed and its users will be shown.

W. Karwowski et al. (Eds.): IHSED 2020, AISC 1269, pp. 186–191, 2021.
https://doi.org/10.1007/978-3-030-58282-1_30

2 Project Development

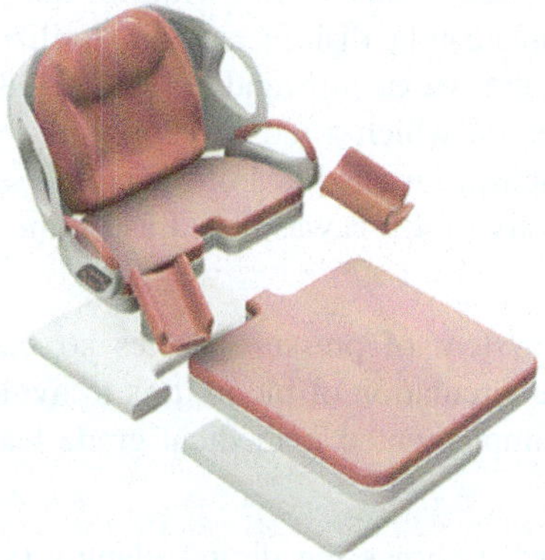

Fig. 1. Final design

2.1 Main Body

The body structure consists of 3 main parts: the backrest, the mother's seat, and the gynecologist/obstetrician seat. Each one of these pieces are made up of two elements for each member: the solid structure made of ABS plastic, and the padding that is made out of memory foam covered with seamless bacteriostatic skin with an epoxy powder finish that helps against corrosion [2].

Backrest: It acquires an enveloping shape on the patient due to the side rails used, while the padding, through an ergonomic shape, provides comfort and support to the lower back; which is the most affected during labor [3]. In addition, it has a removable head that is also adjustable to the prefered height, giving the mother stability to the upper part of the torso. The backrest manages to adapt to different positions according to the progress of the delivery process; or in other cases, according to the requirements of the doctor or the patient.

The Mother's Seat: The seat was ergonomically designed to better support the patient's pelvic section. The pelvic section, being a primordial area during the delivery, requires greater visibility and accessibility for easy maneuvering, by means of which a semicircular cutout included in the design of the seat.

The Gynecologist/Obstetrician Seat: The seat for the doctor, being part of the main structure, has the advantage that the height can be modified independently as desired by means of the hydraulic piston system, it can also be used as support for the patient's legs.

2.2 Accessories

Organic Waste Bin: The waste canister sits within a rail so it can easily be inserted into it, so in this way the area can be disinfected and sterilized with a greater speed and minimal effort. Based on the Mexican Standard (NOM-087-ECOL-SSA1-2002) [4], the bin has small fasteners, on which a red or yellow bag will be placed, as the case may be for the specific biohazard waste [5]. Following the same regulation, the volume requirement within a capacity of 4.5 L was met. Lastly, the material used was surgical steel (stainless steel 304).

Leg Supports: Using a system of pressure valves so they can be adjusted to the patient's legs, favoring the circulation of blood, this to avoid the formation of clots in the arteries. The material implemented is medical grade leatherette with a polypropylene interior [6].

Multiparameter Monitor: It shows the digital display (visual and auditory) of the vital signs of the fetus and the patient. In reference to the audiovisual and physical part, it's an LCD with plastic covers in ABS material. It has the option to modify the brightness and sharpness of the screen; as well as raising or lowering the volume. The technical parameters does not exceed 60 Hz/240 V for the optimal performance. Operates on rechargeable or disposable batteries that work independently of the birthing bed's supply [7].

Patient and Physician Controls: Designed to be intuitive and attractive to the user thanks to its simple iconography system. Within the technical parameters, the IP67 grade stands out (indicator of resistance against the ingress of solids and liquids) [8]. Similarly, it works on the basis of radio frequency, by virtue of its long signal range. It's made with an ABS plastic cover and the buttons are made of elastomer (Dynaflex).

Handrails and Side Railings: They are ergonomic and provide additional support to the mother. Also they can be used as a fulcrum to gain more strength when pushing during labor. They consist of two parts, a more solid part that is made with ABS plastic and a padded part of Memory Foam, which allows the adaptability and support for the mother's hand, which reduces the pressure in the area.

2.3 Available Positions

The functional and ergonomic design of the birthing bed includes the following positions:. [9, 10] (Table 1).

Table 1. Available positions

Name	Description
Squatting position	Position that allows the diameter of the pelvis to open and with the force of gravity, facilitate the labor
Semi lying down	Effective position for labor, the fetus manages to fit into the pelvis in the correct way so that the delivery can proceed safely and normally
Gynecological position (Lithotomy)	The main advantage is the easy access for the obstetrician to the woman's abdomen to monitor the fetus
Lateral position	It will only be on the left side due to the fact that otherwise, the weight and pressure of the body could trap the vena cava, possibly producing an abnormal blood flow that can affect the placenta and the fetus
CPR position	Cardiopulmonary resuscitation should be performed with the patient in a completely horizontal position

2.4 Users

The birthing bed "Odette" presents various advantages since it takes into account the needs and comforts of all participating users. The aspects that were considered for each case are listed below [11, 12] (Table 2).

Table 2. Users

Name	Improvements implemented
Mother	Ease and freedom to choose the right position for the delivery which brings benefits such as comfort, security, feeling of freedom, pain reduction, among others
Obstetrician gynecologist	Easy access to the mother's pelvic area. Use of part of the birthing bed as a seat
Pediatrician	Accessibility to the area to receive the newborn
Companion	Designed so that the companion can have interaction with the birthing bed and participate without causing any inconvenience to other users
Anesthesiologist	Easy access and monitoring of the patient
Maintenance staff	Being able to carry out the cleaning in an adequate way, counting with sanitary curves that allow optimal cleaning of the area
Nurses	Intuitive design that facilitates the correct operation and arrangement of the birthing bed

3 Evaluations

Based on the UCD analysis, the following four aspects evaluated were taken into account, as well as the preferences mentioned above. According to the research carried out in South Korea, these qualities are considered as most important, which relate to the physical dimensions, emotional and social aspects [13] (Table 3).

Table 3. Results of surveys conducted with mothers, gynecologists/obstetricians, pediatricians, maintenance personnel, nurses, anesthesiologists and companions.

Ambit	First design birthing bed	Final design birthing bed
Safety	19% of users consider the birthing bed to be completely safe	56% of users consider the birthing bed to be completely safe
Comfort	14% of users consider the birthing bed to be completely comfortable	43% of users consider the birthing bed to be completely comfortable
Adaptability	17% of users consider the birthing bed to be completely adaptable	52% of users consider the birthing bed to be completely adaptable
Necessary accessories	13% of users consider the birthing bed to have completely necessary accessories	48% of users consider the birthing bed to have completely necessary accessories

4 Conclusions

Based on the information presented in this document, it can be concluded that thanks to the use of user-centered design methodology, it was possible to increase user satisfaction by taking all users involved in the different phases of a natural childbirth into consideration throughout the design process. This allowed us to identify the specific needs of each person and take them into account during the design process to design a product that offers optimal satisfaction to all users.

References

1. Chun, S., Nam, K.: User-centred design approaches for planning inpatient room of geriatric long-term care hospitals: design factors with practical suggestions (2017)
2. Secretaría de Salud Mexicana. Norma Oficial Mexicana (NOM-016-SSA3-2012). https://www.gob.mx/cms/uploads/attachment/file/35887/NOM-016-SSA3-2012.pdf
3. Personal de Mayo Clinic. Dolor de espalda en el trabajo de parto; cómo prevenir el dolor y las lesiones (2019). https://www.mayoclinic.org/es-es/healthy-lifestyle/adult-health/in-depth/%20back-pain/art-20044526
4. Secretaría de Salud Mexicana. Norma Oficial Mexicana (NOM-197-SSA1-2000). https://www.ucol.mx/content/cms/13/file/NOM/NOM-197-SSA1-2000.pdf
5. Tipos de residuos hospitalarios; forma de eliminación de los residuos hospitalarios | Salud Pública y algo más. https://www.madrimasd.org/blogs/salud_publica/2010/05/04/131890
6. ¿En qué consiste la presoterapia? https://muysaludable.sanitas.es/salud/consiste-la-presoterapia/
7. Seguridad Eléctrica de Equipos Médicos. https://www.energymed.com.co/seguridad-electrica-de-equipos-medicos-parte-2/
8. Mi nuevo dispositivo tiene grado de protección IP54, ¿qué significa? https://www.nobbot.com/pantallas/que-es-grado-proteccion-ip/
9. Las diferentes posturas en el parto. https://www.efisioterapia.net/articulos/diferentes-posturas-parto
10. Los 4 tipos de Pelvis. ¡Conócelas! https://eominternacional.com/los-4-tipos-de-pelvis-conocelas/

11. World Health Organization. WHO recommendation on fundal pressure to facilitate childbirth | RHL (2018). https://extranet.who.int/rhl/topics/preconception-pregnancy-childbirth-and-postpartum-care/care-during-childbirth/care-during-labour-2nd-stage/who-recommendation-fundal-pressure-facilitate-childbirth
12. Parto natural (para Padres) - Nemours KidsHealth. https://kidshealth.org/es/parents/natural-childbirth-esp.html
13. Instituto Mexicano del Seguro Social. Vigilancia y Atención Amigable en el Trabajo de Parto en Embarazo no Complicado (GPC-IMSS-052-19) (2019). http://www.imss.gob.mx/sites/all/statics/guiasclinicas/052GRR.pdf

Virtual Learning Objects' of Math Educative Process

Omar Cóndor-Herrera[1(✉)], Janio Jadán-Guerrero[2], and Carlos Ramos-Galarza[2,3]

[1] Maestría en Innovación y Liderazgo Educativo, Universidad Tecnológica Indoamérica, Quito, Ecuador
italo_omar@hotmail.com

[2] Centro de Investigación en Mecatrónica y Sistemas Interactivos MIST, Carrera de Ingeniería en Ciencias de la Computación y Carrera de Psicología, Universidad Tecnológica Indoamérica, Av. Machala y Sabanilla, Quito, Ecuador
janiojadan@uti.edu.ec

[3] Facultad de Psicología, Pontificia Universidad Católica Del Ecuador, Av. 12 de Octubre y Roca, Quito, Ecuador
caramos@puce.edu.ec

Abstract. Nowadays, we are facing rapid educational and technological era, therefore there is increased need to highlight the connection for teachers to rediscover improved teaching-learning processes. The actual educative processes are being developed in virtual learning environments, offering flexible, far reaching distance learning, and better tools for teachers and students in the 2.0 web, this has allowed discovering countless opportunities to improve learning environments, offering students an education that responds to the reality where they can flourish, innovate and develop. The present research brings up as its objective the analysis of the implementation of virtual learning objects in learning math, and as a result, the creation of an innovator technological proposal in the process of teaching-learning.

Keywords: Virtual learning objects · Virtual learning environments · e-learning · b-learning · Math educative process

1 Introduction

The implementation of virtual learning objects (VLO) arises as an alternative to face educative issues such as low academic performance, desertion, and repetition of school years; with these tools given to professors, it is pretended to overcome these situations, proposing an educative innovation in relationship to methodology and didactic resources that are used by teachers, focusing in the area of math through the implementation of VLOs, with the use of the information and communication technologies (ICT), transferring the educative scenario to the virtual learning environment (VLE).

According to the data reported by the Organization for the Cooperation and Economic Development (OCDE) [1], Latin American countries, the context of interest for this investigation, presents a low-performance level in evaluations of math and other

W. Karwowski et al. (Eds.): IHSED 2020, AISC 1269, pp. 192–197, 2021.
https://doi.org/10.1007/978-3-030-58282-1_31

examined areas. Specifically, Colombia has reported an average of 390 in the standardized evaluation in the areas of sciences, reading and math, achieving the place 61st out of 70 evaluated countries. Peru obtained the 62nd place, with an average of 387 points in math. On the other hand, Uruguay holds the 51st place, with an average of 418, which places these countries among the last places in the table, far from countries such as Singapore that counts with an average of 564, representing to the 4.8% of the total of their students presenting low-academic performance in sciences, reading and math (OCDE) [1].

These data allow highlighting a regional level of difficulty in Latin American countries, referring to the development of learning and comprehension of math and a high percentage of students with low academic general performance. This reality is not different in Ecuador, taking into analysis the Report of Education in Ecuador for development PISA (National Institution of Educative Evaluation) [INEVAL] [2], it is evident that the obtained results achieved by the students are not different to those reflected in the region.

The obtained results in Ecuador arises to an average of 377, being 10 points less in comparison to the academic performance evidenced in Peru and with 44% of students presenting low-academic performance, it is mentioned that in reading subject 51% of the students could not reach the level 2, this number increases to 57% in science and an alarming 71% in math [2]. These data show Ecuador students' low-academic performance level, emphasizing math area, which represents a higher issue because of the elevated percentage of students that are placed lower the level of optimal performance.

2 Virtual Learning Objects (VLOs)

In a first approach to the term VLO, authors such as Callejas, Hernández, and Pinzón [3] state that the term Learning Object was named for first time in 1992 by Wayne, who associated LEGO learning normalized blocks, to reuse them in educative procedures, conceptualizing a VLO as a Lego, because it allows in its structure to add many other pieces like in a puzzle [3]. These virtual learning objects could be implemented as educative resources in the E-learning setting (see Fig. 1), as well as in M-learning. According to Colombia's National Ministry of Education (MEN) [4] a virtual learning object could be defined as a set of digital resources that could be used in a variety of contexts, with an educative purpose and to be constituted by at least three internal components: contents, learning activities and contextualization elements. Also, the learning object must have an external information structure (metadata) to facilitate its storage, identification and data recovery.

Therefore, a VLO is the set of digital resources such as materials, videos, activities, images, and diagrams, that once those are linked and organized with a pedagogical objective may be used as resources of learning of a determinate topic, since its development involves a variety of tasks and materials immersed in the virtual learning object, allowing the easier comprehension of any subject. It is important to mention that the VLO is adaptable to the student's learning rhythm and the content could be revised any time and for many times, as it is stated by many other authors, the VLO's goal is to be used as a learning tool, so, students can learn in their own time and rhythm, independently, the bases of a specific topic.

Fig. 1. Students learning with VLOs

3 Implementation of VLOs in Math's Educative Process

As it was mentioned, the VLO may be applied in the teaching-learning process, since it offers many possibilities to students, such as the learning rhythm adaptability, the implementation of different activities that calls the student's attention, through the use of multimedia resources, games, quizzes, etc., with the advantage of being used as many times as the student wants, assimilating the content of the VLO in his/her rhythm. Figure 2 shows an example of an activity containing a VLO: a video quiz.

Fig. 2. VLO's audio-visual activity

As it is stated by Morales et al. [5], the personalization of the different activities, such as exercises, readings, videos and the variety of resources that a VLO offers, allows the student a fluid learning. Through the use of VLOs, and virtual learning environments (VLE), it is possible to work on exercises as part of mechanization processes, applied problems, or as part of an evaluation procedure of any subject, enabling teacher-learning processes. It is necessary to arise the need for a permanent follow-up to the procedures and the corresponding examination for its optimal functioning by the professor [5]. Figure 3 shows an example of an evaluation's question that could be find in a VLO.

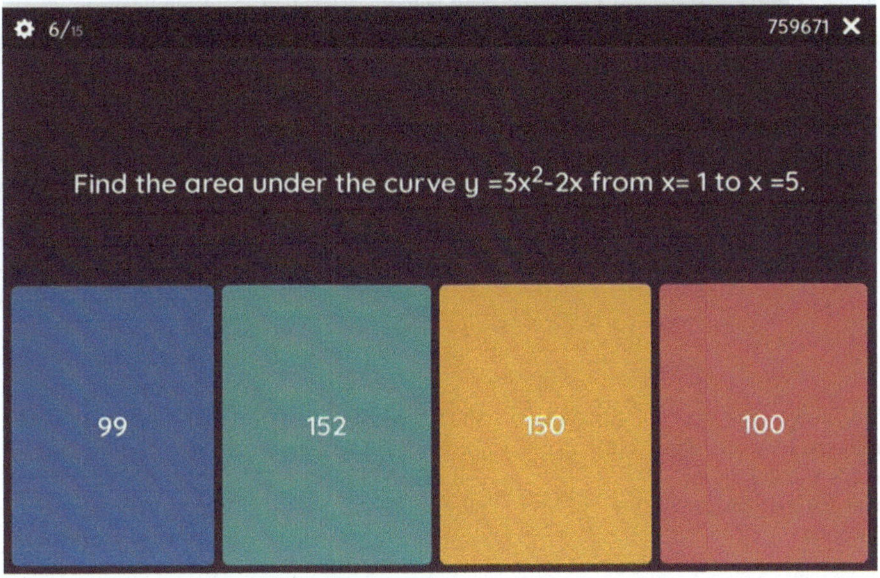

Fig. 3. Example of an examination in a VLO.

Therefore, for future investigation, it is of great interest to implement the VLOs as a didactical resource in the teaching-learning process, to improve students' math academic performance.

4 VLOs Applied in the Teaching-Learning Procedures: Previous Studies

To identify the VLO's use relevance in the educative process, it was made a previous revision of the literature, focusing in the use of VLO in the improvement of math's performance. Table 1 describes the s findings.

Table 1. VLO's use literature

Name	Authors	Investigation	Findings
Construction of virtual learning objects in engineering from a based-problem approach	Parra and Narváez [6]	Construction of objects to be applied in engineerings such as the representation of variables, data input, and output, assignation of expressions	Virtual learning environments involves the organization of learning activities that directs students' self-determination
Interactive math: ¿Another way of teaching math?	Terán [7]	Data was obtained applying the participative observation, interviews, field notes, photographs, and videos	To achieve children's math learning with pleasure and positive attitude that will allow the accomplishment of meaningful learning, teachers must propitiate challenges that will get the whole kid's interest and in this way changing mental schemas
Virtual objects' prototype for exercising math's learning in the first- grade of basic education	Lizcando [8]	VLOs' prototype design for first grade children and the application of a pilot study	It is noted a general improvement after using the VLO, comparing the performance pre and post-test, that could be explained considering that the game exercising requires recalling concepts

5 Conclusions

The implementation of VLOs allows students autonomous learning that is adapted to their own time. Since it is a reusable object, it allows the student to make feedback through this object to the different doubts and questions that might present about a specific point of the addressed topic, as it is the matter of this investigation, math.

The VLOs allow inserting on them a variety of activities such as games. Videos, quizzes, readings, video quizzes, among many others that raise the attention of the student, enabling learning with higher levels of motivation and in a natural way, since it offers to students a technological environment according to the present era.

As a future investigation, we propose research with the Design Thinking Strategies directed to students and professors in order to improve the relationship between the VLOs design and the implementation of new resources on them.

Finally, it is of great interest to the research team to develop virtual learning objects for different subjects and topics, for being applied in the process of teaching-learning, and to evaluate the impact that will have in the digital process.

References

1. OCDE, PISA (2016). https://www.oecd.org/pisa/pisa-2015-results-in-focus-ESP.pdf
2. INEVAL, Educación en Ecuador resultados PISA para el desarrollo (Education in Ecuador, results PISA for the development) Quito (2018)
3. Callejas, M., Hernández, E., Pinzón, J.: Objetos de aprendizaje, un estado del arte (Learning objects, state of art). Entramado **7**(1), 176–189 (2011)
4. Ministerio de Educación Nacional (National Ministry of Education), Banco Nacional de objetos de aprendizaje e informativos (National Bank of learning objects and informatics) (2007). http://portalapp.mineducacion.gov.co/drupalM/
5. Morales, M., Luz, Y., Gutiérrez, L., Ariza, L.: Guía para el diseño de objetos virtuales de aprendizaje (OVA). Aplicación al proceso (Guide for designing virtual learning objects (VLO). Process application). General José María Córdova, pp. 127–147 (2016)
6. Parra, E., Narváez, A.: Construcción de objetos virtuales de aprendizaje para ingeniería desde un enfoque basado en problemas (Construction of virtual learning objects to engineering from a problem-based approach). Revista Virtual Universidad Católica del Norte **31**, 84–104 (2010)
7. Terán, M.: Matemática interactiva: ¿Otra forma de enseñar la matemática? (Interactive math: ¿Another way to teach math?). Educere **6**(21), 88–93 (2003)
8. Lizcando, A.: Prototipo de objeto virtual de aprendizaje para la ejercitación en matemáticas de primer grado de educación básica (Prototype of virtual learning objects in math to exercise on first-grade of basic education). **58**, 96–115 (2010)

Relationship Between Technological Resources and Meaningful Learning in Secondary Students

Johanna Bonilla-Guachamín[1(✉)], Janio Jadán-Guerrero[2], David Rojas-Londoño[1], and Carlos Ramos-Galarza[2,3]

[1] Maestría en Educación, Innovación y Liderazgo Educativo, Universidad Tecnológica Indoamérica, Quito, Ecuador
edjoal513@hotmail.com, davidrojas@uti.edu.ec
[2] Centro de Investigación en Mecatrónica y Sistemas Interactivos MIST, Carrera de Ingeniería en Ciencias de la Computación y Carrera de Psicología, Universidad Tecnológica Indoamérica, Av. Machala y Sabanilla, Quito, Ecuador
janiojadan@uti.edu.ec
[3] Facultad de Psicología, Pontificia Universidad Católica del Ecuador, Av. 12 de Octubre y Roca, Quito, Ecuador
caramos@puce.edu.ec

Abstract. Technological resources have as a goal to facilitate knowledge acquisition, storage, and use, to achieve the results proposed in the teaching-learning process and to obtain meaningful learning. The research proposed as following is based on a correlational methodology, which is oriented towards the improvement of the teaching-learning process and has as aim, to study the relationship between technological resources and meaningful learning of students belonging to high-school level. Results obtained are as following: (a) higher availability of technological resources produce an increase of knowledge acquisition ($r = 0.42$, $p =< 0.001$) and (b) while there is higher scientific literature available, students' skills will be higher too ($r = 0.53$, $p =< 0.001$).

Keywords: Technological resources · Meaningful learning · Social science

1 Introduction

The investigation of the relationship between technological resources and meaningful learning of students belonging to high-school is important since it allows to determine and deep into the causes of an educative latent problem. Also, education at the present faces new demands, such as the use of technological resources for the students to acquire meaningful knowledge through their study years [1].

The profile of the instructor and students has evolved with the digital era and the technological innovations that the XXI century offers. The instructor from the XXI century must be prepared to face the new challenges of education because of these innovations. These challenges implicate to promote a lasting learning attitude, to propitiate critical thinking and curiosity to investigate, as well as being flexible

W. Karwowski et al. (Eds.): IHSED 2020, AISC 1269, pp. 198–202, 2021.
https://doi.org/10.1007/978-3-030-58282-1_32

according to the students' context and their learning environments. Also, integrating technology to the classroom guided by pedagogical strategies that promote collaborative work and active learning, allowing to build an educative framework where students are, as result, independent, self-sufficient and critical, guided by innovator professors with the capacity of connecting technology and pedagogy [2].

Results from the United Nations for the Educational, Scientific and Cultural Organization (UNESCO), through the Second Regional Comparative and Explaining Study, stated that there are a few goals achieved in primary and secondary education in countries form Latin American [3].

2 Technological Resources

Education faces great challenges responding to the new technological changes. Technological resources use has become a nodal point in the teaching-learning process, as well as the knowledge scope, because it facilitates the professor's development, to learn interactively, offering quality education and improving student's learning process.

A technological resource is a means that depends on technology to accomplish its goal. Technological resources could be tangible and intangible. Nowadays, these are essential pieces in schools, businesses, and homes [4]. In this sense, technology has become the key ally to professors when they are capable to integrate it as another resource in the teaching-learning process.

Nowadays, education faces multiple technological challenges and with them, the main one is to give answers to the society about knowledge, the use of ICT implies choosing and to adapt the material, as well as to guide the students in the selection of contents for socially pertinent learning. The challenge is accomplishing the student to differentiate among the positive and negative information, looking for information in the network and being capable to build a valuable judgment about the things they find and read [5].

2.1 Technologies for Learning and Knowledge (TLK)

This term has been introduced in the educative setting, which has allowed its development when looking for new alternatives to insert technology in the learning process, these are the TLK, which are closely linked to the education, thanks to the technological advances and its inclusion into the education. The teaching-learning process allows forming the students in an integral manner, since digital didactics in the educative setting is used, in order to implement higher interactivity in the classroom and to achieve meaningful learning.

The TLK serves to design, implement and to evaluate activities and tasks that go beyond the instrumental use of devices, systems and processes, in order to appropriate a setting that will increase the interest and learning management, to exercise, illustrate propose, interact and exemplify [6]. Although, it is important to address that the transformation to ICT to TLK will be pretty difficult if there is not a change in the teaching practice [6].

3 Meaningful Learning

Within the educative context, meaningful learning implies that knowledge has a special meaning for every person, and it is the one linking new information to a relevant concept already existing in the student's cognitive structure.

This proposal arises since David Ausubel, who, in 1963 when the behaviorism framework gained all the attention, proposed as an alternative for the teaching-learning process a model based on the discovering, which privileged activism, postulating that what has been discovered is learned. Ausubel understood that the human mechanism by excellence to raise and preserve knowledge was receptive meaningful learning, in the classroom setting and in daily life [7].

4 Relationship Between Technological Resources and Meaningful Learning: Investigation Hypotheses

The present investigation formulated its hypotheses concerning two variables, on one hand, technological resources, and, on the other hand, meaningful learning. The approach is directly proportional, meaning as next, H_1 while higher is the availability of technological resources, higher acquisition of meaningful learning will be, and, H_2 while higher scientific literature is available, students' skills development will increase too.

5 Method

Participants
This investigation worked with a randomized sample of 231 high school students, belonging to the private educative system from a Latin American country (Quito-Ecuador).

Research Design
This investigation is quantitative and with a correlational approach.

Instruments and Procedure
Collecting information was realized by applying two instruments to the participants, through Google Forms. The first questionnaire measured the use of technological resources and the second one measured meaningful learning. Table 1 shows the instrument's reliability indexes.

Table 1. Cronbach's Alpha analysis

Questionnaire	Cronbach's Alpha	Elements' No.
Technological resources availability	.71	5
Scientific literature availability	.81	5
Knowledge	.78	5
Skills	.79	5

Data Analyses
Descriptive and inferential statistical techniques to analyze hypotheses and instruments' reliability were applied.

6 Results

Hypothesis 1. According to the correlation analysis, it is possible to determine that, there is a directly proportional relationship, as it is shown in the dispersion diagram, where is possible to observe that when the availability of technological resources in the teaching-learning process is higher (variable A), there is a higher acquisition of meaningful learning (variable B) See Fig. 1.

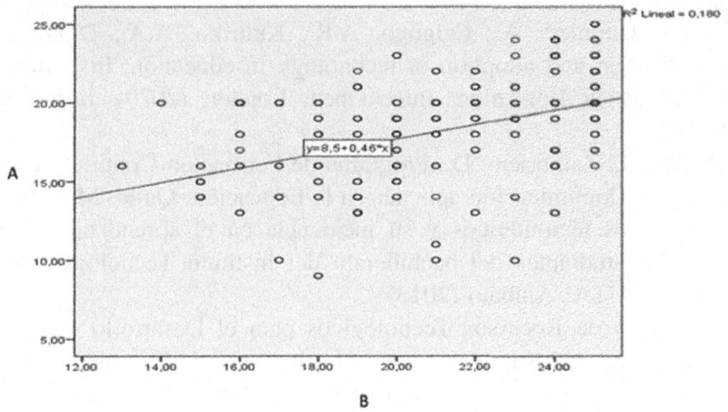

Fig. 1. Dispersion diagram of technological resources and knowledge availability

Hypothesis 2: According to the correlation analysis, it is possible to determine that, there is a directly proportional relationship, meaning that, while higher is the scientific literature availability (variable A), students' skills will increase too (variable B) See Fig. 2.

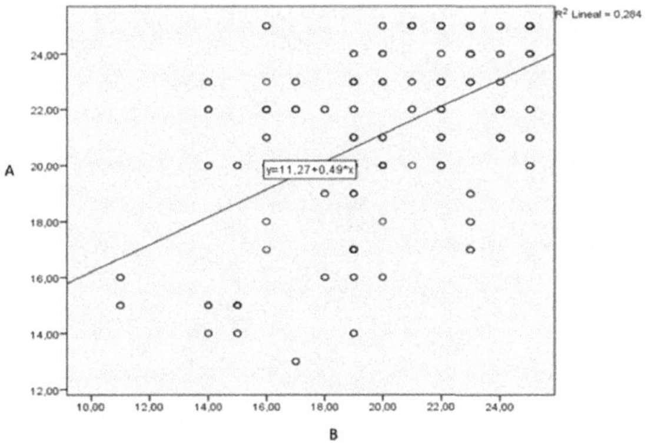

Fig. 2. Dispersion diagram of scientific literature and students' skills.

7 Conclusions

In this correlational study, it is possible to evidence from the hypotheses analyzed that, there is a directly proportional relationship between technological resources and meaningful learning. Thus, while more technological resources are used in the teaching-learning process, more significant knowledge will be obtained by students.

From these results, it is possible to propose the future research of the presented variables for its projection for the execution of studies about the implementation of technological resources and which should be evaluated through experimental studies to determine the effect on the actual teaching-learning processes.

References

1. Mustapha, A., Mohammed, A., Egigogo, A.R., Kutiriko, A.A., Dokoro, A.H.: Factors affecting the utilization and adoption of technology in education. In: Altlnay, F. (ed.) The Role of Technology in Education. IntechOpen, London (2020). https://doi.org/10.5772/intechopen.85712
2. Scoggin, J., Wills, N., Zambrano, D.: Programa de Formación Continua. In: de Tecnologías de la Información y Comunicación aplicadas a la Educación, Quito, Mineduc, p. 199 (2012)
3. Paredes, C.: Recursos tecnológicos y su incidencia en el aprendizaje significativo de la matemática de los estudiantes del bachillerato del Instituto Tecnológico Rumiñahui de la Ciudad de Ambato. UTA, Ambato (2013)
4. Parra, C.: Aplicación de Recursos Tecnológicos para el Desarrollo de la Comprensión y Expresión del Lenguaje de los niños y niñas del sub nivel 2 de Educación Inicial de la Unidad Educativa San Francisco del Alvernia durante el año lectivo 2016–2017. Universidad Central del Ecuador, Quito (2017)
5. UNESCO, Texto 2 Formulación de Proyectos, Lima: ARTOLAN E.I.R.L
6. Pérez-Zúñiga, R., Mercado-Lozano, P., Martínez-García, M., Mena-Hernández, E.: La sociedad del conocimiento y la sociedad de la información como la piedra angular en la innovación tecnológica educativa. Revista Iberoamericana para la Investigación y el Desarrollo Educativo **8**(16), 847–870 (2018)
7. Rodriguez, M.: La teoría del aprendizaje significativo: una revisión aplicable. Revista Electrònica d'Investigació i Innovació Educativa i Socioeducativa **3**(1), 29–50 (2011)

Virtual Assistants and Its Implementation in the Teaching-Learning Process

Omar Cóndor-Herrera[1(✉)], Janio Jadán-Guerrero[2], and Carlos Ramos-Galarza[2,3]

[1] Maestría en Innovación y Liderazgo Educativo, Universidad Tecnológica Indoamérica, Quito, Ecuador
italo_omar@hotmail.com

[2] Centro de Investigación en Mecatrónica y Sistemas Interactivos MIST, Carrera de Ingeniería en Ciencias de la Computación/Carrera de Enfermería/Carrera de Psicología, Universidad Tecnológica Indoamérica, Av. Machala y Sabanilla, Quito, Ecuador
janiojadan@uti.edu.ec

[3] Facultad de Psicología, Pontificia Universidad Católica del Ecuador, Av. 12 de Octubre y Roca, Quito, Ecuador
caramos@puce.edu.ec

Abstract. Nowadays, technological advances encompass realities that in the last century were possible just as fantasies. Society lives now in a technological era where education must look for the path and tools to get the benefits that these advances offer, as well as the technological innate abilities that have been developed in students of this generation, who have been nominated as digital natives. For 2017 the capacities and use of virtual assistants have been significantly expanded, new products are offered in the market, although, these assistants are sub-used with basic functions solely, not taking into account the offered benefits. The present work aims to describe the different virtual assistants that could be used as an educational tool contributing to the teaching-learning process.

Keywords: Virtual assistants · Artificial intelligence · ICT · M-learning · Teaching-learning process

1 Introduction

From the emergence of ICT and the incorporation of them into the educational field there have been new challenges to be created and not just for the production, representation, diffusion, and access to knowledge, but also in innovating constantly and generate new learning conditions in environments which are not the conventional ones, offering chronic and asynchronous communication [1].

The availability of technological devices variety inserted in the teaching-learning process offer a high-rage of possibilities to the professor of the XXI century, to generate learning that responds to the actual reality and where students develop [2]. Unquestionably, technological improvement will keep continuing faster [3], proof of that is the

W. Karwowski et al. (Eds.): IHSED 2020, AISC 1269, pp. 203–208, 2021.
https://doi.org/10.1007/978-3-030-58282-1_33

invention and technological boom of the virtual assistants that work with virtual intelligence. There are many developers which have caused the use of these assistants and the significant expansion worldwide [4].

Thousands of users, between them, students, are familiarized with the usage of different applications, platforms, and technological educative devices since virtual assistant's availability goes from the ones that are incorporated to mobile devices such as SIRI, which is incorporated to Apple's devices, Google Assistant in mobile devices, Cortana, incorporated to computers and recently, intelligent speakers such as Google Home, Google home mini, and the Amazon's Echo dot which counts with Alexa, and many more others.

In that sense, it is proposed a theoretical revision of the principal technological developers that has been created to improve educative processes at different levels, as following. This work contributes significantly to professors to count with a range of possibilities about technological resources that might be used in their daily practice.

2 Virtual Assistants Contributing to Education

A virtual assistant is an intelligent agent, capable of perceiving its environment, processing those perceptions and answering or acting in its environment rationally [5], a software-based on artificial intelligence (AI) that could realize tasks or offering services to a user, being capable of realizing some actions common to humans, taking into account two characteristics: reasoning and behavior [6], this assistant could be set and adapted to users' needs, even being used as an educational tool in a class since these allow a constant interaction user-assistant.

This interaction catches children, teenagers, and adults' attention, since there is the possibility to talk to a virtual assistant, obtaining an answer which generates users' surprise and curiosity [7], producing an inner motivation, this fact could be useful to improve the learning experience, being a novel and advanced way that integrates ICTs into the education field. As it is affirmed by Escorcia and Triviño [8], an advanced integration is evidenced in management strategies for ICT usage, as well as designing learning environments or designing projects which imply technological use. Figure 1 shows how the interaction with the assistants generates emotional responses from users.

2.1 Virtual Assistants in Mobile Devices and Computers

Siri

One of the pioneers of virtual assistants is SIRI, which has been created in 2007 [9]. Apple describes it as artificial intelligence with functions of a personal assistant, it is present in devices iOS, macOS, tvOS, and watchOS, within its functions are answering questions, making recommendations using natural language processing (NLP), contrasting information, makes calculations, becoming a type of mobile encyclopedia, which may be used as another learning tool in the teaching-learning process.

Cortana

This assistant was developed by Microsoft, appeared in 2014, at the beginning it was designed for Windows 10. Between its functions are natural voice recognition without

Fig. 1. Interaction with an intelligent speaker

the need to use the keyboard and answering questions using the information from the searching motor Bing [10]. Figure 2 shows a representation of the Cortana virtual assistant.

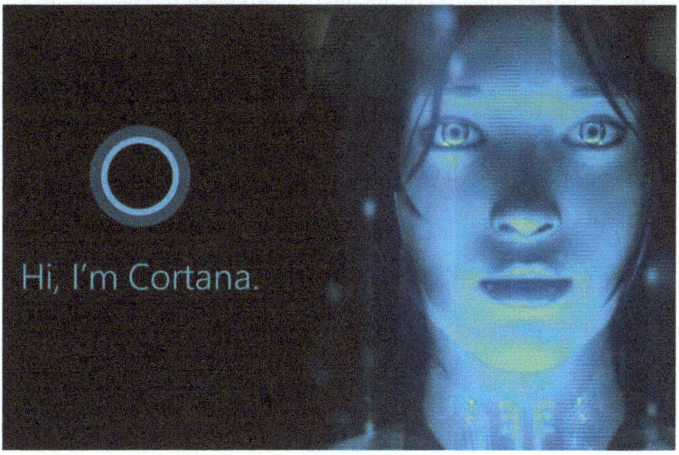

Fig. 2. Cortana's representation

Google Assistant

This is a virtual assistant developed with artificial intelligence by Google, it is present in domestic and mobile intelligent devices [11], appeared in 2016 and it is possible to interact with it through voice, admitting the input to the keyboard and its principal educative uses are voice commanders or fast answering to any topic, students interact as a game with basic knowledge in a range of areas [12] (see Fig. 3).

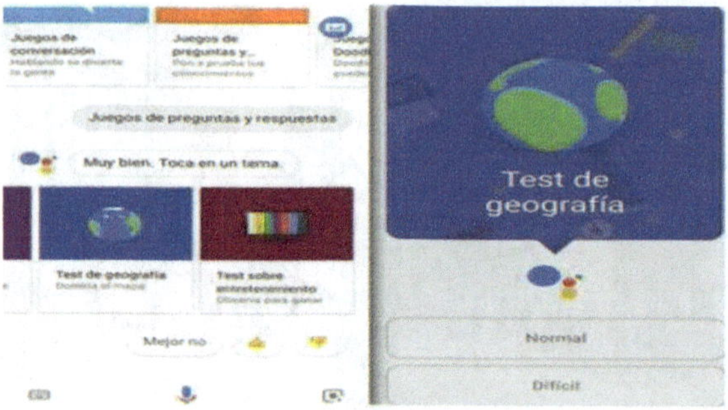

Fig. 3. Educational interaction with Google assistant.

2.2 Virtual Assistants in Intelligent Speakers

The evolution of virtual assistants drew strength when intelligent speakers Siri and Cortana were launched, as well as its mobile devices and computers' version. On the other hand, Google's assistant counts with an intelligent speaker denominated Google home. Samsung counts with its intelligent speaker Bixby, and Amazon counts with the Echo dot, which is integrated with Alexa's assistant and will be detailed in the next paragraph. Figure 4 shows intelligent speakers from each developer already explained.

Fig. 4. Intelligent speaker

Alexa

Alexa was launched in 2014, it is a virtual assistant developed by Amazon, using the intelligent speaker named Echo dot, within its educative functions are answering questions, contrasting and looking for information of any topic, translating to a variety of languages, solving math exercises, makes conversions, calculations, looks up for synonyms, reads fairy-tales, etc.

Also, when it is synchronized to a mobile device, it offers the opportunity to incorporate different abilities, these are nominated by Amazon as «Skills», is equivalent to the applications that are known in mobile devices. With these abilities inserted, Alexa becomes more intelligent allowing users to realize more functions just by telling Alexa what to do. It uses a set of tools, creating natural experiences by voice [13]. Nowadays, there are a variety of skills designed with educative goals, which would be used in the teaching-learning process o even in the autonomous students' learning.

2.3 Usage of Virtual Assistants Contributing to Education Field

Even when the usage field of the assistants in learning environments are not studied deeply, it is possible to point out some options, from which we will mention as following.

Virtual assistants could work as virtual professors, since they could keep an interactive conversation with the user, as it happens in the scholar environment, acting as an interactive professor or guide capable of solving questions, explaining some matters and also realizing automatic and personalized evaluations for each student [14].

Another option is the possibility of enhancing the learning of a second language since Alexa translates words and uses Skills to teaching languages and using answering corrections allowing to verify them and improving into the wanted expression [6]. In this sense, it offers the students the opportunity to acknowledge their mistakes, learning from them and being evaluated from their hits.

3 Psychological Benefits from Virtual Assistants in Learning Processes

Inner motivation is one of the benefits of using virtual assistants in the teaching-learning process, as it was previously analyzed, the usage of virtual assistants results novel and innovating for students, that is why learning became pleasant and improve results. According to Anaya and Anaya [15] when executing a task and there is joy, exists an inner positive motivation, since positive emotions influence inner motivation, thus, students' significant learning is possible.

4 Conclusions

Technological innovation is necessary to the present era education, reducing the gap between education and reality results imperative nowadays, that is why this study presents a closer look to the virtual assistants developed by different technological businesses, analyzing its characteristics and possible usefulness in the educative field.

As a contribution to education, it is proposed from the technology area to implement virtual assistants in the teaching-learning process and being able to evaluate the impact on students' academic achievements.

As for delimitations of this study, it is important to address the accessibility to the different devices and the internet connection that the schools count with, depending on their location, as well as the risk of not being well-used with an educative objective within a set already planned and being converted on distractors agents, instead of educative, so, it is important to consider the curricula planning integrated to virtual assistants to improve learning processes.

It has been affirmed that studying the application of virtual assistants and its usage in the educative field is scarce, that is why this investigation took place, opening a new research line for the future in this area. Also, it is of great interest to the research team to corroborate the effects of the usage of these technological resources with experimental studies.

References

1. Pérez de A, M., Tellera, M.: Las TIC en la educación: nuevos ambientes de aprendizaje para la interacción educativa. Revista de Teoría y Didáctica de las Ciencias Sociales **18**, 83–112 (2012)
2. Borba, M., Askar, P., Engelbrecht, J., Gadanidis, G., Llinares, S., Sanchéz, M.: Aprendizaje mixto, e-learning y aprendizaje móvil en la educación matemática. ZDM **48**, 589–610 (2016)
3. Villarreal, M., Esteley, C., Mina, M.: Modeling empowered by information and communication technologies. ZDM **42**, 405–419 (2010)
4. Wikipedia: WIKIPEDIA la enciclopedia libre. https://es.wikipedia.org/wiki/Asistente_virtual. Accessed 20 Mar 2020
5. Wikipedia: WIKIPEDIA la enciclopedia libre. https://es.wikipedia.org/wiki/Agente_inteligente_(inteligencia_artificial). Accessed 22 Mar 2020
6. Saavedra, E.: Vermislab, 19 Junio 2019. https://www.vermislab.com/inteligencia-artificial-educacion/. Accessed 24 Mar 2020
7. Penalva, J.: XATAKA, 29 Enero 2019. https://www.xataka.com/entrevistas/mis-hijos-hablan-alexa-assistant-siri-me-pregunto-debo-ensenarles-a-ser-amables-preguntarles-favor. Accessed 22 Mar 2020
8. Escorcia, L., Triviño, C.: Tendencias de uso de las TIC en el contexto escolar a partir de las experiencias de los docentes. Educación y educadores **18**(1), 137–152 (2015)
9. Apple: Apple (2020). https://www.apple.com/es/siri/. Accessed 22 Mar 2020
10. Microsoft: microsoft.com (2020). https://www.microsoft.com/en-us/cortana. Accessed 22 Mar 2020
11. Wikipedia: Wikipedia la enciclopedia libre (2020). https://es.wikipedia.org/wiki/Asistente_de_Google. Accessed 22 Mar 2020
12. Google: Google. https://assistant.google.com/learn/. Accessed 22 Mar 2020
13. AMAZON: Amazon Alexa. https://developer.amazon.com/es-ES/alexa. Accessed 22 Mar 2020
14. INSPIRATICS: InspiraTICs. https://www.inspiratics.org/es/recursos-educativos/5-usos-que-ya-tiene-la-inteligencia-artificial-en-el-aula. Accessed 22 Mar 2020
15. Anaya, A., Anaya, C.: ¿Motivar para aprobar o para aprender? Estrategias de motivación del aprendizaje para los estudiantes. Tecnología Ciencia Educación **25**(1), 5–14 (2010)

Motion Capture and Virtual Reality Application in the Interactive Exhibition of Chinese Traditional Furniture

Dehua Yu(✉)

Beijing Institute of Technology, No.5 Zhongguancun South Street, Haidian District, Beijing, China
yudehuabit@163.com

Abstract. The aim of this interaction exhibition of Chinese traditional furniture is to increase visitors' experience of Chinese traditional furniture and its mortise and tenon structure. Cinemagraph effect, virtual reality and motion capture sensors technology are used. First, when visitors go across the virtual ancient painting "Western Garden" drew by Xie Huan from Ming dynasty on the wall, the cinemagraph effect used in the painting would attract more interest of visitors. Second, with more interest, some visitors can approach the real mortise and tenon structure of Chinese traditional furniture, and can fix or unfixed the structure directed by virtual installation instructions on the wall. Third, some volunteers would be chosen to participate the interaction activities wearing the motion and movement capture sensor system. The volunteers would be consistent with the virtual characters in ancient paintings, experiencing the ancient life scenario with ancient furniture and furnishings around.

Keywords: Motion capture · Virtual reality · Chinese traditional furniture · Mortise and tenon

1 Introduction

Cultural heritage study and protection is a vital research field, and many research findings are just published in academic journals and books. However, the display and communication of these research findings for public is not enough. With the help of motion capture and virtual reality technology, the public will have the chance to interact with real mixed with virtual cultural heritage, touching, fixing, folding, and even immersing in the mixed reality display environment [1]. The paper would show the interactive exhibition of Chinese traditional furniture with the technology of motion capture and virtual reality, and study how visitors interact with the real mixed with virtual works.

The aim of the interactive exhibition of Chinese traditional furniture is to show the ancient usage scenarios of Chinese traditional furniture, and display the artful and ingenious mortise and tenon structures [2] to inspire people's imagination and creation.

A famous painting "Western Garden" drew by Xie Huan [3] from Ming dynasty would be used to illustrate an ancient scenario of Chinese refined scholars and their

W. Karwowski et al. (Eds.): IHSED 2020, AISC 1269, pp. 209–214, 2021.
https://doi.org/10.1007/978-3-030-58282-1_34

furniture around. The painting would be shown in the exhibition hall with virtual animation on the wall and real Chinese traditional furniture and furnishings on the exhibition floor.

2 Interaction Ways

The painting has all been digitized, and many details have 3D models and have independent animations to play in proper time, including the characters, trees, animals, furniture and furnishings around.

Three ways would be used to interact with visitors and participants who would like to visit the exhibition.

First is for ordinary visitors. When visitors go across the restored ancient living scene, the position sensor would capture the route of their visiting, and the cinemagraph of the painting would play accordingly in line with the specific location of visitors. From this way, visitors would be attracted by the cinemagraph to observe the virtual ancient living scenario and the real ancient furniture and furnishings.

Second, if the visitors show more interest, they might have interest to take part in the fix and unfix process of the mortise and tenon of Chinese traditional furniture shown on the floor of the exhibition, which also matches with the ancient virtual scenario in the painting. The sensor could capture the action of the visitors, and the animation would show more details and illustration to direct the visitors to finish the process. The whole process would be more interesting and interactive.

Third, if the visitors show great enthusiasm and have time, they could carry on the action capture sensors system to experience the interaction with the virtual characters and the ancient scenario in the painting. After wearing the action capture sensors system, the participants would control the action of one character in the virtual painting, and the character would follow all the actions made by the participants. In other words, the participants would be one member of the ancient scenario in painting, and they can go around and interact with the virtual characters in the painting through capturing the motion of them. Of course, most important task that the participants would be directed to experience is using of ancient furniture and furnishings in the painting. Some furniture can be fixed and unfixed, and some furnishings can be played with. Furthermore, the participants can also sit on the real furniture on the floor of the exhibition, and fix and unfix the real furniture directed by the virtual animation of installation instructions.

Through all of above, the participants would experience the ancient living environment surrounding by real and virtual scenario and furnishings.

2.1 Cinemagraph Effect for Ordinary Visitors

The painting is in the environment of outdoor, surrounding by beautiful natural scenery like trees, hills, pavilion, and even animals. Many furniture and furnishings are placed in outdoor garden, which is called "west garden", and this is also where the name of the painting comes from. Many refined scholars are gathering together to write poems, appreciate paintings and enjoy the garden.

The painting contains series of furniture, for example tables, chairs, stands, beds, and so on, which is suitable to show the ordinary life of Chinese ancient refined scholars. The exhibition assigned proper real ancient furniture and furnishings to match the historical environment of the painting, which mixed the real and virtual environment together. When people visit the exhibition, they would immerse in the life scene of the ancient people and the use state of these furniture vividly (Fig. 1).

Furthermore, when visitors go across the painting on the wall, the position sensor would capture the exact position of the visitor. In the meantime, some vivid cinemagraph [4] would played when then go by, for example, the figures wave, the branches shake, the animals move, and even the incense burner on the incense table curls up.

The purpose of all the interactive details is to attract the visitors' attention to stay for longer time and pay more attention to experience, feel and observe the ancient living environment, furniture and furnishings inside.

Fig. 1. The scene of visitors go across the interaction exhibition

2.2 Touching and Fixing the Ancient Furniture with More Interest

If the implicit tips above all attract the visitors' attention successfully, the visitors might stay for a while to observe the exhibits of furniture matching the virtual painting. The second interactive way would carry out.

The exhibits are real Chinese traditional furniture, some chairs are stable and complete which visitors can sit on directly, and some show the mortise and tenon structure of Chinese traditional furniture which visitors can fix and unfixed directed by corresponding animation of installation instructions played on the wall integrated with the painting.

For example, when visitors touch the fixable leg of the table (Fig. 2), the sensor would capture the action and trigger the switch of play the animation to direct the fix and unfix process, and also show the mortise and tenon structure from different perspectives to help visitors understand the structure. Once the visitors make some mistakes by accident, the mistake would be monitors, the animation would remind the mistake points and direct them to modify them.

Fig. 2. The interaction scene of fix and unfix process

It is easy to study even slightly more complicated mortise and tenon structure without any pressure, which is easier than ordinary exhibition without interaction. Most of all, more participants would have more interest on Chinese traditional furniture and its mortise and tenon structure because of the interactive experience.

2.3 The Interaction by Means of Motion Capture Sensor Technology

The exhibition would also invite some volunteers to wear motion capture sensor hardware [5] to interact with the virtual figures in the painting. The motion capture sensor hardware would record the process of the action and movement of the volunteers. One virtual character who plays the role of ancient people in the painting would follow the full-body action and movement of the volunteers (Fig. 3). It seems that the volunteers have already been in the virtual scenario of ancient life.

Furthermore, the volunteers would be directed to sit on the chair by the table exhibited on the hall by the painting wall, where there are also virtual chair by virtual table in the painting. When the volunteers sit, the virtual character in the painting also sit, and they can communicate with each other, sharing the hot tea, enjoying the fragrance in the censer, and studying about the table they are sitting by.

There would be a Chinese traditional high and low dual-purpose table nearby [6], the volunteers would be directed to fix and unfix the dual-purpose table, and experience two conditions of high table and low table. In the meantime, the virtual character also follow all the motion and movement, keeping pace with the volunteers. The experience of involvement not only make the volunteers increase interaction, but also infect visitors surrounding, and encourage them to participate.

Fig. 3. The Interaction wearing motion capture system

3 Conclusion

The study focused on the digitization display and spread of Chinese traditional furniture and its culture it contains. Mortise and tenon of Chinese traditional furniture is exact and artful, which is the most important structure used in Chinese traditional furniture. However, mortise and tenon structure is hidden inside the furniture, and we cannot observe it from outside. With the help of motion capture and virtual reality technology, people have more perspective to observe, touch, fix, and even immerse in the mixed reality display environment, interacting with real and virtual furniture, which add more possibility to know about the charm and secret of the structure and Chinese traditional furniture itself.

The interaction exhibition of Chinese traditional furniture is a positive attempt to combine the motion capture technology with cultural heritage display and spread.

Acknowledgements. This paper is supported financially by MOE (Ministry of Education in China) Project of Humanities and Social Sciences Research Youth Fund (Project No. 19YJC760144).

Reference

1. Yu, D.: Virtual reduction and interaction of Chinese traditional furniture and its usage scenarios. In: 10th International Conference on Intelligent Human System Integration, vol. 1131, pp. 552–558. Springer, Heidelberg (2020)
2. Yang, Y.: Study of Ming-style Furniture, p. 40. Beijing Construction Industry Press, Beijing (2002)
3. Chinese Portrait Paintings, vol. 2, pp. 153. Beijing Press, Beijing (2004)
4. Niewland, M.: Framed in Time: a Cinemagraph Series of the Everyday & Grounded Theory of Cinemagraphy (2012). https://www.niewland.com
5. Roetenberg, D., Luinge, H., Slycke, P.: Xsens MVN: Full 6DOF Human Motion Tracking Using Miniature Inertial Sensors. https://www.researchgate.net/publication/239920367
6. Robert, D.J.: Classical Chinese Furniture in the Minneapolis Institute of Arts, pp. 136–137. Art Media Resources Ltd., Chicago (1999)

Exploratory Approach to Performance of Smart Components of Intelligent [Smart] Buildings

Lekan Amusan[1(✉)], Emetere Moses[2], and Ojelabi Rapheal[3]

[1] Lekan Amusan Building Technology Department College of Science and Technology, Covenant University, Ota, Ogun State, Nigeria
lekan.amusan@covenantuniversity.edu.ng
[2] Moses Emetere Department of Physics, Covenant University, Ota, Ogun State, Nigeria
moses.emetere@covenantuniversity.edu.ng
[3] Ojelabi Rapheal Building Technology Department College of Science and Technology, Covenant University, Ota, Ogun State, Nigeria
rapheal.ojelabi@covenantuniversity.edu.ng

Abstract. Intelligent building is gaining access to the construction industry all over the world in recent times. Building functionality and the need for occupant comfort has been the major drive toward advent of intelligent Buildings. In the light of this, there is a need to carry out continuous evaluation of performance of the building to ensure value for money. In this study therefore, post-occupation building managers at different levels of types of buildings were censored within the scope of the following parameters; awareness on current state of practice in office intelligent building application, the level of satisfaction of managers and the users of intelligent buildings' performance, effectiveness of the security system and access control in building and performance enhancement measures to be adopted and factors that influences effective performance of intelligent building accessories. Survey method was used with the aid of questionnaire the resultant data was analyse using relative agreement index (RAI), Mann-Whitney U-Test, Pearsons's Chi-square test and Student's T-test. The study presented factors that influences integration of automated accessory in building and their performance, effectiveness of security system and access control, conventional automation application in building products among others. The study recommend constructive innovative approach to intelligent building management, strategic plan should be instituted in maintenance operation, proactive management building accessory, and collaborative stake holder inclusive management, as panacea to forestalling negative attendant consequence of facility negligence in intelligent building.

W. Karwowski et al. (Eds.): IHSED 2020, AISC 1269, pp. 215–224, 2021.
https://doi.org/10.1007/978-3-030-58282-1_35

1 Introduction

The concept of intelligent buildings has become an important area of focus in construction and real estate practice all over the world. Intelligent building is used to describe the total embodiment of building component that tends to operate in an automatic way with little or no input of human effort. It is a wind of change in the construction sector that tend to sweep off the art of getting work done mechanically with automatic or learned machine approach. The development started from Europe, Asia and other developed countries and into Africa. Construction industry has witnessed various applications of intelligent systems that cut across various application during construction process. However, there have been researches that focus on various aspect of intelligent buildings. There are researches that focused on design, installation and management of components of an intelligent building. In [1] [Kubba 2017]; [2, 3], it was pointed out that researches have been focused on design and installation and maintenance while little had been done on carrying out post occupancy evaluation of the buildings.

In a study carried out by [4] G intelligent building was described from international perspective considering the international context of the component of an intelligent building. An intelligent building according to [4] is a building that has the ability to combines the best available ideas, designs, materials and technologies in order to produce and maintain an interactive, adaptive, responsive integrated and dynamic intelligent environment to achieve the users' objectives over the lifespan of the building. [5] shared similar view with [4–6] in submitting that, an intelligent building acts as a productive workspace that is also cost effective and user-friendly through an integrated system of structure, building services and management with smooth running within the components Also, [6, 7] opined that in Intelligent buildings are becoming adaptive to our environment and contribute to the cyber-physical aspect of ecosystem, to corroborate this fact [5], Modern houses are now sources of data along with our mobile devices while taking note and imitating behaviors found in hardware, software and artificially intelligent systems [8] and [9]. In line with the above review, sustaining the intelligent systems in a building is of importance as the choice of the intelligent systems, and that user compatibility is the key. Also, [5] and [3] stressed the importance of integrating an intelligent building components in a way that sustainability of the components is maintained throughout the life cycle of the building.

This study therefore aims at comparing the levels of satisfaction of the end users of the traditional buildings as compared to the fast-developing intelligent building systems also measure the levels of comfort and adaptation to the occupants' needs. This would help provide feedback to the manufacturer of the building components for improvement on the manufacturing process for better building performance. This is supported in the approach in adopted in [5, 6] and [7].

2 Materials and Methods

Primary data was obtained from post occupation managers and users of intelligent buildings that are on ground at the selected locations of the research while Survey materials adopted structured questionnaire design in a closed structure manner as carried out in similar studies such as in [13, 14] and [16].

2.1 Material and Tools

In this study different materials and tools were used, part of the materials used are A-4 papers for the drafting of questionnaire, audio equipment, markers, pencils and biro. Also, Statistical tools of SPSS was engaged in the processing of data collated from the respondents. Some of the tools include Relative Agreement Index (RAI), Mann-Whitney U-Test, Pearsons's Chi-square test and Student's T-test.

The Relative Agreement Index was calculated using the following relation.

$$\mathrm{RAI} = \sum \frac{\mathrm{Wi}}{\mathrm{A} \times \mathrm{N}} \quad (1)$$

Where RAI = Relative Agreement Index, Wi = Weighted Sum, A = The number of items on Likert scale of 1–5. N = individual weight of the scale item on Likert scale 1–5. The component of the Likert Scale include (SA: Strongly Agree (5), A: Agree (4); SD: Strongly Disagree (2); D: Disagree (1); N: Neutral (3).

Similarly, survey design method was used in the study with population comprised of 17 Intelligent buildings types that spans across Commercial building, Office building, Health care and Residential buildings. The buildings were classified in the following order; 8 Commercial building which is equivalent to 40% of total building types. 4 Office building types of 20% of total building while 3 Residential of 15% and 5 Health-care building based on availability of such buildings [17] and [18].

3 Results

In this section results of the survey was presented, it include description of the experimental results, cross validation of facts and conclusion on results presented. The parameters presented include the following: Break down of Respondents; Diversity of Post Occupation Manager Representation; Break down of Respondents Construction Experience; Companies Represented by Managers; Intelligent Building Types and Post Occupation Manager; Intelligent Building Types and Post Occupation Managers in Charge; Perception on Awareness of Intelligent Building System; Post Occupancy Evaluation of Performance Indicator[KPI] of Intelligent Building; Post Occupancy Evaluation of Performance Indicator[KPI] of Intelligent Building; Effectiveness of Components of Intelligent Buildings; Satisfaction Level of Facility Managers on Intelligent Building System Performance and Issues and Challenges involved in Integration of Components of Intelligent. The processed data was presented in charts and tables. The breakdown of respondents that participated in the survey was as presented in Fig. 1, 30 Electrical control system managers which constitute 30% of total respondents and 40 facility users, participated in the survey, also, 20 Door access control managers which formed 26.67% of total respondent was found, 15 Mechanical system control managers constituting 20% of total respondents participated while 10 Heating and Ventilation access control managers that constitute 13.33% was sampled. The implication of the structure above is that majority of the works that the professional would engage in the building are majorly electrical-based in nature. Similarly, Mechanical based duty that is available in the building is also next in quantity hence the high quantity of mechanical engineering system managers sampled among the respondent. Summarily, the trend indicated the bulk of the intelligence system components to be of mechanical and electrical based [21].

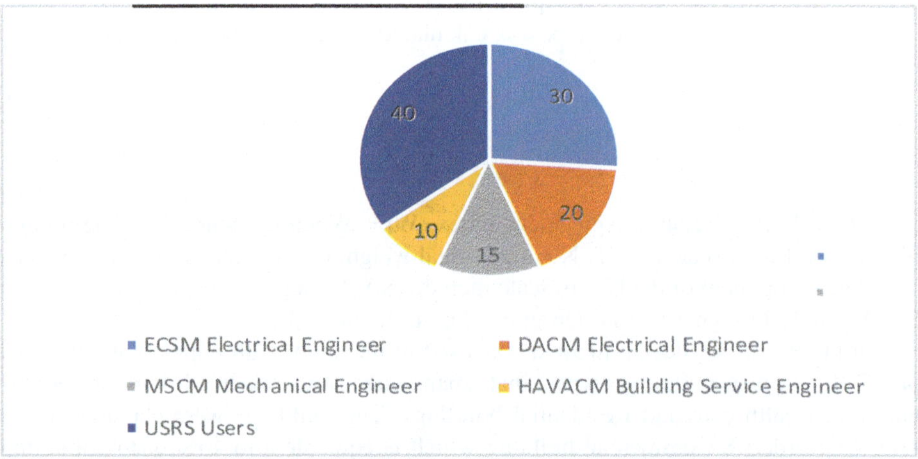

Fig. 1. Break down of respondent [post occupation manager and users]

Figure 2 presents diversity of post occupation managers experience representation. The survey indicated that Electrical control system managers [ECSM] which constitute 32% of total respondents had 28 years of general facilities management experience. Door access control managers [DACM] who are basically electrical engineer constitute 28% of managers sampled and has 25 years of general construction experience, also, 22% of mangers are Mechanical engineering control system managers [MSCM] and has 18 years of professional experience while the remaining 20% are Heating and Ventilation control system managers [HAVACM] with 18years of professional experience. Electrical and Mechanical engineer appeared to have more professional experience. The reason could be linked to the existence of Electrical based items in building which predates the advent of building access control systems thus, electrical engineer has been maintaining electrical items before the advent of other mechanical and electromechanical components of building. Introduction of automatic heating system as part of intelligent building system was fairly new in the construction industry, hence

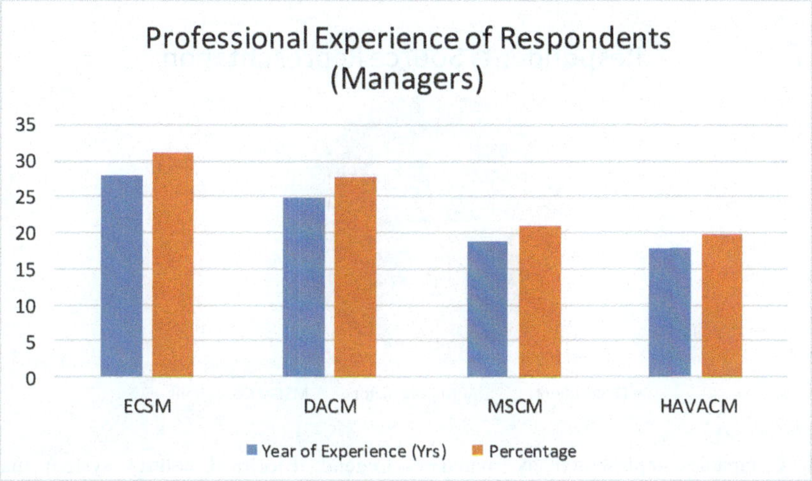

Fig. 2. Diversity of post occupation manager professional experience representation Legend: Electrical control system managers [ECSM]; Door access control managers [DACM], Mechanical services control manager [MSCM] and Heating and Ventilation access control managers [HAVACM].

reason for the least year of experience by the Heating and ventilation control system managers [14, 15] and [13].

Figure 3 illustrates the breakdown of the post occupation managers and the respective companies that they represents. ECSM occupies 45% of the total sample of managers and represents clients of the intelligent buildings, DACM takes 30% of the total respondents and are forwarded to the duties of intelligent building manager by contractors while MSCM have 25% of total managers figure and represents consultant firms. The trend of the presentation could be linked to the procurement system of intelligent building that involves the clients, contractor and consultant's initiative. This enables each stakeholder to have specific lines of impact which would still continue to the maintenance stage. Each of the clients that is responsible for the installation of a specific component would still be engaged in the management of such component. Sometimes the procurement style could be build-operate-transfer for instance, in this scenario, the component is being managed after installation and certain professional are trained who might likely take over the management of the facility, in this way continuity is established [4, 8, 14].

In Fig. 4 Category of Intelligent building used in this study was illustrated. 8 Commercial building which is equivalent to 40% of total building types. 4 Office building types of 20% of total building while 3 Residential of 15% and 5 Health-care building of the total buildings sampled was used. The ECSM that are electrical based were found on Commercial building, since the main component in them is the intelligent electrical system and lighting, DACM engaged the office buildings and MSCM are on Residential buildings while HAVACM are engaged on Health-care buildings [11].

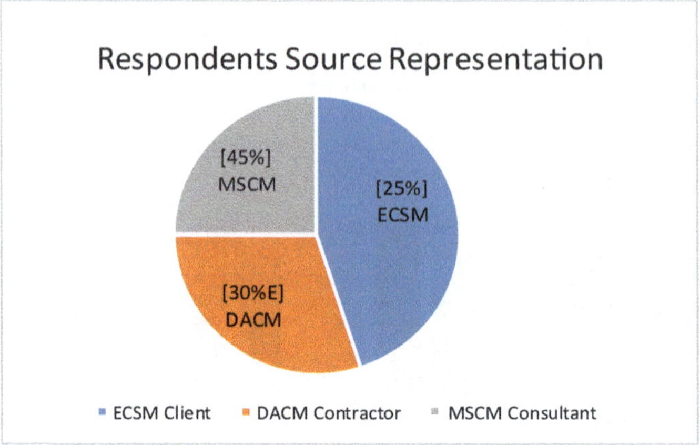

Fig. 3. Companies represented by managers Legend: Electrical control system managers [ECSM]; Door access control managers [DACM], Mechanical services control manager [MSCM] and Heating and Ventilation access control managers [HAVACM].

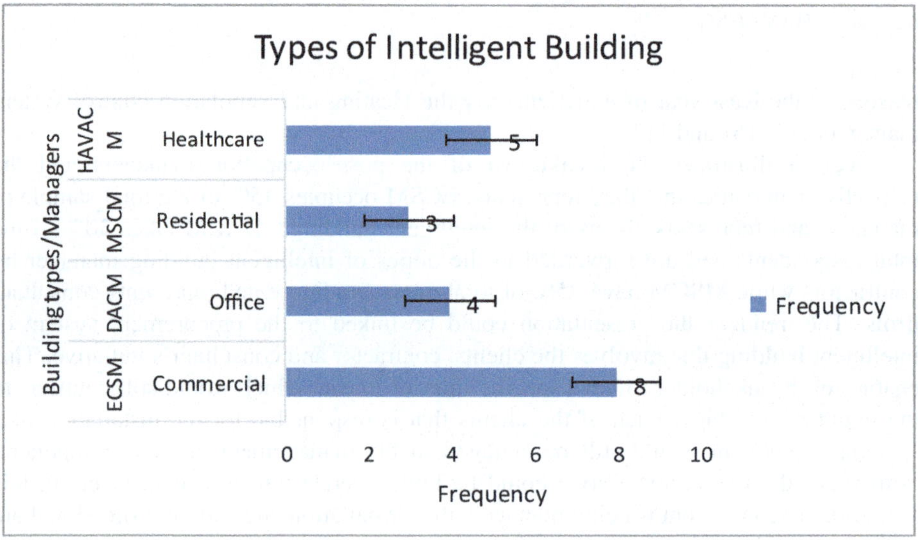

Fig. 4. Types of Intelligent building and cadre of managers in control Legend: Electrical control system managers [ECSM]; Door access control managers [DACM], Mechanical services control manager [MSCM] and Heating and Ventilation access control managers [HAVACM].

3.1 Awareness on Intelligent Buildings Application

Survey was conducted generally about awareness of intelligent building system. The survey results is presented in Table 1.

Table 1. Perception of Building managers on extent of awareness of intelligent building System

Awareness about intelligent buildings	RAI	RANK
I have participated in the configuration and design of some intelligent buildings	0.873	1st
Understand Intelligent building involves incorporating advanced technical components	0.864	2nd
Encourages machine enabled replacement of functions	0.827	3rd
There are components of vertical and horizontal movement in building that are being controlled automatically	0.794	4th
In intelligent building components can be remotely controlled	0.786	5th
Design of intelligent buildings encourages the use of 3Dand 4D systems	0.732	6th
Understand that Intelligent buildings encourages real time surveillance and monitoring	0.729	7th

Table 1 illustrates cross section of opinion of respondents while censoring the extent of their understanding of intelligent building system. Participant agreed to the fact that they have been involved in the configuration, design and construction of intelligent building at one time or the other. In addition, majority agreed to the fact that intelligent buildings involves incorporating advanced technical component in building, this tend to reveal their exposure to some components of intelligent buildings. Substantial part of the respondents has also participated in replacement of components in intelligent building considering their involvement in the replacement or and maintenance functions. Similarly, respondents are aware about the existence of intelligent vertical and horizontal transportation system in the buildings, intelligent chutes, ramps and lifts are among the components that are found in the buildings. Finally, respondents attested to the fact that they are knowledgeable about remote controlled building components and real time control of operation within and without the building through intelligent access control monitoring system [11, 12].

3.2 Analysis of Effectiveness of Building Component

Table 2. Effectiveness of smart components of intelligent buildings

Smart building parameters	ECSM		HAVACM		MSCM	
Smart building component	RAI	RANK	RAI	RANK	RAI	RANK
Smart building security system	0.87	1st	0.85	4rd	0.82	1st
Smart HVAC system	0.88	2nd	0.89	1st	0.82	1st
Intelligent automated lightning	0.79	3rd	0.86	2nd	0.76	3rd
Smart Occupant energy monitoring	0.75	4th	0.86	2nd	0.75	4th
Smart water supply system	0.73	5th	0.72	5th	0.71	5th
Intelligent fire system	0.73	5th	0.73	6th	0.60	6th
Smart Access control and monitoring	0.70	7th	0.70	7th	0.62	7th
Smart Automated parking system	0.67	8th	0.67	9th	0.56	8th
Smart Indoor air quality system	0.63	9th	0.78	8th	0.52	9th
Smart Internet infrastructure	0.61	10th	0.62	10th	0.51	10th

Legend: Electrical control system managers [ECSM]; Door access control managers [DACM], Mechanical services control manager [MSCM] and Heating and Ventilation access control managers [HAVACM].

Table 2 presents survey on effectiveness of components of intelligent buildings. The main area of strength in intelligent building is the component of the building. Ideally, any intelligent building should contain electrical, mechanical, and electromechanical components, however classification of the content is often on account of component functionality. The component of the building in this context is group in term of electrical based component, mechanical based and electro-mechanical component. The measurement of effectiveness could be explored based on the components' performance. In line with the results presented in the table, the Intelligent building security system was ranked first by ECSM and MSCM with RAI values 0.87 and 0.82 respectively, while intelligent HVACM system was also rated first with RAI values 0.89 and 0.82 by HVACM and MSCM managers. It could be observed that the response tend to skewed towards the area of competence of the professionals. For instance, be the HVACM was able to ranked HVAC systems first because that is their area of competence and on account of wealth of experience should be able to give a valid submission on the effectiveness of performance of the component. Intelligent automatic lighting system was ranked third by ECSM and MSCM managers with RAI values 0.79 and 0.76. It could be deduced that basic components that are working effectively are electrical based components, for instance the security system works based on electrical circuit control likewise the Heating and the ventilation systems. Automatic lighting is also as important as other similar components, this could be linked to the dependency of various spaces in building on internal and external lighting. In support of the view presented. [17] stressed the importance of illumination and energy supply to the internal space of building and that there is correlation between effective energy supply in building relative to the general green building performance.

4 Conclusion

One of the cardinal objective in post occupancy evaluation all over the world is carrying out satisfaction test and analysis of the users of a facility or building. It would help in identifying areas that requires attention. Also, satisfaction test could serve as a means of generating information that could be used in trouble-shooting, test running and calibration of building components. Therefore, Table 8 presents the outcome of satisfaction test conducted on the different types of building using the managers' experience as the focus. It was discovered that the respondents indicated high level of satisfaction as regards the following intelligent building components.e.g. smart energy efficient lift control system, automated lighting system, and automatic temperature regulator, solar energy generating system and fire systems among others. The components were ranked as highly satisfied by four types of managers, that is, the ECSM, DACM, MSCM and HAVACM.

Similarly, the following mechanical and electro-mechanical based components were ranked second by the by three groups of the managers, they include ECSM, MSCM and HAVACM. The components combines the operation of mechanical and electrical components of building system. For instance, they are the components that are always in use, and in a study carried out by [2, 18, 19] and [20] air quality processing systems was described as one of the necessary component in building, similarly plumbing system, door access control system and burglar alarm were rated as one of the essential component of users preference in building, therefore [18] advocated water supply system as one of the essential items in building. The trend could be attributed to criticality of user need in a building as a determinant of the pattern of satisfaction over an item.

References

1. Kubba, S.: Green project commissioning. In Handbook of Green Building Design and Construction LEED, BREEAM, and Green Globes, Butterworth –Heinemann, Oxford (2017)
2. Altomonte, S., Saadouni, S., Kent, M.G., Schiavon, S.: Satisfaction with Indoor environmental quality in BREEAM and Non-BREEAM certified office buildings. Archit. Sci. Rev. **60**, 343–355 (2017). https://doi.org/10.1080/00038628.2017.1336983
3. Bhavani, R.G., Mohen, M.: Awareness on lighting energy saving and user satisfaction in residential and office buildings. In: International Conference on Biological, Civil and Environmental Engineering (BCEE-2014), Dubai (UAE), 17–18 March 2014 (2014)
4. Hoseini, G., Dahlan, N.D, Berardi, U., Hoseini, G.A., Makareni, N.: Sustainable energy performance of green buildings: a review of current theory. Renew. Sustain. Energy Rev. **25**(1), 1–7 (2013)
5. Sherif, K.: The Use of Intelligent Buildings to Achieve Sustainability through Architectural Proposal (2011)
6. Brooks, D.J.: Intelligent buildings: an investigation into current and emerging security vulnerabilities in automated building systems using an applied defeat methodology. (2012). https://doi.org/10.4018/978-1-4666-2659-1.ch001

7. Ai-Abbas, S.H.: Energy conservation in residential building using green architecture principles. Master Thesis. World SB 14 Conference, Cairo University, Egypt (2011)
8. Lekan, A., Dolapo, D., Joshua, O.: Cost and time performance of construction projects. Int. J. Mech. Eng. Technol. (IJMET) **8**(10), 918–927 (2017)
9. Manic, M., Rodriguez-Andina, J.J., Amarasinghe, K., Reiger, C.G.: Intelligent buildings of the future: cyberaware, deep learning powered, and human interacting. IEEE Ind. Electron. Mag. (2016)
10. Brooks, D.J., Coole, M., Haskell-Dowland, P., Griffiths, M., Lockhart, N.: Building Automation & Control Systems: An Investigation into Vulnerabilities, Current Practice & Security Management Best Practice. ASIS Foundation Project (2017)
11. Osawaru, F., Amusan, L., Awotinde, O., Akanya, C., Asiyanbola, O., Akinbo, F.: Planning materials supply for construction works. In: 22nd International Conference on Science and Sustainable Development, ICSSD 2018: The Role of Science in Novel Research and Advances in Technology; Center for Research, Innovation and Discovery, Covenant University (CUCRID), Nigeria, 14 May 2018 through 16 May 2018, Code 138212 (2018)
12. Lekan, A., Clinton, A., Blossom,O., Babatunde, O.: Building informatics: post occupancy narratives of automation component of intelligent and non intelligent buildings. Int. J. Pharm. Res. **6**(5) (2020)
13. Clements-Croome, T.D.J.: What do we mean by intelligent buildings. Autom. Constr. **6**(5/6), 395–400 (1997)
14. Yiu, C.Y., Ho, C.W., Lo, S.M.: Weathering effects on external wall tiling systems. Constr. Build. Mater. **21**(3), 594–600 (2007)
15. Harris, S.Y.: Building Pathology: Deterioration, Diagnostics, and Intervention. John Wiley & Sons Inc., New York (2001)
16. BSI (1991 BS4778-3.1: 1991) Quality Vocabulary: Part 3: Availability, Reliability and Maintainability Terms, Section 3.1 Guide to Concepts and Related Definitions, British Standards Institution, UK (1991)
17. Seeley, I.H.: Building Maintenance, 2nd edn. Macmillan, Basingstoke (1992)
18. Mozer, M.C.: An intelligent environment must be adaptive. IEEE Intell. Syst. Appl. **14**(2), 11–13 (1999)
19. Harrison, A., Loe, E., Read, J. (eds.): Intelligent Buildings in South East Asia. E & FN Spon, London (1998)
20. Raouf, A.M., Al-Ghamdi, S.G.: Framework to optimize cost of quality in delivering and operating green buildings. In: Proceedings of International Conference on Sustainable Infrastructure 2019, Los Angeles, CA, USA, 6–9 November 2019, pp. 338–347 (2019). https://doi.org/10.1061/9780784482650.035

Systems Design and Human Diversity

Future Trends in Education for a More Sustainable Human Systems Design: The CREATION Project

Ana Margarida Ferreira[1,2(✉)], Stefania Savva[3], and Nicos Souleles[4]

[1] IADE, Universidade Europeia, Lisbon, Portugal
Ana.Margarida.Ferreira@universidadeeuropeia.pt
[2] UNIDCOM/IADE, Unidade de Investigação em Design e Comunicação, Lisbon, Portugal
[3] The Hub Nicosia, Nicosia, Cyprus
Stephania.hubnicosia@gmail.com
[4] Art + Design: E-Learning Lab, Cyprus University of Technology, Limassol, Cyprus
Nicos.Souleles@cut.ac.cy

Abstract. Europe faces various challenges towards sustainable development. The influx of refugees, youth and gender unemployment, frequent global health threats, spiraling conflict, cultural intolerance, socially marginalized groups, and others are some of the diverse and large-scale social challenges that cannot be ignored. Such reality requires innovative and sustainable responses to address them and puts the economy, the design education sectors and, cultural and creative industries in the spotlight. This paper will reflect upon these ideas and present the CREATION PROJECT, the ReACt Model and the newly created Consortium of seven European organizations that joined to address these issues. It envisages to develop an entrepreneurial skills framework for adult education for the creative and cultural sectors across Europe. The main focus is on bridging the gender gap in entrepreneurship and promoting cultural dialogue. The project will promote the learning of entrepreneurial skills enabling the underrepresented groups to succeed in cultural and arts entrepreneurship. It aspires also contribute to their socioeconomic and cultural sustainability and wellbeing.

Keywords: Co-CREATION · Social design education · Entrepreneurship education · Social innovation · Gender equity · Cultural diversity · Sustainability

1 Introduction

As presented in the United Nations' report, "Transforming our World: The 2030 Agenda for Sustainable Development" [1], Europe faces many challenges to sustainable development. Large populations still live below poverty levels due to systematic inequality and enormous disparities in opportunity, wealth, and power around the world. The influx of refugees, youth and women unemployment, frequent global health

W. Karwowski et al. (Eds.): IHSED 2020, AISC 1269, pp. 227–233, 2021.
https://doi.org/10.1007/978-3-030-58282-1_36

threats, spiraling conflict, cultural intolerance, socially marginalized groups, and others are some of the diverse and large-scale social challenges that cannot be ignored. To reverse this constant reality requires innovative and sustainable responses to address the challenges posed. The EU Eurostat report indicates that [2], in 2019, asylum seekers increased again being the first time that the number of asylum applications did so since 2015. Germany, France, Spain, Cyprus, Greece and Italy are the main countries of destination. These stark facts require new and clear strategies, collaboratively designed and multidisciplinary in science, research, innovation, education or policy-making domains to address issues such gender equality, reduction of inequalities, quality and diversity of education, decent work and economic growth, as listed in the 17 Sustainable Development Goals (SDGs) far from the desirable global targets [3].

2 European Commission's Megatrends and New Challenges and Scenarios in Science, Research, Innovation and Education

Considered as enablers and drivers for Europe's needed transition to a greener and more sustainable place, science, research and innovation are considered key areas to address. To achieve the defined goals, the European Union (EU) [4], recognizes the importance of five main pillars for the reframing exercise towards a safer and just space for Humanity. Acting synergistically, the five (5) main pillars are: a) co-CREATION, working and acting together; b) diffusion, sharing knowledge; c) uptake, turning research into sustainable solutions with social and economic value; d) transformation, changing the way we consume and produce; and last but not least e) directionality, leaded by research and innovation. These pillared actions take place in a context whereby global and long-term forces, or megatrends shape our world and greatly influence our future [5]. Education, in general, and Higher Education (HE) in particular, have a critical role to play in helping achieving the intended process of sustainable social change in a context of rapid technological pace. HE contributes to the development of research and human capital, mandatory for innovation, productivity, and economic growth. According to OECD [6], by developing technical, professional, and discipline-specific knowledge and skills, cognitive and information processing skills, and social and emotional skills, learners are better prepared for active citizenship and the world of work in the context of 4th Technological Revolution.

3 Creative and Cultural Industries, Social Design and Artistic and Cultural Entrepreneurship

In the past two decades, there has been increased evidence on the rise of the creative economy in general and the cultural and creative industries in particular to spur innovation and knowledge transfer development, and sharing both commercial and sociocultural value [7]. Global, technological, and social changes are rapidly expanding the scope, potential and interest in creative industries. As a result, there are increasingly blurred lines between industries and sectors, tools and communication channels [8]. As we move towards a more connected, integrated, and creative future, few scenarios arise. A strategic one is linked to the idea of creative economy as a route to economic diversification and gaining critical soft skills such as creative thinking and problem solving, of importance for employees in smart environments. The shift from Science, Technology, Engineering and Mathematics (STEM) to 'STEAM' education and the reinforcement of HE and Business cooperation are highlighted as a way to reduce the mismatch between education and the job market [9]. Figure 1 shows the state-of-art and the opportunity spaces to cooperation improvement.

Fig. 1 State of cooperation from the higher education and business viewpoints (*code: 0: Not at all; 1–4: Low; 5–7: Medium; 8–10: High*) [10]

As pointed out by NESTA [11], central to the envisioned and paradigmatic sustainable change are the local qualification networks and collaborative learning processes in which different social agents share their perspectives, expectations, and experiences. Through these joint efforts, new insights, creative strategies, experiences or new business models are discovered, completing the open, inter, and cross-disciplinary character of innovation practice, research and learning [8]. Such understanding and conceptualized action is even stronger in adult and lifelong learning. Its practice becomes an instrument for employers to prepare their workforce for the introduction of innovative technologies and creative processes or, in a more social perspective, to provide access to upskilling and opportunities for learners willing to find new employment opportunities [12].

4 Social Design and Entrepreneurship Education, the CREATION Project and the ReACt Model

Although in recent years there have been increased attempts to address some of these challenges, the demand for the skills and competencies that foster design for social change and innovation surpasses the needs [13]. In Europe, HE approaches in this area remain fragmented, and inconsistent, comprising mostly small-scale and isolated academic initiatives as articulated by Souleles [14] and Rocha [15]. This awareness sparked the Creation Project and framed the Multi-Dimensional Research-Action-Change Pedagogical Model [ReACt Model] evolutionary design. The CREATION project is a European ongoing project, comprised by a Consortium of seven partners from Ireland, Cyprus, Greece, Germany, Italy and Portugal, and envisaging the development of an entrepreneurial skills framework for adult education for the creative and cultural sectors across Europe. As a central aspect is the highlighting of the power of co-design in cultural entrepreneurship and creative and cultural industries towards sustainable development and social inclusion. The process of moving from the CREATION theoretical framework, to an operational level through the ReACt Model, is showed below in Fig. 2.

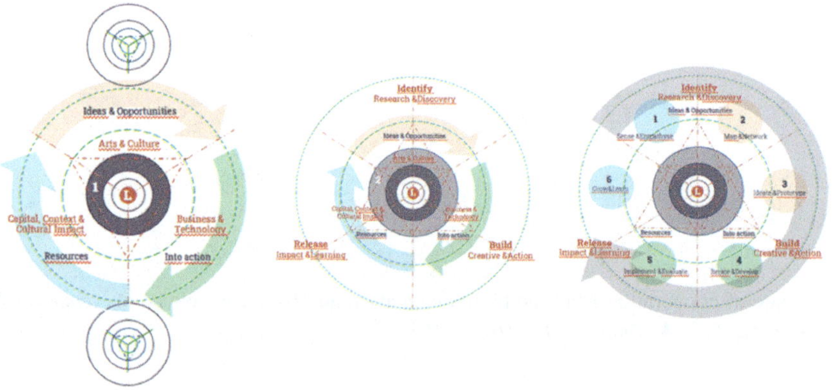

Fig. 2 ReACt Model: Synthesizing, in a learner-centric system, a continuum and integrative flux of learning activities and outcomes in three domains of scientific action, designing a new and open education framework to social change

The main references for the ReACt Model, include Moore's 'Concentric Circles Model of Sustainable Development' [17] and Savva's ECAS Framework [18], and linked to social innovation [11], entrepreneurship and 'theory of change' mindsets [19, 20], as well as design thinking and participatory practices [21, 22]. In this way a multilayered and learner-centric ecosystem is created, framing in a dynamic and coherent way, the integration of Arts, Culture, Business and Technology and Capital, Context and Impact domains. The intention is to synthesize through a holistic and integrative learning experience, the scientific knowledge, learning practices and outcomes of those domains. The evolutionary dimension of the model is displayed in Fig. 3.

Evolving in complexity and interacting gradually with the economic, social, cultural, and environmental outer systems, CREATION ReACt model intends to contribute to fill the still existing gap in the field of theory, pedagogy, and practice of cultural entrepreneurship and in the training and competence development of women entrepreneurs from diverse backgrounds, which is acknowledged by Savva et al. [17].

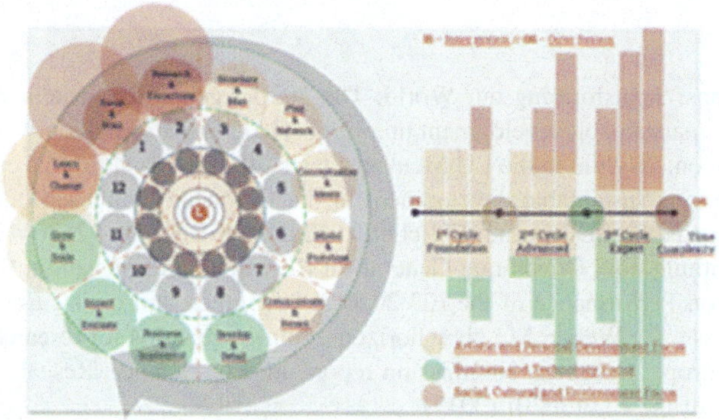

Fig. 3 ReACt Model, in an evolutionary perspective (*from inner to outer systems*), reflecting changes over time, the historical dimension of learning experience and the sustainability of the educational model

5 Conclusions and Further Research

In a 2018 review of the European Reference Framework of Key Competences for Lifelong Learning [23], a new recommendation on key competences was proposed by the European Council highlighting the rapidly changing and highly interconnected world and the need for a wide range of skills and competences, continually developed throughout life, for a more equal, inclusive, sustainable and democratic society. Literacy, science, technology and engineering, personal, social and learning to learn, citizenship, entrepreneurship and cultural awareness and expression are some of the eight key competences highlighted. Moreover, it was stressed that, in support of competence-oriented education, training and learning in a lifelong learning context, three challenges were to be addressed: a) the use of a variety of learning approaches, contexts, and environments; b) the support for teachers and other educational staff and, last but not least, c) assessment and validation of competence development and its impact. With regards to the approach to design, the importance of an educational paradigmatic shift and a broader understanding of the ecosystem was already identified by Findeli [24], in his paper "Rethinking Design Education for the 21st Century". The CREATION project and the ReACt model are still early contributions to these societal and learning challenges. Empirical data will revise and inform its design, reinforcing the exploratory and top down research approach. This strategy will allow further reflection upon it and a better adaptation and response to real social domains and educational needs, aiming for sustainable and inclusive solutions.

Acknowledgements. This research is part of the project Cultural and Arts Entrepreneurship in Adult Education (CREATION 2019-1-PT01-KA204-061315), award for 24 months Erasmus + Key Action 2, Adult Education Sector (KA204) under the National Call 2019, National Agency of Portugal.

References

1. United Nations: Transforming our World: The 2030 Agenda for Sustainable Development (2015), https://sustainabledevelopment.un.org/post2015/transformingourworld/publication
2. European Union: Asylum in the EU Member States (48) (2020). https://ec.europa.eu/eurostat
3. European Union: Sustainable development in the European Union - Monitoring report on progress towards the SDGs in an EU context (2020). ISBN 978–92–76–17443–1
4. European Commission, Directorate-General for Research and Innovation: Science, Research and Innovation Performance of the EU 2020 A fair, green and digital Europe (2020)
5. Ricci, A., Sessa, C., Weber, M.: New horizons: future scenarios for research and innovation policies in Europe, European commission report. In: EC: Science, Research and Innovation Performance of the EU (2020) (2017)
6. OECD: Benchmarking higher education system performance: Conceptual framework and data, Enhancing Higher Education System Performance, OECD Paris (2017)
7. UNCTAD: Creative Economy Outlook - Trends in international trade in creative industries 2002–2015 & Country Profiles 2005–2014 (2018). https://unctad.org/en/PublicationsLibrary/ditcted2018d3_en.pdf
8. Ferreira A.M., Souleles N., Savva, S.: Social design, innovation and ergonomics: reflections on education, transdisciplinarity and new blurred models for sustainable social change. In: Goossens, R., Murata, A. (eds.) Advances in Social and Occupational Ergonomics. AHFE 2019. Advances in Intelligent Systems and Computing, vol. 970, pp. 37–48. Springer, Cham (2020).
9. O'Shea, M., et al.: Supporting Key Competence Development: Learning approaches and environments in school education, European Union (2020)
10. Davey, T., et al.: The state of university-business cooperation in Europe, European Union (2018). In: The Education and Training Monitor 2019, Directorate-General for Education, Youth, Sport and Culture (DG EAC) (2019). ISSN 2363–1988
11. Murray, R., Caulier-Grice, J., Mulgan, G.: The Open Book of Social Innovation, Social Innovator Series: Ways to Design. NESTA & Young Foundation, Develop and Grow Social Innovation (2010)
12. Sekmokas, M., et al.: Workforce skills and innovation diffusion: trends and policy implications, EU Working paper (2020)
13. Souleles, N., Ferreira, A.M., Savva, S.: Threshold concepts and design for social change. In: Goossens, R., Murata, A (eds.) Advances in Social and Occupational Ergonomics. AHFE 2019. Advances in Intelligent Systems and Computing, vol. 970, pp. 80–89. Springer (2020)
14. Souleles, N.: Design for social change and design education: social challenges versus teacher-centred pedagogies. Des. J. **20**(suppl1), S927–S936 (2017). https://doi.org/10.1080/14606925.2017.1353037
15. Rocha, H., Ferreira, A.M., Azevedo, J.M.: Paradigm shift in design education: an overview on issues and possibilities for change. In: Proceedings of DRS 2018 International Conference: Catalyst, vol. 7, pp. 2914–2926. Limerick, Ireland, June 2018
16. Moore, M.: Concentric circles model of sustainable development (2000). In: Thatcher, A. (ed.) Defining Human Factors for Sustainable Development working paper (2013)
17. Savva, S., et al.: Social design for inclusive local learning ecosystems: an entrepreneurial cultural affinity approach. EAI Endorsed Trans. Creative Technol. **6**(20), e3 (2020)
18. Reeler, D.: A three-fold theory of social change and implications for practice, planning, monitoring, and evaluation. Community Dev. Res. Assoc. 1–35 (2007). https://www.eccnetwork.net/sites/default/files/media/file/1469709548wpdm_Reeler.pdf
19. Lackéus, M. & OECD: Entrepreneurship in Education (2015)

20. IDEO: HCD - Human Centered Design Toolkit, 2nd edn. (2011)
21. Drain, A.: Sanders, E: A Collaboration System Model for Planning and Evaluating Participatory Design Projects. Int. J. Des. **13**(3), 39 (2019)
22. European Commission: Council Recommendation on Key Competences for Lifelong Learning, Official Journal of the European Union, C 189/1 (2018). https://eur-lex.europa.eu/legal-cotent/EN
23. Findeli, A.: Rethinking design education for the 21st Century: theoretical, methodological, and ethical discussion. Des. Issues 17(1), 5–17 (2001). Massachusetts Institute of Technology

Technology in Favor of Disability: Prevalence Study in Ecuador

Hugo Arias-Flores[1(✉)], Pamela Acosta-Rodas[2], Janio Jadán-Guerrero[1], and Carlos Ramos-Galarza[1,2]

[1] Centro de Investigación en Mecatrónica y Sistemas Interactivos MIST/Carrera de Administración de Empresas/Carrera de Ingeniería en Ciencias de la Computación/Carrera de Psicología/Carrera de Administración de Empresas, Universidad Tecnológica Indoamérica, Av. Machala y Sabanilla, Quito, Ecuador
{hugoarias, janiojadan}@uti.edu.ec, caramos@puce.edu.ec

[2] Facultad de Psicología, Pontificia Universidad Católica del Ecuador, Av. 12 de Octubre y Roca, Quito, Ecuador
{mpacosta, caramos}@puce.edu.ec

Abstract. People with disabilities and special needs are very important part of the society, their needs have not been covered over time. Globalization and technological development helped identify the importance of various groups with disabilities. In Ecuador, this identification allows this social group to be visible and particularly people with visual disabilities, who represent approximately 11% of the population with disabilities. The objective of this research is to identify Ecuador's cities that maintain a higher prevalence of people with disabilities. To comply with the study, a descriptive analysis was conducted from public information about people registered by the authorized government control body. Obtaining this information makes it possible to establish a baseline of attention with real data, which allows getting a greater focus on the development of this social group and the attention they should receive.

Keywords: Disability · Prevalence · Studies · Latin America · Visual disability · Physical disability

1 Introduction

The International Classification of Functioning, Disability, and Health was approved in 2001 [1], in which, disability is an evolutionary concept that results in social interaction, contemplating deficiencies, activities' limitations, and participation restrictions. Disorders that affect corporal or structures' functionalities are understood as deficiencies; on the other hand, difficulties to accomplish tasks are activity's limitations, and participation restrictions meaning there are problems to participate in vital situations [2, 3]. Creating a classification about the types: physical disability, sensorial disability, intellectual disability, and mental disability (see Fig. 1).

W. Karwowski et al. (Eds.): IHSED 2020, AISC 1269, pp. 234–238, 2021.
https://doi.org/10.1007/978-3-030-58282-1_37

Fig. 1. Graphic representation of disability types approached in the study

Certainly, disability is related to social factors, such as poverty and social exclusion [4, 5], these disadvantages affect the most to women [4], and, families counting with a member presenting any disability have an increasing tendency to poverty, since the same income level represents a lower lifestyle in homes with people presenting disability in comparison with families that do not count with a member with disability [6]. This fact highlights a strong relationship between poverty and disability.

In this context, the number of people presenting disabilities increases as the result of the aging population, chronic health problems, poverty, armed conflict, and urban and gender violence [7]. In this respect, 15% world's wide population lives with any type of disability [8], and, in Latin America it is estimated that 70 millions of people live with any kind of disability [9], referring to Ecuador, the 5,6% of the total population presents a disability of any type [10].

According to this reality, it is possible to visualize the importance of developing a study about disability in Ecuador, to create verified information and analyze disabilities' prevalence. The research carried out by Cambois et al. [11] affirms that there are social differences in the prevalence of disability in the 26 European countries studied, in comparison to the higher prevalence reached in lower social level groups and lower prevalence in the higher social level, establishing that this analysis would help identifying determinant facts and the efficiency of national policies implemented to approach the social differences in health.

2 Method

Disability data was obtained from the Ecuadorian National Board for Disability Equality, which is open to be searched in its webpage [12].

The sample counted with a total of 473652 valid data, from which, 208122 (43.9%) belonged to female gender, and, 265508 of male gender (56.1%), and, 22 (.00%) belonging to LGBTI gender. Taking into account the age of this sampling, 2744 (0.6%) were aged between 0 to 3 years, 6078 (1.3%) between 4 to 6 years, 24208 (5.1%), between 7 to 12 years, 27533 (5.8%) between 13 to 17 years, 69203 (14.6%) between 18 to 29 years, 222335 (46.9%) between 30 to 64 years, and 121551 (25.7%) between 65 years and above. Data was recruited from each province of Ecuador.

3 Results

The first step was to analyze the frequency of the type of disability that is present by province, to which the participants belong. These results are shown in Table 1.

Table 1. Type of disability by province

	Type of Disability					Total
	1	2	3	4	5	
Azuay	14655	3754	1555	3873	6136	29973
Bolivar	2877	1236	271	1548	1571	7503
Cañar	3313	1205	562	1280	2413	8773
Carchi	2075	611	205	1096	992	4979
Chimborazo	5728	1717	451	3505	3691	15092
Cotopaxi	4545	1436	382	2401	2755	11519
El Oro	9348	2154	910	2264	5660	20336
Esmeraldas	7483	2020	635	1675	4138	15951
Galapagos	329	68	49	88	153	687
Guayas	58230	12465	6949	14188	26440	118272
Imbabura	5327	1340	566	3162	2894	13289
Loja	5834	1753	882	2046	4587	15102
Los Ríos	13358	2310	861	2267	4652	23448
Manabí	25694	6370	2902	4487	7920	47373
M. Santiago	2359	989	256	710	1438	5752
Napo	1765	559	125	636	954	4039
Orellana	3443	1061	197	864	924	6489
Pastaza	1474	373	152	507	747	3253
Pichincha	32135	8803	4896	12349	16845	75028
Santa Elena	5406	906	354	1417	2287	10370
Sto. D. Tsachilas	6255	1644	569	1490	2850	12808
Sucumbíos	2668	816	312	788	1442	6026
Tungurahua	5037	1440	596	3440	3103	13616
Zamora Chinchipe	1640	565	137	562	1070	3974
Total	220978	55595	24774	66643	105662	473652

Note: 1. Physical. 2. Visual. 3. Psychosocial. 4. Auditory. 5. Intellectual.

Afterward, the percentage of disability according to the province to which participants belong. In Fig. 2 results found are presented. There were conducted association analyses among participant's sociodemographic variables, where it was found that the gender to which they belonged would have an association with the type of disability, where male gender has a higher risk probability of presenting a disability $x^2_{(8)} = 317.14$, $p = <.001$. Likewise, it was found a significant association between the disability and participants' age x^2 (464) = 98668.05, $p = <.001$ where it was found that the age between 30 and 64 years is the age range with higher probabilities to present a disability.

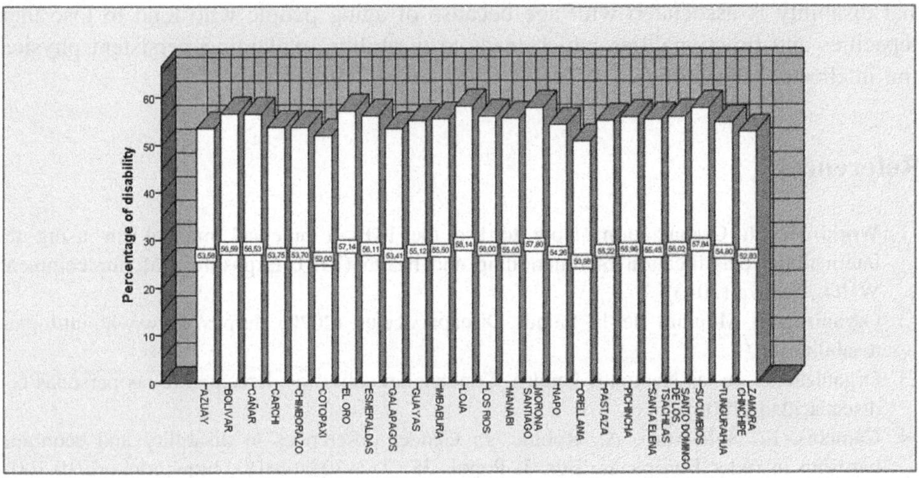

Fig. 2. Disability level according to the province they belonged

Finally, it was analyzed the type of disability according to the percentage of presence. See Table 2.

Table 2. Type of disability

Type of disability			
		Frequency	Percentage
Valid	Physical	220978	46,7
	Visual	55595	11,7
	Psychosocial	24774	5,2
	Auditory	66643	14,1
	Intellectual	105662	22,3
	Total	473652	100,0

4 Conclusions

In this report, it has been analyzed the prevalence of disability present in Ecuador through the technological development that has been built from the national board in charge of the attention of human beings presenting these conditions.

In this respect, Ecuador is divided into 25 provinces, from which, Guayas and Pichincha are the ones with higher density population, on one hand, Guayas counts with the principal port of the country, centering the highest economical movement, and, on the other hand, Pichincha is the capital.

This would explain the higher concentration of people with disabilities in these provinces. Living in Guayas and Pichincha represents better educational, labor and health opportunities. An important fact to take into account from the results presented is that disability is associated with age because of aging people who tend to lose their capacities and functionalities which leads to disability, explaining persistent physical and intellectual disabilities.

References

1. World Health Organization: How to use the ICF: a practical manual for using the international classification of functioning and health (ICF). Exposure draft for comment. WHO, Geneva (2013)
2. Organización Mundial de la Salud. Discapacidades (2020). https://www.who.int/topics/disabilities/es/
3. Organización de las Naciones Unidas. Convención sobre los derechos de las personas con discapacidad (2006)
4. Cambois, E., Solé-Auró, A., Robine, J.: Gender differences in disability and economic hardship in older Europeans. Eur. J. Popul. **35**, 777–793 (2018). https://doi.org/10.1007/s10680-018-9504-2
5. Parodi, G., Sciulli, D.: Disability and social exclusion in Italian households. Soc. Indic. Res. **144**, 767–784 (2019). https://doi.org/10.1007/s11205-019-02068-1
6. Vu, B., Khanam, R., Rahman, M., et al.: The costs of disability in Australia: a hybrid panel-data examination. Health Econ. Rev. **10**, 6 (2020). https://doi.org/10.1186/s13561-020-00264-1
7. World Health Organization: World report on ageing and health. World Health Organization, Geneva (2015)
8. World Health Organization and World Bank: World report on disability. WHO, Geneva (2011)
9. Comisión Económica para América Latina y el Caribe: Informe regional sobre la medición de la discapacidad. Una mirada a los procedimientos de medición de la discapacidad en América Latina y el Caribe. Santiago de Chile (2014)
10. Ortíz, J.: La discapacidad en el Ecuador en cifras, año 2010. Revista de la Facultad de Ciencias Médicas de la Universidad de Cuenca **31**(1), 74–81 (2013)
11. Cambois, E., Solé-Auró, A., Brønnum-Hansen, H., Egidi, V., Jagger, C., Jeune, B., et al.: Educational differentials in disability vary across and within welfare regimes: a comparison of 26 European countries in 2009. J. Epidemiol. Community Health **70**, 331–338 (2016). https://doi.org/10.1136/jech-2015-205978
12. Consejo Nacional para la Igualdad de Discapacidades: Estadísticas de discapacidad (2020). https://www.consejodiscapacidades.gob.ec/estadisticas-de-discapacidad/

Community Management Model of Water Resources. Case Study: Urban Planning of the Vinces Canton, Ecuador

Arturo Cadena(✉), Felipe Espinoza, Gabriela Vega, and Jesús Hechavarría

Faculty of Architecture and Urbanism, University of Guayaquil, Cdla. Salvador Allende, Av. Delta y Av. Kennedy, Guayaquil, Ecuador
{arturo.cadenaca, felipe.espinozao, gabriela.vegag, jesus.hechavarriah}@ug.edu.ec

Abstract. This article deals with community management applied to water resources, from the urban planning and public-community model to mitigate the effects of water mismanagement in Vinces canton. The objective is to propose a community management model of water resources to mitigate the negative effects of the current management system in the Vinces canton through the use of public-community models and guidelines of the 2030 Agenda. A descriptive methodology of model studies is used management system with its success models and objective six of the 2030 Agenda. Using the qualitative method, the current conditions of Vinces' water management are analyzed using surveys and interviews with public actors and focus groups. It is concluded that the model acts according to the needs of a satisfactory water management for the users of Vinces canton.

Keywords: Community management · Hidric resource · Urban planification · Community public model · 2030 Agenda

1 Introduction

Currently, there is no adequate water management in the Vinces canton, so there are problems related to the collapse of the sanitary infrastructure, floods and drinking water deficits, with rural areas being the least favored [1].

Although the greatest use of water is for agricultural activity, the most critical aspects of availability are related to the supply of drinking water for industrial processes and for the generation of electrical energy. Excess water generates floods, and/or landslides that affect the water supply by contaminating the water sources with sediments and causing destruction of the supply and distribution systems, causing direct impacts on the availability and quality of the water for the different users [2].

Environmental management by communities plays an important role in river management, since it is proven that the best decision-makers are those who live with the problem.

Under these circumstances, community water management studies have been explored. With these efforts it has been possible to maintain the water resource and thus ensure that water can reach each of their homes [3].

W. Karwowski et al. (Eds.): IHSED 2020, AISC 1269, pp. 239–244, 2021.
https://doi.org/10.1007/978-3-030-58282-1_38

2 Methodology

The methodology implemented to carry out this article is based on the collection of primary information through the review of information from various bibliographic sources such as: publications, studies, projects, theses from various universities at national and international level on public management models. and community. To create a database that allows the management of water resources in the canton to be described as a future model of community management through the creation of public-community alliances [4], and to emphasize successful experiences in management models.

In the real context of Vinces canton, through qualitative methods such as the semi-structured interview, the needs and problems will be described. In order to obtain important data to help design the environmental management model applied to the water resource, which responds to the users of Vinces canton. Finally, the results obtained in the field will be synthesized through an analysis to determine the strengths and threats that the water resource has in the Vinces canton, which will allow the researcher to correct the failures in public water management.

Through the participant observation method [5] which serves to verify and follow up on how the inhabitants use the water resource, the levels of use will be described as domestic, industrial and of irrigation, for the elaboration of a matrix of use.

From the data obtained in the qualitative methods, the management model will be developed from public and community alliances.

The goals and indicators of the sixth objective of sustainable development of the 2030 Agenda will be studied in order to evaluate each one of the points, preparing a check list of positive and negative results valued using a traditional weighting method.

3 Results

3.1 Scientific Review

In this section, through scientific documentation search platforms such as Scopus, Science Direct, Scielo and repositories of already published documents, an updated database (2015–2020) was obtained. Five documents were obtained referring to the different water management models under the community development approach with public-community alliances and an analysis document of the 2030 Agenda from which the following main ideas were obtained from each of these scientific documents:

Community participation: exploring legitimacy in socio-ecological systems for environmental water governance [6] - Scientific article: This article explores the relationship between legitimacy and participatory forms of decision-making in relation to responsible public institutions of the environmental governance of water in Australia. These can obtain greater flexibility and choice, since it is the management of water from citizen participation as agitator of public processes.

Global water policy and local payment choices in rural Africa [7] - Scientific article: In order to achieve "basic" drinking water for all by 2030. Efficient models of public, private and community services are analyzed for the supply of drinking water through incentives at its action level.

Are urban informal communities capable of co-production? The influence of community - public partnerships on water access in Lilongwe, Malawi [8] - Scientific article: Alternative models of community-state co-production in water management are studied. Surveys, focus group interviews were used to examine the impact of co-production on access to water and service provision in informal settlements in Lilongwe, Malawi.

The Community Management Model for Rural Drinking Water in Chile: Institutional Context, Regulations and Reform Intentions [9] - Scientific article: The Rural Drinking Water Program implemented by the State of Chile For more than 50 years, the organized rural community has been successful in supplying drinking water to localities in its sector. Then, the subsidiary role of the State and community management by local actors have been key elements for the validity of this program. A bill is currently being discussed that seeks to reform and endow the sector with institutions and regulations - incorporating sanitation - through the "Rural Sanitary Services".

Strengthening of the Community Management Model of Water for Human Consumption in the rural sectors of the Espejo canton - Case of the "La Libertad" Drinking Water Management Board [10] - Thesis (PUCE Repository): The research aims to develop tools to strengthen the management model of the Administrative Board for Drinking Water and Sanitation in La Libertad, Espejo canton, Carchi province. Analyzes the models of community organization for the management of drinking water for 600 families in their administrative and technical - operational aspects established in the current legislation.

The Sustainability index of the provision of clean water treatment plants (IPAB) in supporting SDG 2030 programs for the availability and management of sustainable clean water in Soppeng Regency, South Sulawesi Province, Indonesia [11] - Conference paper: To achieve goal six of the 2030 Agenda this document took a sustainability measurement of the sustainability of water treatment provision in the Soppeng district. The sustainability analysis was carried out using a multidimensional scaling approach that weights the results and shows whether the project is sustainable or not.

As a synthesis of the review of these 6 documents, it is obtained that the community management models are much more efficient in time and in actions to improve the water system, while the state public system is important in the provision of infrastructure for sanitary operation. This means that they must cooperate as a public administrative model and community management.

On the other hand, complying with a measurable sustainability index as expressed in the last document will make the models last over time and improve the quality of life of users.

3.2 Qualitative Method

Through surveys and interviews with the Municipal GAD of Vinces canton, focus groups and representatives of the urban and rural community, a sample of 50 participants was taken and the perception of the sanitary system of rainwater, wastewater and the provision of drinking water was evaluated, evaluating its service, endowment and conditions. Qualitative research seeks a representative sample, expressing user experiences (Table 1).

Table 1. Percentage weighting of water systems management opinion

Variables	Indicators		
	Good	Regular	Bad
AA.LL. system sewer service	4%	28%	68%
Provision of the AA.PP. system	30%	20%	50%
Conditions of the AA.SS: infrastructure	26%	24%	50%

The following are the general opinions collected in the interview:

1. The sewer service for the discharge of rainwater is terrible with two rains since the system collapses and my house is flooded.
2. The provision of the drinking water system is intermittent, but usually arrives between 8 to 12 h a day.
3. Every month the manhole covers overflow, sometimes you have to wait 10–15 days for them to solve the problem (Table 2).

Table 2. Water use levels per component

Water use levels	Real consumption	Deficit
Domestic	15%	85%
Industrial	70%	30%
Irrigation	7%	93%

Using this table, we can determine that 69% of the water provided for these services is wasted and does not meet the need.

3.3 Guidelines of the Community Management Model Proposed

- Preserve and avoid contamination of water wells - community management.
- Provide sewerage infrastructure to combat the deficit of 85% of provisioning - public management.
- Construction of a wastewater treatment plant to supply the 93% deficit of the irrigation system - public management.
- Participatory action group for repairs in the sanitary system - community management.
- Use of green infrastructures to improve the agricultural irrigation system - community management.

- Maintain the average levels in obtaining well water and the treatment of its surpluses - community management.
- Maintaining natural water courses in rural area - community management.
- Maintain Integrated Management of Water Resources through multi-level cooperations and alliances together with the Secretary of Water - public-community management.
- Reforest the banks of the Vinces river and preserve the riverside flora - community management.
- Manage a budget for community water management - public management.
- Creation of the Community Water Management Association of the Vinces canton - public-community management.
- Establish a policy of cooperation by competences between public and community management - public management.

3.4 Territorial Community Management Model

This territorial model is structured according to the 12 guidelines obtained from the community management model of the 2030 Agenda. The model consists of a treatment plant located in the northern part of the main canton in which, all the tributary of the Vinces river goes down, in the city center of Vinces, the community management association and the green infrastructure poles will be located. They will be located in the Antonio Sotomayor Parish, in the southern area in the La Soberana, Bagatela and Matecito areas and in the northern area in the San Antonio and San Francisco areas, these being the most productive areas with the greatest lack of green infrastructure for irrigation (Fig. 1).

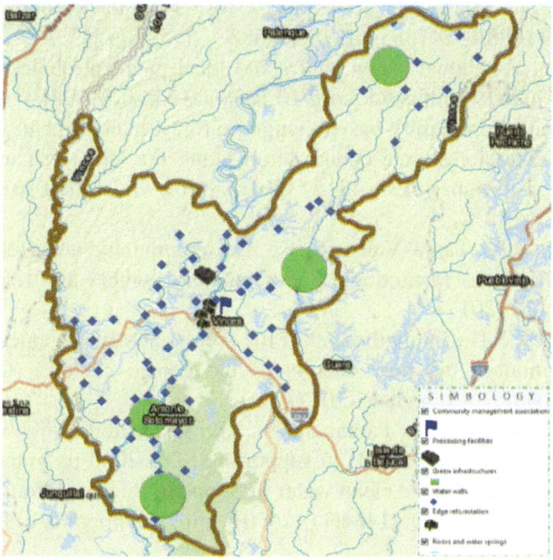

Fig. 1. Desired territorial model of community water management. Vinces canton.

4 Conclusion

It is concluded that the community management model enhances the decision-making process on water management through participatory actions by the control entity. The application of the proposed model will reduce the risks of floods and unsanitary conditions at the cantonal level. Decentralization is in line with the current sustainability framework and its guidelines comply with the agreement of the 2030 Agenda. The presented water management is aligned with a sustainable and lasting future of the resource.

References

1. Valero Fajardo, C.L., Hechavarría Hernández, J.R.: PEST analysis based on fuzzy decision maps for the ordering of risk factors in territorial planning of the Vinces Canton, Ecuador. In: Advances in Intelligent Systems and Computing, February 2020. AISC, vol. 1131, pp. 1190–1194 (2020). https://doi.org/10.1007/978-3-030-39512-4_181
2. Byakagaba, P., Egeru, A., Barasa, B., Briske, D.D.: Uganda's rangeland policy: intentions, consequences and opportunities. Pastoralism **8**(1) (2018). https://doi.org/10.1186/s13570-017-0111-3
3. Adissi, P., Pinheiro, F., Cardoso, R.: Gestão ambiental de Unidade Produtivas, 1a. Elsevier, Rio de Janeiro (2013)
4. García, D.: El agua: patrimonio y derecho (2007)
5. Bottin-Rousseau, S., Perrut, M., Picard, C., Akamatsu, S., Faivre, G.: An experimental method for the in situ observation of eutectic growth patterns in bulk samples of transparent alloys. J. Cryst. Growth **306**(2), 465–472 (2007). https://doi.org/10.1016/j.jcrysgro.2007.05.035
6. Godden, L., Ison, R.: Community participation: exploring legitimacy in socio-ecological systems for environmental water governance. Aust. J. Water Resour. **23**(1), 45–57 (2019). https://doi.org/10.1080/13241583.2019.1608688
7. Hope, R., Ballon, P.: Global water policy and local payment choices in rural Africa. npj Clean Water **2**(1) (2019). https://doi.org/10.1038/s41545-019-0045-y
8. Aheto, D.W., et al.: Community-based mangrove forest management: implications for local livelihoods and coastal resource conservation along the Volta estuary catchment area of Ghana. Ocean Coast. Manag. **127**, 43–54 (2016). https://doi.org/10.1016/j.ocecoaman.2016.04.006
9. Schuster Olbrich, J.P., Tapia Valencia, F.: The community management model of rural drinking water in Chile: Institutional and regulatory context and reform intentions. Leg. Forum **16**, 110–120 (2017)
10. Yazán Montenegro, P.D.: Strengthening of the community management model of water for human consumption in the rural sectors of the Espejo canton - Case: drinking water administration board 'La Libertad' (2017)
11. Azmanajaya, E., Paulus, C.A., Paranoan, N.: The sustainability index of the provision of clean water treatment plants (IPAB) in supporting SDG 2030 programs for the availability and management of sustainable clean water in Soppeng Regency, South Sulawesi Province, Indonesia. J. Phys. Conf. Ser. **1464**(1) (2020). https://doi.org/10.1088/1742-6596/1464/1/012052

Ecological Waste Planning. Case Study: Comprehensive Waste Management Plan at the Simón Bolívar Air Base, Guayaquil, Ecuador

Santiago Tisalema(✉), Jesús Hechavarría, Gabriela Vega, and Marcial Calero

Faculty of Architecture and Urbanism, University of Guayaquil, Cdla. Salvador Allende Av. Delta y Av. Kennedy, Guayaquil, Ecuador
{santiago.tisalemas, jesus.hechavarriah, gabriela.vegag, marcial.caleroa}@ug.edu.ec

Abstract. The objective of this work is to propose a comprehensive waste management plan at the Simón Bolívar air base under ecological criteria. An exploratory research methodology is used through the review of scientific articles, the adequate management of waste considering ecological corridors as a reference to improve the criticism of these guidelines and establish a comprehensive ecological environmental management. In the research carried out, the conditions of the existing and endemic vegetation are analyzed, and new proposals are made. The workshops carried out were developed with focus groups using the participatory method to propose the intelligent waste system. It is concluded that the application of intelligent systems in waste management obtains economic and environmental benefits and that the designs of ecological corridors improve the urban image of the Simón Bolívar air base.

Keywords: Comprehensive waste management · Environmental management · Urban planning

1 Introduction

Waste management currently plays a fundamental role in caring for the environment [1]. That is why this work focuses on proposing an improvement to environmental management at the Simón Bolívar air base. The planning that is done is limited to locating garbage containers and organizing the collection schedule. Solid waste is deposited in the unclassified collection center where it can accumulate up to 5 tons per day, in a total of 30 containers distributed by the air base. The collection center for the 3 sectors receives non-hazardous solid waste from canteens, administrative offices and public housing. There is no presence of hospital or maintenance waste (airplanes and land vehicles), which are collected by environmental administrators certified by the Municipality of Guayaquil. The frequency of collection of non-hazardous solid waste is carried out 3 times a week at night, but it does not fulfill the function of classifying or properly managing this waste.

W. Karwowski et al. (Eds.): IHSED 2020, AISC 1269, pp. 245–250, 2021.
https://doi.org/10.1007/978-3-030-58282-1_39

About 340 families live permanently in the air base, although they have recreational areas, these lack adequate vegetation to mitigate the impact of the high temperatures of Guayaquil. This unfavorable situation is further confirmed by climate change, repeatedly causing heat waves that trigger thermal stress on the inhabitants. The proper use of vegetation has resulted in good practices in caring for the environment [2]. However, at the air base there are initiatives to plant trees by the residents but for the most part they do not generate enough shadows to avoid the effects of the sun. This work proposes a new management plan to remediate the damage caused by human action throughout the years of history of the air base with the initiative of reconnecting fragmented ecosystem spaces through ecological conservation corridors [3] and an intelligent system of waste collection and classification.

2 Methodology

An exploratory research methodology will be used through the review of scientific articles, historical research and books on scientific bases that take intelligent waste management as a reference and propose ecological corridors to analyze their good practices in comprehensive environmental management [4]. Useful contributions based on stated objectives are extracted from this review.

In the second phase of the methodology, a survey was conducted to analyze data on the presence of introduced plants and suggested endemic plants according to the locality. Through a weighted matrix of adaptability, durability and benefits, the plant species to be implanted in the ecological corridor were chosen, complemented by new species that will improve the ecosystem services of the comprehensive management plan.

In the third phase of the methodology, through workshops with focus groups within the study area of the air base, a design proposal for the waste system was presented, based on good scientific review practices, to result in a participatory urban planning process under a GIS graphic scheme [5] of the comprehensive environmental management plan.

3 Results

3.1 Detection, Monitoring and Alert

In this section, 3 of the systems that have been implemented in smart containers and waste management systems were analyzed as an example of good practices and sustainable waste management for their application in the comprehensive management plan. As the Smart Dual Dustbin model for waste management in SmartCities [6] in the work, a Dual Dustbin model is proposed. The proposed methodology used two trash cans (trash cans A and B) placed together, where the sensor systems in the solid waste management systems are analyzed to verify the level and availability of the landfill. We also propose an intelligent container (ITB) with optimization of the efficiency of garbage collection through IoT detection [7], where a real-time detection system is carried out with additional calls to the collection service for its next collection in the place. The system detects three levels of garbage, such as empty, medium, full. For its geolocation it uses GPS systems that work with a battery based on solar panels. The efficient IoT-based smart container for a clean environment [8] provides a system that covers the monitoring of the state of the garbage container and includes new systems such as climate detection and mishandling of the container when garbage is thrown out of the container. of garbage.

3.2 Urban Articulation

Also considered is the capacity of an ecological corridor to articulate fragmented spaces and improve the urban image is considered, considering the social and ecological functions of urban green space. The Gravity Model Toolbox (an automated, open source ArcGIS tool) was used for building and prioritizing green corridors in cityscapes [9]. The green corridors connecting isolated remnant habitat patches (parks) can increase connectivity and provide eco-systemic services.

The least cost route model [10] was used to identify possible links between patches, based on network structures and cost area, which measures the theoretical energy cost of traveling between elements of the landscape. This process allowed applying concepts and guidelines for solving various impairments in physical and environmental quality problems in green corridors using a research method with a qualitative approach with descriptive research [11].

3.3 Green Planning

In the analysis, several plant species were considered, including the 5 recommended by "The guide to the flowery sky of Guayaquil", to provide ecosystem services within the study area, such as: Neem, Red Mangrove, Jacaranda, Guayacan and Gramillon. Among the benefits of these plant species stands out the improvement of air quality and the production of wide shading radii. In addition, 9 plant species were analyzed for the ecological corridor as shown in Table 1.

Table 1. Results and averages of the weighting matrix

Vegetal species	Results			Average
	Adaptability durability benefits			
Almond	3	3	2	2,66
Mango	2	3	2	2,33
St. Augustine grass	2	2	2	2,00
Olive	2	2	2	2,00
Veranera	2	1	3	2,00
Palm tree	2	2	1	1,66
Dwarf palm	1	1	1	1,00
Zebra tree	1	1	1	1,00
Mother-in-law's tongue	2	1	1	1,33

According to these results, the choice trend prefers endemic plant species such as: Almond (2.66), Mango (2.33) and San Agustin Grass (2). As well as some introduced species such as: Olivo and Veranera (2). These species will also be included in the proposed ecological corridor.

3.4 Participatory Design

The workshops were prepared jointly with the administrative area of the air base and the residents to determine the conditions of the container design and the technological system to be applied (Fig. 1).

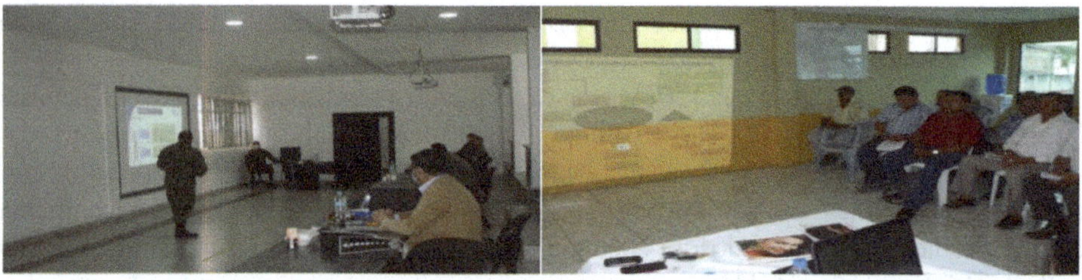

Fig. 1. Participatory workshops with the focus groups of the Simón Bolívar air base

The design proposal was based on three criteria provided by the focus groups: safety, odor control and efficient handling, so the design obeys a closed container that handles several separate interior compartments and that has efficient management thanks to solar panels.

3.5 Comprehensive Management Plan for Environmental Management

This comprehensive management plan articulates the underlying spaces with green ecological corridors taking into account the selected vegetation: Neem, Red Mangrove, Jacaranda, Guayacan and Gramillon that will allow the improvement of air quality and the production of wide shading radii. Soil protection between St. Augustine grass and the Gramillon as well as ornamental shrubs like Olive and the Veranera and fruit trees like the Almond and Mango. Figure 2 shows the outline of the comprehensive management plan proposed.

For the evacuation of the 5 tons per day of waste produced by the air base, 22 containers were installed, for a cumulative capacity of 9.68 tons. Another 4 tons could be additionally evacuated in 2 containers in the collection center, which allows the removal of garbage efficiently 3 times a week. Each of these containers includes fill sensors, alerts, and trash separators (Fig. 3).

Fig. 2. Outline of the comprehensive management plan proposed.

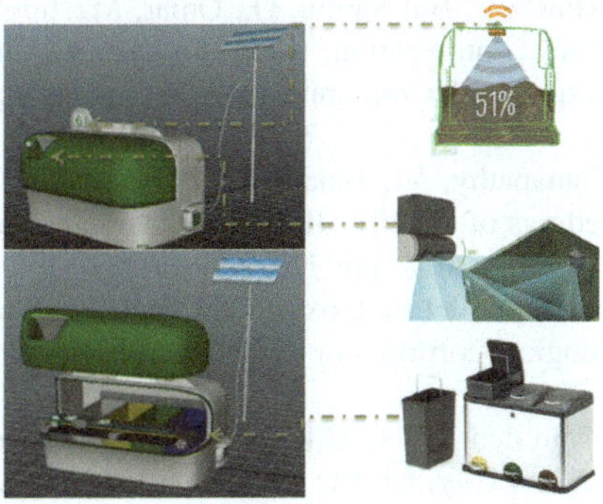

Fig. 3. Result of participatory design for the smart container.

4 Conclusion

It is concluded that applying ecological corridor designs through reforestation and choosing species according to the environment will diminish the negative impacts of the high temperatures of Guayaquil in the air base, which will allow the mitigation of air and soil contamination as well as an increase in shadows from the trees, greater thermal comfort and better quality of life for the inhabitants. The operation of the waste collection and classification system with the use of intelligent systems in the containers will give greater economic and environmental efficiency in the long term.

References

1. Sow, I.: Wastes management in african cities. In: Advances in 21st Century Human Settlements, pp. 201–218. Springer (2019)
2. Aheto, D.W., et al.: Community-based mangrove forest management: Implications for local livelihoods and coastal resource conservation along the Volta estuary catchment area of Ghana. Ocean Coast. Manag. **127**, 43–54 (2016)
3. Jongman, R.H.G.: Homogenisation and fragmentation of the European landscape: Ecological consequences and solutions. Landsc. Urban Plan. **58**(2–4), 211–221 (2002)
4. Bernal, R., Sánchez, E., Mauricio, D., Raymundo, C.: Comprehensive management model for solid waste collection and transportation in peruvian urban municipalities. In: Ahram, T., Taiar, R., Colson, S., Choplin, A. (eds.) IHIET 2019. AISC, vol. 1018, pp. 959–966. Springer, Cham (2020).
5. Rodríguez-Espinosa, V.M., Aguilera-Benavente, F., Gómez-Delgado, M.: Green infrastructure design using GIS and spatial analysis: a proposal for the Henares Corridor (Madrid-Guadalajara, Spain). Landsc. Res. **45**(1), 26–43 (2020)
6. Rohit, G.S., Chandra, M.B., Saha, S., Das, D.: Smart dual dustbin model for waste management in smart cities. In: 2018 3rd International Conference for Convergence in Technology, I2CT 2018 (2018)
7. Sohail, H., Ullah, S., Khan, A., Bin Samin, O., Omar, M.: Intelligent trash bin (ITB) with trash collection efficiency optimization using iot sensing. In: 2019 8th International Conference on Information and Communication Technologies, ICICT 2019, pp. 48–53 (2019)
8. Murugaanandam, S., Ganapathy, V., Balaji, R.: Efficient IOT based smart bin for clean environment. In: Proceedings of the 2018 IEEE International Conference on Communication and Signal Processing, ICCSP 2018, pp. 715–720 (2018)
9. Wanghe, K., et al.: Gravity model toolbox: an automated and open-source ArcGIS tool to build and prioritize ecological corridors in urban landscapes. Glob. Ecol. Conserv. **22**, June 2020
10. Xiu, N., Ignatieva, M., van den Bosch, C.K., Zhang, S.: Applying a socio-ecological green network framework to Xi'an City, China. Landscape Ecol. Eng. **16**(2), 135–150 (2020).
11. Sari, D.P., Alhamdani, M.R.: The urban design guidelines of Sungai Pinyuh street corridor. In: IOP Conference Series: Earth and Environmental Science, vol. 409, no. 1 (2020)

Safety Engineering and Systems Complexity

Investigating the Measurement of Resilience Engineering for Improving Organisational Safety

Manikam Pillay[1(✉)], Ishanka Weerasekara[1], Udara C. R. Ranawalage[2], and Emmanuel B. Boateng[1]

[1] School of Health Sciences, The University of Newcastle, Callaghan, NSW 2308, Australia
manikam.pillay@newcastle.edu.au, {ishanka.weeresekara, emmanuel.boateng}@uon.edu.au
[2] School of Architecture and Built Environment, The University of Newcastle, Callaghan, NSW 2308, Australia
udarachathurang.ranasingheranawalage@newcastle.edu.au

Abstract. This project investigates the measurement of resilience engineering. A growing body of peer-reviewed studies continues to be published on resilience engineering, demonstrates its recognition and importance to safety across a range of industrial contexts. However, little attention has focused on developing an understanding of how it has been conceptualized and measured. This is a significant gap which can limit its operationalization, benchmarking and evaluation n for research and practice. This paper presents an integrative review project currently underway which seeks to address this gap. After completing a systematic search and selection strategy seventeen articles were selected for analysis. Initial findings suggest fifteen survey instruments have been used in these studies.

Keywords: Resilience engineering · Integrative review · Organisational safety PRISMA

1 Introduction

This paper is a work-in-progress and is part of a broader research project on advancing organisational health and safety through Resilience engineering (RE). RE was introduced as an approach for improving organisational safety following the Columbia space shuttle disaster [1]. The seminal work on the topic was published in 2006 [2] and included extended versions of papers presented and discussed at the first International Symposium on RE in 2004. Since then it has gained traction in a range of industries such as healthcare, nuclear power plants, petrochemical facilities, electricity distribution and railways [3, 4]. As such a growing body of knowledge on the topic is available. Despite this, progress on its adoption across contemporary high-risk industries such as construction, healthcare, manufacturing, and mining have been relatively slow. In part, this is due to the difficulty in operationalising RE [5]. There is no uniformly accepted definition of RE, and no common characteristics for evaluating RE

W. Karwowski et al. (Eds.): IHSED 2020, AISC 1269, pp. 253–257, 2021.
https://doi.org/10.1007/978-3-030-58282-1_40

quantitatively [3]. Key proponents have argued for the need to focus on learning, responding, monitoring and anticipating [6]; while others have suggested behaviours [7], cognition [8], buffering, flexibility, margins and tolerance [9]; safety culture [1, 10]; safety trade-offs alongside production, quality and efficiency targets [9, 10], gap between work as imagined and work as performed [11, 12], or broader processes that assist organisations to achieve both safety and production targets [12]. This research seeks to address these gaps by investigating how RE has been measured in the published literature.

2 Method

An integrative review was adopted for this research. This method involves a structured approach and can be used to evaluate the strengths of evidence, identify gaps in the current research, build the bridge between related areas of published works, generate research question(s), identify theoretical or conceptual frameworks, and explore methods used [13]. The Preferred Reporting Items for Systematic Reviews and Meta-Analyses (PRISMA) [14] as adapted for a previous review [3] were used. Articles were selected based on a search of six electronic databases (CINAHL, Google Scholar, PsycINFO, PubMed, Scopus, and Social Sciences) using resilience engineering in the Titles and abstracts were searched and supplemented with additional ones from reference list of papers. Papers published in English between January 2003 and December 2019, which focussed on safety and described quantitative methods and approaches were included in the final selection. Key information relating to research aims, theory/model, research design, characteristics measured, and instruments used were extracted into an Excel sheet. The selected articles were critically appraised using a using an eight-item questionnaire adapted from the Critical Appraisal Skills Programme (CASP) [15] and the quality assessment tool of diverse study designs [16].

3 Results

The search across the six electronic databases and supplemented with hand searches generated 3884 articles from which a final set of seventeen was selected for analysis and synthesis. The backgrounds of these are summarized in Table 1. Fifteen surveys were reported across these studies. As previously indicated, the analysis and synthesis of this work are still in progress; however, some general statement can be made. Quantitative studies of RE did receive much attention until 2010, with the first two studies were published in 2013. Only four studies were supported with a theoretical framework such as with drift-towards failure [17] and organisational behaviour [18]. Two studies used an integrated RE theory [19, 20], while one safety culture [21].

Table 1. Background details of studies selected for final review

Context and Country	Purpose	Instrument
Gold Mining Australia	Introduced a theoretical framework for RE and a toolkit for investigating RE	Structured safety climate survey
Aviation Austria	Developed and validated an Inventory for assessing resilience through behaviour	Structured questionnaire surveys
Process industry Iran	Examined the validity of a survey method for measuring RE	Structured questionnaire surveys
Petrochemical plants Iran	Assessed factors affecting the resilient levels using Fuzzy cognitive maps	Experts' Views and Structured questionnaire survey
Petrochemical departments, Iran	Evaluated performance of Integrated RE through questionnaires and data envelopment analysis	Structured questionnaire survey
Public hospitals Iran	Designed a questionnaire to assess crisis management based on RE principles	Structured questionnaire survey
MSWM Companies Italy	Assessed Resilience in Municipal Solid Waste Management Companies	Structured questionnaire survey
Petrochemical plant Iran	Proposed framework to assess human resources productivity considering RE, motivational factors and health, safety, environment, and ergonomics	Structured questionnaire survey
Process industries Poland	Examined whether RE concept is related to the implementation of OSH management systems and safety performance	Structured questionnaire survey
Hospitals Iran	Present a new framework for assessing the crisis management based on RE principles	Structured questionnaire survey
Process industry Iran	Assessed RE factors based on system properties	Semi-structured interviews with MCQs, analysis of documents
Aluminium factory Iran	Evaluated Integrated Resilience Engineering (IRE) using mathematical programming	Structured questionnaire survey
Construction industry Canada	Developed and validated a Safety Climate Resilience Model	Structured questionnaire survey
Petrochemical Plants Iran	Designed a validated instrument to Measure Resilience Safety Culture	Structured questionnaire survey
Oil and gas industry Kuwait	Used resilient safety culture model to measure impact of remoteness and mental health	Structured questionnaire survey
Steel industry Iran	Quantified and determined priorities of RE dimensions	Structured questionnaire survey
Petrochemical Plants Iran	Evaluated impacts of RE on integrated health, safety, environment, and ergonomics management system	Structured questionnaire survey

3.1 General Characteristics of Studies and RE Measured

Eleven articles were published from Iran, with the remaining ones from Australia, Austria, Canada, Italy, Kuwait, and Poland. Industries from which these were published from included petrochemical plants, process industries, public hospitals, aviation, aluminum manufacturing, construction, gold mining, solid waste management and steel manufacturing. Fifteen (15) survey instruments were reported across these studies, suggesting this can be used for advancing quantitative evaluation of RE. As most of the published are from Iran, care needs to be taken when drawing from these for other countries. Similar care needs to be taken when proposing an instrument for the general industry because of the heterogeneity of the studies.

The number of RE characteristics measured in these studies differed considerably, with a minimum of three (3) and a maximum of thirteen (13). The most common characteristics measured included six of seven themes suggested by Wreathall [12]. Behaviours [7], buffering, flexibility, margins and tolerance [9] were included in two instruments; while cognition [7, 8], or the gap between work as imagined and work as performed [11, 12] were mentioned in one study but did not feature in any empirical investigations. What was not clearly spelt out, however, was what these characteristics specifically represented i.e. independent variables, mediating variables, or outcomes. This is something that needs to be investigated further.

4 Conclusion

This research is work-in-progress so any definitive conclusions regarding how RE has been measured in the literature cannot be made at this stage. However, the preliminary findings suggest that questionnaire surveys are the most common tools used for investigating RE quantitatively, so this can be used to conduct benchmarking studies in the field. Some key characteristics for measuring RE have been identified, but further analysis is required to specify what they represent. Findings from such analysis are useful in developing an appropriate theoretical framework and a supporting survey instrument for advancing quantitative RE studies. Future work will report on the development, pilot testing and validation of such surveys.

Acknowledgements. This research was supported through an internal University of Newcastle School of Health Science Strategic Pilot Grant Awarded to the first author in 2018.

References

1. Woods, D.D.: Resilience engineering: Redefining the culture of safety and risk management. Hum. Factors Ergon. Soc. Bull. **49**, 1–3 (2006)
2. Hollnagel, E., Woods, D.D., Leveson, N.: Resilience Engineering: Concepts and Precepts. Ashgate, Aldershot (2006)
3. Pillay, M.: Resilience engineering: an integrative review of fundamental concepts and directions for future research in safety management. Open J. Saf. Sci. Technol. **7**, 129–160 (2017)

4. Righi, A.W., Saurin, T.A., Wachs, P.: A systematic literature review of resilience engineering: research areas and a research agenda proposal. Reliab. Eng. Syst. Saf. **141**, 142–152 (2015)
5. Anderson, J.E., Ross, A.J., Back, J., Duncan, M., Snell, P., Walsh, P., Jaye, P.: Implementing resilience engineering for healthcare quality improvement using CARE model: a feasibility model. Pilot and Feasibility Studies 2 (2016)
6. Hollnagel, E.: Safety -I and safety-II: The past and future of safety management. Ashgate Publishing Ltd., Farnham, England (2014)
7. Back, J., Furniss, D., Blandford, A.: Cognitive resilience: reflection-in-action and on-action. In: Resilience engineering workshop, 25–27 June. MinesParis Tech (2007)
8. Furniss, D., Back, J., Blandford, A., Hildebrandt, M., Broberg, H.: A resilience markers framework for small teams. Reliab. Eng. Syst. Saf. **96**, 2–10 (2011)
9. Woods, D.D.: Essential characteristics of resilience. In: Hollnagel, E., Woods, D.D., Leveson, N. (eds.) Resilience Engineering: Concepts and Precepts. Ashgate Publishing Ltd., Aldershot (2006)
10. Flin, R.: Erosion of managerial resilience: From Vasa to NASA. In: Hollnagel, E., Woods, D.D., Leveson, N. (eds.) Resilience Engineering Concepts and Precepts, pp. 223–233. Ashgate Publishing Ltd., Aldershot (2006)
11. Dekker, S.W.A.: Resilience engineering: chronicling the emergence of confused consensus. In: Hollnagel, E., Woods, D.D., Leveson, N. (eds.) Resilience Engineering Concepts and Precepts, pp. 77–92. Ashgate Publishing Ltd., Aldershot (2006)
12. Wreathall, J.: Properties of resilient organizations: an initial view. In: Hollnagel, E., Woods, D.D., Leveson, N. (eds.) Resilience engineering concepts and precepts, pp. 275–285. Ashgate Publishing Ltd., Aldershot (2006)
13. Whittemore, R.: The integrative review: updated methodology. J. Adv. Nursing **52**, 546–553 (2005)
14. Liberati, A., et al.: The PRISMA statement for reporting systematic reviews and meta-analyses of studies that evaluate health care interventions: explanation and elaboration. PlOS Med. **6**, W1–W30 (2009)
15. Critical Appraisal Skills Programme. http://www.casp-uk.net/casp-tools-checklists
16. Sirriyeh, R., Lawton, R., Gardner, P., Armitage, G.: Reviewing studies with diverse designs: the development and evaluation of a new tool. J. Eval. Clin. Pract. **18**, 746–752 (2012)
17. Pillay, M., Borys, D., Else, D., Tuck, M.: Safety culture and resilience engineering: Theory and application in improving gold mining safety. In: Dominy, S. (ed.) Gravity Gold 2010, pp. 129–140. AusIMM, Ballarat, Australia (2010)
18. Heese, M., Kallus, W., Kolodej, C.: Assessing behaviour towards organizational resilience in Aviation. In: Fifth Resilience Engineering Symposium Managing Trade-offs. Resilience Engineering Association (2013)
19. Azadeh, A., Salehi, V., Ashjari, B., Saberi, M.: Performance evaluation of integrated resilience engineering factors by data envelopment analysis: the case of a petrochemical plant. Process Saf. Environ. Prot. **92**, 231–241 (2014)
20. Azadeh, A., Salmanzadeh-Meydani, N., Motevali-Haghighi, S.: Performance optimization of an aluminum factory in economic crisis by integrated resilience engineering and mathematical programming. Saf. Sci. **91**, 335–350 (2017)
21. Garg, A., Alroomi, A., Tonmoy, F., Mohamed, S.: Quantitative assessment of resilient safety culture model using relative importance index. In: 11th International conference on construction in the 21st century (CITC-11). Springer Nature (2019)

An Operational Model for Developing Process Operator Students' Safety Competence in on-the-Job Learning

Susanna Mattila, Noora Nenonen, Sari Tappura(✉), and Sanna Nenonen

Unit of Industrial Engineering and Management, Tampere University, 33014 Tampere, Finland
{susanna.mattila, noora.nenonen, sari.tappura, sanna.nenonen}@tuni.fi

Abstract. In the safety-critical process industry, safety competence is emphasized. It is essential to ensure that process operator students adopt good safety competence during their studies. Due to Finnish vocational education and training (VET) reform, on-the-job learning has increased, inducing a need for new methods. In this study, an operational model for developing process operator students' safety competence in on-the-job learning was constructed and evaluated. The model was compiled in cooperation with six process industry companies, five VET organizations, and eight expert organizations. The main data collection methods were interviews and workshops. The model consists of eight phases and provides instructions, tools, and good practices for implementing on-the-job learning. Companies and VET organizations considered the model useful and identified several purposes of use for the model.

Keywords: On-the-job learning · Process industry · Vocational education and training · Vocational student · Workplace safety

1 Introduction

New technologies, materials, extreme conditions, and the lack of young professionals are challenges in the process industry, where the potential for major hazards and accident risks is always present [1, 2]. Industrial operators have a significant role in process safety [3]. The education of operators should include proper safety education, as young workers typically have a higher injury rate than more experienced ones [4].

The new legislation on vocational education and training (VET) in Finland emphasizes the role of on-the-job learning in vocational education [5]. On-the-job learning is an essential part of developing process safety competence in process operator students. Successful learning at a workplace needs effective cooperation between a VET provider, a company, and a student and the integration of students' different learning environments [6]. The reconciliation of workplace demands and educational goals is important for the student to achieve the desired learning outcomes [7]. For the cooperation of process industry companies and vocational institutions, new

W. Karwowski et al. (Eds.): IHSED 2020, AISC 1269, pp. 258–264, 2021.
https://doi.org/10.1007/978-3-030-58282-1_41

methods were needed. The aim of this study is to construct and evaluate an operational model for developing process operator students' safety competence in on-the-job learning.

2 Materials and Methods

This study utilized a constructive research approach, which aims to solve problems through the construction of a practical solution [8]. In this study, an operational model was developed to find solutions that support the development of students' safety competence in the context of on-the-job learning in the process industry. The compilation of the model was based on the main phases typical of constructive research: setting of requirements, construction of alternative solutions, and validation of the constructions [8].

First, the development needs concerning students' safety competence and on-the-job learning in the process industry were charted to define the requirements for the model. Second, the design and content of the model were constructed by synthesizing the data collected for the study. Third, the model was revised according to the review comments of the study participants, and its utilization/usefulness was evaluated.

The model was compiled in cooperation with chemical and process industry companies (n = 6), their collaborator VET organizations (n = 5), and other expert organizations (n = 8) including, for example, labor unions, trade associations, and safety authorities. The main data collection methods were interviews and workshops (Table 1). In addition, a risk assessment exercise with observation, an accident analysis, a questionnaire for students, and the literature are used.

Table 1. Methods and materials used in the construction of the model

Phase	Methods	Participants
Phase 1 Setting of requirements	Interviews	All participating organizations (18 interviewees)
	Workshop 1 Development needs	All participating organizations (19 participants)
Phase 2 Construction of design and content	Company interviews	Company representatives (n = 46) from all companies (workplace instructors, superiors, safety and HR specialists, students)
	Workshop 2 Process safety competence	All participating organizations (20 participants)
	VET provider interviews	Process industry teachers (n = 7) from all VET institutions
	Risk assessment exercise with observation	Process industry teachers (n = 9) and process operator students (n = 35) from three VET institutions
	Accident analysis	Accident (n = 1381) data and reports (n = 17) involving process operators and industry from the Finnish Workers' Compensation Center
	Safety outlook questionnaire	Process operator students (n = 117)
	Literature	Previous research, legislation, other guidance related to on-the-job learning
Phase 3 Validation	Workshop 3 Review of the model	All participating organizations (16 participants)
	Feedback interviews	All participating organizations (19 interviewees)

Interview and workshop participants included persons involved in on-the-job learning in companies and VET organizations and other experts in the field of on-the-job learning and safety. The interviews were mainly semistructured individual interviews and were recorded and transcribed. The workshops were facilitated by the researchers and consisted of common discussions, group works, and presentations of research results. The interviews and workshops covered several topics relevant for the study, for example, good practices and development needs related to on-the-job learning and safety, safety competence requirements, and the usefulness of the model [e.g., 9–11].

The risk assessment exercise aimed to discover how well process operator students can assess risks [12]. The accident analysis provided a summary of the circumstances and consequences typical of accidents at work involving process operators and industry and recommendations for accident prevention. The aim of the questionnaire was to study whether the students had the safety attitude and ability desired by the process industry companies. The data were collected through an e-survey directed to the process operator students of the cooperative VET institutions. The questionnaire was based on previous studies and consisted of items covering perceptions toward accidents, safety instructions, communication, the role of safety, safety competence and training, and safety performance, responsibilities, and risks.

3 Results

3.1 The Operational Model

The operational model describes the process of on-the-job learning. The model consists of eight phases realized either while preparing for or during on-the-job learning (Fig. 1) [13]. The model describes in more detail each phase and its significance and implementation. In addition, safety issues relevant for each phase are emphasized.

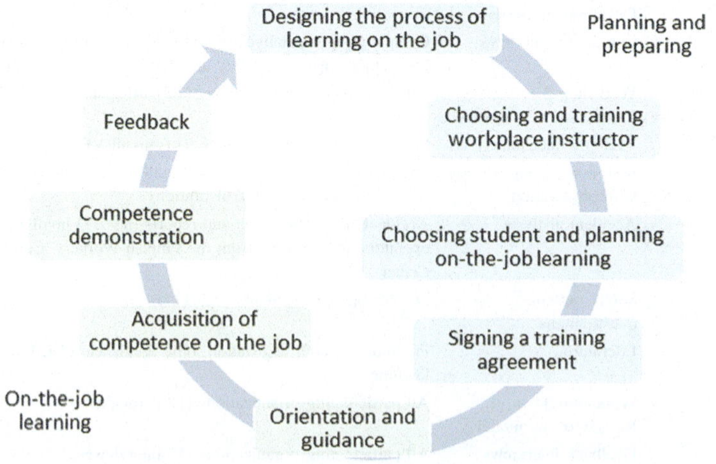

Fig. 1. Phases of the on-the-job learning process

For every phase of the process, the model contains examples of good practices on how to implement the phase. Moreover, the operational model provides tools and instructions for helping workplaces, VET providers, and students apply it in cooperation (Table 2). The tools and instructions can be used for several purposes, for example, defining practices related to on-the-job learning, self-assessment of safety competence, and training teachers and workplace instructors.

Table 2. Tools and instructions for applying the operational model

Tool or instruction	Description
Everything okay? -checklist	The most important tasks of the VET provider, workplace, and student at each phase
Features of a good workplace instructor	List of features associated with a good workplace instructor
Topics for workplace instructor training	Topics to be covered when training workplace instructors
Students' essential safety competence	Safety competence requirements essential for process operators
Safety outlook questionnaire	Questionnaire on students' outlook on safety and considering safety matters in education
Occupational safety responsibilities	Instruction on safety responsibility distribution between VET provider organization, workplace, and student
Topics for orientation and guidance	Topics that should be included in orientation and guidance
Risk assessment exercise	Tips for giving risk assessment exercises to students
Tips from companies to students	Companies' general expectations for student attitudes and behaviors during on-the-job learning
Accident information	The overview of accidents in the process industry
Topics for teacher visits during on-the-job learning	Topics to discuss between the teacher, student, and workplace representative when the teacher visits the workplace during on-the-job learning
Exercises for students during on-the-job learning	Examples of exercises that teachers or workplaces can give students during on-the-job learning
Tips for planning and conducting competence demonstrations	List of things to consider when planning a competence demonstration, tips on how to prepare for a competence demonstration, and questions and subjects to cover during the competence demonstration
Examples of feedback questions	Examples of questions for collecting feedback for development from a student to a company or VET organization, from a company to a VET organization, or vice versa

3.2 Usefulness of the Model

The cooperative organizations of the study considered the model useful and identified several purposes of use for it. The main reason why the model was considered useful was because it defines and documents the process for on-the-job learning. The study participants further explained that defining and documenting this process ensures the quality of the on-the-job learning and guidance given to students, unifies related practices, and clarifies the parties' roles. The model can be used to define, develop, benchmark, and train the processes, practices, and responsibilities related to on-the-job learning. These were rarely documented in the companies and VET organizations, and related information was mainly scattered and tacit.

The model was considered a useful practical guide with good examples for VET institutions and workplaces. Good practices in the operational model serve as checklists, diminishing the need to remember things and making operations smoother. Moreover, companies and vocational institutions can utilize the model when developing and unifying their practices in cooperation. Vocational institutions become familiar with companies' practices and vice versa. For example, the operational model can be used when drawing up an agreement on cooperation between a company and a vocational institution. The model can also be used to train teachers and workplace instructors.

4 Discussion

As a result of the study, an operational model for developing students' safety competence in on-the-job learning in cooperation with companies and vocational institutions was compiled. The operational model defines the general process of on-the-job learning through eight phases and related safety issues. Its purpose is to offer an example and good practices to companies and vocational institutions, who can define and document their own process of planning and carrying out on-the-job learning with the help of the model.

At least in Finland, a common documented model for on-the-job learning has not been available. Dang [14] considers the relationship between vocational education training providers and enterprises and proposes a model to link those in Vietnam. The paper does not define the process, but it emphasizes the importance of the cooperation between the government, the school, and the industry.

Good communication has been suggested as an essential but often inadequately functioning factor in successful safety management, particularly in situations where several parties are involved [15]. Poortman et al. [16] have reported ineffective communication between school and work in dual higher professional education. In this study, it was noticed that the cooperation and development of students' safety competence will benefit from using similar practices in companies and vocational institutions, for example, performing safety observations.

This study has some limitations. It covered only a limited number of companies and vocational institutions. The model should still be further tested in practice in different sectors. Nevertheless, the study participants considered the model useful in ensuring the overall quality of the on-the-job learning process and suitable for multiple purposes. The fluency of the on-the-job learning process and the focus on safety competence can increase companies' and VET organizations' attractiveness to students. Moreover, developing the process safety competence of vocational students increases their prospects of finding employment and provides a qualified workforce for the process industry in the future.

Acknowledgments. The authors thank the research funding provided by the Finnish Work Environment Fund and participating process industry companies. Furthermore, the authors appreciate the cooperation of the participating organizations.

References

1. De Rademaeker, E., Suter, G., Pasman, H.J., Fabian, B.: A review of the past, present and future of the European loss prevention and safety promotion in the process industries. Process. Saf. Environ. **92**, 280–291 (2014)
2. Rodríguez, M., Díaz, I.: A systematic and integral hazards analysis technique applied to the process industry. J. Loss Prev. Process Ind. **43**, 721–729 (2016)
3. Nazir, S., Sorensen, L.J., Øvergård, K.I., Manca, D.: How distributed situation awareness influences process safety. Chem. Eng. Trans. **36**, 409–414 (2014)
4. Salminen, S.: Have young workers more injuries than older ones? an international literature review. J. Saf. Res. **35**, 513–521 (2004)
5. Ministry of Education and Culture: Vocational Education and Training Will Be Reformed. Press Release, 10 August 2017
6. Virtanen, A., Tynjälä, P., Eteläpelto, A.: Factors promoting vocational students' learning at work: study on student experiences. J. Educ. Work **27**, 43–70 (2014)
7. Fjellström, M., Kristmansson, P.: Learning as an apprentice in Sweden: a comparative study on affordances for vocational learning in school and work life apprentice education. Educ. Training **58**(6), 629–642 (2016)
8. Kasanen, E., Lukka, K., Siitonen, A.: The constructive approach in management accounting research. J. Manag. Acc. Res. **5**, 243–264 (1993)
9. Tappura, S., Nenonen, S., Nenonen, N., Kivistö-Rahnasto, J.: Process safety competence of vocational students. In: Arezes, P.M. (ed.) AHFE 2019. AISC, vol. 969, pp. 383–392. Springer, Cham (2020)
10. Tappura, S., Nenonen, S., Nenonen, N.: Developing safety competence process for vocational students. In: Ahram, T., Karwowski, W., Taiar, R. (eds.) Human Systems Engineering and Design, IHSED 2018. Advances in Intelligent Systems and Computing, vol. 876, pp. 668–674. Springer, Cham (2018)
11. Nenonen, N., Tappura, S., Nenonen, S., Mattila, S.: Teachers' perceptions of the safety competence of process operator students. In: Arezes, P.M., Baptista, J.S., Barroso, M.P., Carneiro, P., Cordeiro, P., Costa, N., Melo, R.B., Miguel, A.S., Perestrelo, G. (eds.) Occupational and Environmental Safety and Health II. SSDC, vol. 277, pp. 141–149. Springer, Cham (2020)
12. Nenonen, N., Nenonen, S., Tappura, S.: Process operator students' abilities to assess OSH risks. In: Ahram, T., Karwowski, W., Pickl, S., Taiar, R. (eds.) Human Systems Engineering and Design II. IHSED 2019. Advances in Intelligent Systems and Computing, vol. 1026, pp. 566–572. Springer, Cham (2019)

13. Tappura, S., Nenonen, N., Mattila, S., Nenonen, N.: Prosessialan ammatillisten opiskelijoiden turvallisuusosaamisen ja yritys-oppilaitosyhteistyön kehittäminen. Tampere University, Tampere (2019). (in Finnish)
14. Dang, V.H.: The relationships between the vocational education training providers and enterprises: theory and practice. Int. J. Educ. Literacy Stud. **4**, 47–53 (2016)
15. Nenonen, S., Vasara, J.: Safety management in multi-employer worksites in the manufacturing industry – opinions on cooperation and problems encountered. Int. J. Occup. Saf. Ergon. **19**, 167–183 (2013)
16. Poortman, C.L., Reenalda, M., Nijhof, W.J., Nieuwenhuis, L.F.M.: Workplace learning in dual higher professional education. Vocations Learn. **7**(2), 167–190 (2014). https://doi.org/10.1007/s12186-014-9111-2

Measuring the Outcomes of Safety Training

Sari Tappura(✉) and Aki Jääskeläinen

Tampere University, 33014 Tampere, Finland
{Sari.Tappura,Aki.Jaaskelainen}@tuni.fi

Abstract. Safety training has been identified as one of the most important safety management practices. Previous research has suggested several models for evaluating the effectiveness of safety training. Evaluation is crucial to implement evidence-based training interventions. However, the outcomes of safety training are difficult to demonstrate. Studies better specifying and measuring the outcomes of training are needed. The aim of this study was to design and test a model for evaluating the outcomes of safety training. The model addressed 1) safety knowledge, 2) safety attitudes, beliefs and motivation, 3) safety behavior and 4) safety performance. A survey (n = 25) was used to measure the employees' perceptions after the safety training. Based on the results, some improvements in safety knowledge, attitudes, behavior and safety performance can be found. The model was found as suitable for its purpose. The results can be utilized in identifying factors related to the effectiveness of safety training.

Keywords: Occupational health and safety · Safety training · Safety performance · Performance measurement

1 Introduction

Safety management practices aim to improve working conditions at the workplace and positively influence employees' attitudes and behaviors with regard to safety, thereby improving safety performance [1]. Safety training has been identified as one of the most important safety management practices that predicts safety knowledge, safety motivation, safety compliance and safety participation which are key factors of safety performance [1–4]. Appropriate safety training could improve organizations' safety culture [5].

Previous research has suggested models for evaluating safety training effectiveness. These models address several themes, such as safety knowledge, safety attitudes and beliefs, safety behavior and health [6–8]. Based on the Ricci et al. [8] meta-analysis, the most effective training in terms of improving safety knowledge and attitudes are the individual self-learning modality, supplied by learning sessions no longer than 1 h and not compulsory for the worker. The best effects on increasing safety behaviors were obtained by behavioral, not compulsory training led by a researcher or an expert, individually supplied and for a longer duration. On the whole, the effectiveness was higher in voluntary learning conditions.

W. Karwowski et al. (Eds.): IHSED 2020, AISC 1269, pp. 265–270, 2021.
https://doi.org/10.1007/978-3-030-58282-1_42

According to Ricci et al. [8], a questionnaire is the most common method for evaluating the effectiveness of safety training. However, the most significant measures of efficacy are practical tests and questionnaires (administered immediately after the intervention) for knowledge training and only practical tests (in a follow-up time from immediately until three months) for attitudes training. The most significant measures were obtained by questionnaires, administered up to three months after the intervention.

Evaluation is crucial to implement evidence-based training interventions and improving the content and focus of training [8]. Previous studies [7, 8] highlight a strong support for the effectiveness of training on worker safety attitudes and beliefs and, to a lesser extent, on worker's knowledge. Further, safety training has found to have a medium effect on employee behavior and only minor effect on health. Different kinds of training are effective with regard to safety knowledge and attitudes than with regards to safe behavior [8]. Previous research suggests that classroom training does not reveal itself very effective [8]. More research is needed about the effectiveness of training, for example, applying different methods, in a small group, and regarding vulnerable employees [8]. Furthermore, training should be continuous since the training outcomes can be significantly reduced over three months [8].

Outcomes of safety training are difficult to demonstrate. There is a need for studies better specifying the outcomes of safety training and presenting and testing instruments for evaluating the outcomes of safety training.

Safety performance indicators could be better utilized in evaluating the safety training effectiveness. We investigate how the outcomes of safety training can be measured and whether safety training has positive effects on safety performance Therefore, this paper presents a framework and method for measuring safety training outcomes and discusses the employees' perception on the effectiveness of training on safety performance.

2 Materials and Methods

This study designed and tested a model for evaluating the outcomes of safety training. Testing was carried out in a large company providing support services for facilities management. This study utilizes a design science approach in which the intention is to develop scientific knowledge and solve practical problems. The six main phases of design science process [9] can be found in Fig. 1. This study follows the first four steps while deploying and maintaining are not in the scope. First, the scope of the model is defined. Second, the design and content of the model is designed based on earlier literature. Third, the model is tested with the potential users of the model.

Fig. 1. Main phases in constructing a model [9].

The scope of the model was defined based on the company's needs to find out how effective is the specific training intervention. The aim of the training was to increase hazard awareness, promote employees' safe behavior and support the development of safety culture. A model and its content were designed based on the literature addressing safety knowledge, safety attitudes and beliefs, safety behavior and safety performance with regard to safety training effectiveness [6–8]. The model was reviewed by safety experts of the case company and revised accordingly. Finally, the model was tested in a case company via paper survey.

Company-specific safety training was carried out as participatory classroom lessons with practical examples and discussions moderated by an expert. The whole work team including the supervisors participated in one training session. Training session were arranged during November and December 2019, and total amount of the participants was about 300.

A survey was used to measure the employees' perceptions after the safety training. The survey was carried out during February (2 to 4 months after the training sessions), and 25 responses were received. Only one respondent represented supervisor position. Around half of the employees had worked more than 5 years in their current position, and almost all had at least 1 years experiences in the position. The survey consisted of 48 questions; a Likert-scaled evaluation included 43 questions, and 5 open questions were asked. The respondents evaluated the effects of the safety training with relation to the four themes of the survey. Data was captured from the response sheets using Webropol survey tool. Descriptive analysis of the gathered data was carried out. The utilization of the model was evaluated.

3 Results

The model for measuring the outcomes of safety training was designed based on the existing literature on safety management and models related to the effectiveness of safety training. The literature-derived content was further revised by the safety experts in case company in order to capture realistic outcomes in the training in question. The model addressed: 1) safety knowledge, 2) safety attitudes, beliefs and motivation, 3) safety behavior and 4) safety performance (Table 1). The mean scores of the items are shown in Table 1.

Table 1. The contents of the model and the main results (n = 25)

Perspective	Sections	No of questions	Mean
Safety knowledge	Working environment	4	3.73
	Ability to work safely	4	3.74
Safety attitude, beliefs and motivation	Understanding the significance of safety	5	3.82
	Risk perception	4	3.81
	Safety Motivation	3	4.01
Safety behavior	Adhesion to the instructions	4	3.76
	Feedback	3	3.53
	Active safety participation	5	3.64
Safety performance	Effects on occupational injuries	1	3.84
	Effects on workplace climate	1	3.80
	Supervisory performance	4	3.66
	Employee performance	5	3.86

1 = disagree, 2 = partly disagree, 3 = neither agree nor disagree, 4 = partly agree, 5 = agree

Most of the respondents (80%) agreed that the safety training has positive effects on occupational safety by decreasing the number of occupational injuries (mean 3.84). Similarly, most of the respondents (64%) agreed that the training improves the workplace climate (mean 3.80). Some of the respondents stated that the training stirs themselves to think risks before starting the job. Most of the respondents (74%) also felt that the training improved their safety knowledge and ability to work safely in different kind of situations (mean 3.73). Most of the respondents (70%) perceived that the training had positive effects on their safety attitude, beliefs and motivation (mean 3.88). Interestingly, the training seemed to have greatest effect on the safety motivation of the employees (mean 4.01), such as more active tackling of the problems and caring for each other. The training had also some positive effects on employee behavior (mean 3.65), such as more active use of personal protective equipment. Safety climate was perceived quite positive (mean 3.77), for example, the majority of the employees stated that they take the fellow workers' safety suggestions seriously (mean 4.04) and can discuss safety matters openly (mean 4.00). Moreover, supervisors ensure that every employee can influence his/her own occupational safety (mean 3.80) and encourage employees to work always according to the safety instructions (mean 3.76).

The case company representatives evaluated that the model is suitable for its purpose, as it is comprehensive and compiles the major objectives for the effective safety training. The testing demonstrated that it includes relevant criteria for evaluating training outcomes and reveals differences between different perspectives to safety performance. Evaluation can also be conducted with reasonable efforts by employees.

4 Discussion

Based on the results, some perceived improvements in safety knowledge, attitudes, behavior and safety performance can be found. According to Ricci et al. [8] meta-analysis, OHS training induced positive effects on employees' safety knowledge, attitudes and beliefs, but minor evidence of training effectiveness on employees' safe behavior and health. On the contrary, Robson et al. [7] found strong evidence of the effectiveness of training on employees' safe behaviors. However, changes in behavioral routines are difficult to self-evaluate. They could be better evaluated by, for example, observations in workplaces [8].

Training adult workers is often a challenge because they are all motivated in different ways, have different sets of experiences, different expectations and different sets of skills and knowledge [8]. Safe behavior depends not only on perception and knowledge but also on safety climate at the workplace [10, 11]. Co-workers' orientation and social pressure may influence workers' intention to act in a safe way [8]. Nevertheless, safety training remains inadequate if it does not support behavioral changes.

This study focused on safety training aiming at behavioral and cultural changes. Raising safe behaviors is typically linked to in-depth and organization-wide interventions and not only to knowledge extension or occasional interventions [6, 8]. Previous research [8] suggests that classroom training does not reveal itself very effective. Behavioral training and hands-on practice with greater effect on individual training and provided by experts may promote safe behavior. Moreover, training should be in-dept and continued in order to have lasting outcomes. More research is needed about the effectiveness of training, for example, applying different methods, in a small group, and regarding vulnerable employees.

The initial phase of the model testing shows positive results and implications. The model was found as suitable for its purpose. In a further study, safety performance indicator data will be gathered to complement the survey data and to evaluate whether different sources of information provide conforming message. In the future, a second survey round could be carried out to study the permanence of the effects. During the same period, the safety performance quantitative data could be examined to cross-check the effects.

The study has some limitations. The study used self-report measures which may affect the reliability of results. The use experiences of the new model are still limited. The number of respondents to the survey could have been higher. In the future, the model should be further tested with more companies in different kind of industries with regard to different kind of safety training interventions.

This study contributes to the identification of a training intervention that could be efficient in improving safety performance. The results of the study can be utilized in identifying and measuring the outcomes of training and planning adequate interventions to improve safety.

The resulting model will benefit both the research and practice of safety management. The researchers can utilize the model in survey research. The practitioners can utilize it in evaluating safety training interventions and their (cost-)effectiveness. This study also presents a model for evaluating safety training effectiveness which combines safety knowledge, safety attitudes, safety behavior and safety performance. These perspectives can be used in linking different indicators and data sources to the evaluation of safety training effectiveness. By capturing safety performance, the new model highlights the link between training effectiveness and safety performance. Moreover, the developed new framework may be utilized as a leading indicator of safety performance.

Acknowledgments. The authors sincerely appreciate the contributions of all the respondents who participated in the study. The authors gratefully acknowledge the research funding provided by the Finnish Work Environment Fund, participating companies, and Tampere University.

References

1. Vinodkumar, M.N., Bhasi, M.: Safety management practices and safety behaviour: assessing the mediating role of safety knowledge and motivation. Accid. Anal. Prev. **42**, 2082–2093 (2010)
2. Hinze, J., Gambatese, J.: Factors that influence safety performance of specialty contractors. J. Constr. Eng. Manag. **129**(2), 159–164 (2003)
3. Wu, T.-C., Chen, C.-H., Li, C.-C.: A correlation among safety leadership, safety climate and safety performance. J. Loss Prev. Process Ind. **21**(3), 307–318 (2008)

4. Neal, A., Griffin, M.A., Hart, P.M.: The impact of organizational climate on safety climate and individual behavior. Saf. Sci. **34**(1), 99–109 (2000)
5. Bahn, S.: Moving from contractor to owner operator: impact on safety culture: a casestudy. Empl. Relat. **35**(2), 157–172 (2013)
6. Burke, M.J., Sarpy, S.A., Smith-Crowee, K., Chan-Serafin, S., Salvador, R.O., Islam, G.: Relative effectiveness of worker safety and health training methods. Am. J. Public Health **96** (2), 315–324 (2006)
7. Robson, S., Stephenson, C.M., Schulte, P.A., Amick III, B.C., Irvin, E.L., Eggerth, D.E., Chan, S., Bielecky, A.R., Wang, A.M., Heidotting, T.L., Peters, R.H., Clarke, J.A., Cullen, K., Rotunda, C.J., Grubb, P.L.: A systematic review of the effectiveness of occupational health and safety training. Scand. J. Work Environ. Health **38**(3), 193–208 (2012)
8. Ricci, F., Chiesi, A., Bisio, C., Panari, C., Pelosi, A.: Effectiveness of occupational health and safety training: a systematic review with meta-analysis. J. Workplace Learn. **28**(6), 355–377 (2016)
9. Holmstöm, J., Ketokivi, M., Hameri, A.-P.: Bridging practice and theory: a design science approach. Decis. Sci. **40**(1), 65–87 (2009)
10. Quick, B.L., Stephenson, M.T., Witte, K., Vaught, C., Booth-Butterfield, S., Patel, D.: An examination of antecedents to coal miners' hearing protection behaviors: a test of the theory of planned behavior. J. Saf. Res. **39**(3), 329–338 (2008)
11. Reason, J.: The Human Contribution – Unsafe Acts. Accidents and Heroic Recoveries. Ashgate Publishing, Farnham (2008)

Remote Inspection

Lee T. Ostrom(✉) and Cheryl A. Wilhelmsen

Idaho Falls Center Campus, University of Idaho, Idaho Falls, ID, USA
{Ostrom, cherylw}@uidaho.edu

Abstract. Inspection of complex mechanical systems can be costly and require expert inspectors to perform the work. These systems include bridges, power plants, aircraft, and even factories. In many cases, these systems are in remote places globally and are difficult for qualified inspectors to visit. Inspections can be done remotely using sophisticated cameras, set up to take three-dimensional images. We showed that LIDAR and a three-dimensional, immersive environment could be used successfully for these types of inspections. However, there are human factors issues and risks associated with remote inspections. This brief paper discusses aircraft damage, inspection techniques, and the benefits, human factors issues, and risks of remote inspection.

Keywords: Remote inspection · LIDAR · 3-D immersive environment

1 Background

We have worked on aircraft maintenance and inspection, human factors, ergonomics, and risk since the early 1990s. Many of the tools we use were initially developed in the 1950s and 1960s by defense contractors to analyze complex weapons systems [1]. In the 1970s and 1980s, these same tools were updated and adapted to analyze the risk associated with nuclear power and weapons production [2]. We began using risk assessment tools to examine aircraft maintenance tasks in the early 1990s. When we initially started this work, we focused on aircraft maintenance [3]. The vast majority of the structural components of commercial aircraft in service during the 1990s were made from aircraft-grade aluminum alloys. Aluminum alloys are perfect for aircraft because it withstands the extreme environments commercial and military airplanes are subjected [4]. At the request of NASA and Boeing Commercial Aircraft Company, we shifted our focus to aircraft inspection [5]. Aircraft inspection, in general, is categorized as General Visual Inspection (GVI), Detailed Visual Inspection (DVI), and Directed Inspections [6]. An obvious statement is that inspecting aircraft made from aluminum alloys is much different than inspecting aircraft primarily composed of composite materials. We developed risk models of aluminum aircraft inspection before shifting over to performing inspection studies on composite inspection [7].

There is a big difference in the types of damage metal aircraft experience and the kinds of damage composite aircraft experience [8]. Aluminum alloys primarily crack and tear in response to an impact, for example. Composite aircraft experience small dents that can lead to delaminations. Delaminations can go unseen, whereas a crack in metal is much more visible. Both materials can suffer catastrophic damage. However,

W. Karwowski et al. (Eds.): IHSED 2020, AISC 1269, pp. 271–276, 2021.
https://doi.org/10.1007/978-3-030-58282-1_43

most of the time, the initial damage begins as small cracks or delaminations. For example, Aloha Airlines Flight 243, a large part of the fuselage separated from the Boeing 737 in mid-flight (Fig. 1). The failure mechanism resulted from multiple site fatigue cracking of the skin adjacent to rivet holes along with the lap joint upper rivet row and tear strap disbond [9]. Aluminum alloy aircraft are still being produced, and Alcoa Aluminum claims they have an alloy that can compete with composites for weight and strength [10].

Fig. 1. Aloha airlines flight 243 damage (Source federal aviation administration)

The Boeing Commercial Aircraft Company's (Boeing) 787 was the first commercial airliner with many of its structural components made of composite material [11]. The 787 Dreamliner was flown for the first time in December 2009. 50% of the structure is made up of composite.

The Airbus A350 was the second mostly composite airliner to enter commercial service. Composite materials have been incorporated in planes for many years. Composite materials were first used on commercial aircraft in the 1950s. Boeing's 707 was made of 2% fiberglass, and in the '80s, Airbus was using 5% composite on the A310-300. Gradually both companies have increased the use of composite through the years. The Airbus A340 was using 10% of composite materials in the 1990s, and the Boeing 777 was using 12%. By the turn of the century, the advance made in composite manufacture allowed the aeronautical industry to increase its use of composite significantly. The Boeing 777 moved from 12 to 50% on the 787, while Airbus A340 moved from 10 to 25% on the A380 and finally to 53% on the A350XWB. Composites have a high strength to weight ratio, and the resulting aircraft weighs significantly less, producing fuel savings [12].

2 Remote Inspection Human Factors

Human inspectors have a relatively tricky task determining what is and what is not damage, whether aluminum or composite [13]. There is a wide range of non-destructive testing equipment that is used to help find metal damage. These inspection modalities include eddy current testing, boroscope examinations, dye penetrant testing, and ultrasonics, to name a few [14]. These systems do not lend themselves to remote inspection. Remote visual inspection on aluminum aircraft can be performed for cracks, but many small cracks could be missed.

Similarly, composite aircraft damage is challenging to find. Figure 2 shows minor composite damage on the surface. However, thermography [15], a non-destructive testing method, reveals the underlying damage (Fig. 3). The damage shown in Fig. 4 cannot be missed. Catastrophic accidents have occurred as well. An accident involving American Airlines flight 587, the tail fin of an Airbus A300-605R came off the aircraft after takeoff from JFK Airport [16]. A similar accident happened to the Air Transat flight 961 in 2005, when a large part of the composite rudder had disintegrated [17].

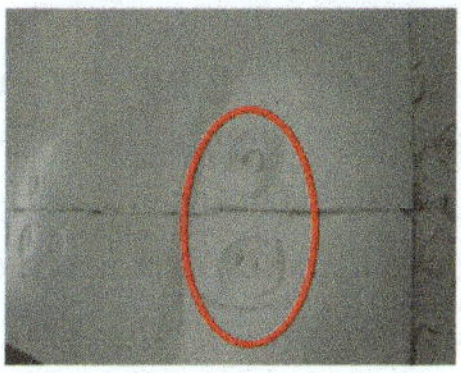

Fig. 2. Photograph of dents around rivets in a composite aircraft part (Source [2])

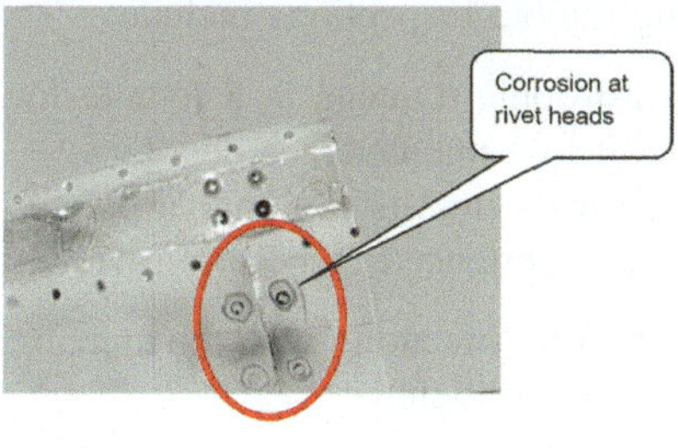

Fig. 3. Thermographic image of composite damage (Source [2])

Fig. 4. American airlines, airbus 319 DAMAGED by hail (Source, courtesy of American airlines)

Remote inspection can be performed in several different ways. In conjunction with a 3D Immersive environment, we showed that LIDAR could be used to perform remote inspections [18, 19, and 20]. Various companies have also developed camera systems for remote inspections, along with Virtual Reality systems, for example, the AVA-TOUR System [21]. The human factors and logistical related issues associated with using these processes for inspection include:

- Training of the inspectors, both at the remote site where the data are collected and at the analysis site where the data are interpreted.
- The lighting of the aircraft for the camera systems.
- Cleaning of the surfaces for inspection. Dirty surfaces can be interpreted as damage.
- Placement of the data collection devices to ensure coverage of all areas of the aircraft. The installation of these systems could be a big issue. Even in person, GVI and DVI inspections frequently miss damage on the top of aircraft surfaces [22].
- The expense of the data collection and data display technology.

The benefits of performing remote inspections include:

- Finding potentially catastrophic damage at a remote location, instead of flying a potentially flawed aircraft to a maintenance depot.
- Reducing the number of inspectors at regional airports.
- Having potentially more experienced inspectors review inspection data.

The risks of remote inspection follow along with the human factor/logistical issues and are contrary to the benefits. Risks include:

- Missing damage that a live inspector would be able to find.
- Falsely indicating damage that keeps a healthy aircraft at a remote location.
- Certifying a damaged aircraft for flight because the damage was not correctly identified.

3 Summary

This brief paper discussed some of the issues with aircraft damage and the potential catastrophic accidents that can happen if structural damage is not detected. Remote inspection can be performed using a variety of techniques. When displayed in a 3D immersive environment, we proved that LIDAR scans are beneficial for remote inspection purposes. There are other systems that can be used, as well. However, there are human factors issues and risks associated with isolated inspection, as there is with all technologies.

References

1. Keller, W., Modarre, M.: Reliab. Eng. Syst. Saf. **89**(3), 271–285 (2005)
2. Perkins, J.H.: Development of risk assessment for nuclear power: insights from history. J. Environ. Stud. Sci. **4**, 273–287 (2014). https://doi.org/10.1007/s13412-014-0186-8
3. Ostrom, L.T., Wilhelmsen, C.A.: Ergonomic assessments of commercial aircraft maintenance tasks: line maintenance. In: Advances in Industrial Ergonomics and Safety VII (1995)
4. Quest for Performance: The Evolution of Modern Aircraft. NASA History (2012)
5. Ostrom, L.T., Wilhelmsen, C.A.: Task and risk analysis of aviation maintenance and inspection tasks. In: Presented at the Tenth International Symposium on Aviation Psychology, May 1999
6. Federal Aviation Administration, Volume 3 General Technical Administration, 8900.1 CHG 569. Accessed 11 Nov 2018
7. Ostrom, L., Wilhelmsen, C., Valenti, L., Hanson, E.: Probability of crack detection: a unique approach to determining inspector reliability. In: Presented at the System Safety Society Meeting, Denver, Colorado, August 2002
8. Mraz, S.: Basics of aerospace materials: aluminum and composites. Mach. Des. Thu. **10**, 40 (2014). Accessed 19 June 2014
9. Aircraft Accident Report, Aloha Airlines Flight 243, Boeing 737–100, N73711, Near Maui, Hawaii, April 28, 1998 (PDF). National Transportation Safety Board. June 14, 1989. NTSB/AAR-89/03. Accessed 5 Feb 2016
10. Paur, J.: New Aircraft Aluminum Set To Compete With Composites. https://www.wired.com/2011/06/new-aircraft-aluminum-set-to-compete-with-composites/, Accessed July 2020
11. Cummins, N.: Simply Flying, Airbus vs Boeing – Which Plane Manufacturer Is Best?. Accessed February 2019, https://simpleflying.com/airbus-vs-boeing/2019
12. Worries about new composite made airplane, 1001 Crashes.com (2011). Accessed February 2019, https://www.1001crash.com/index-page-composite-lg-2.htm
13. Erhart, D., Ostrom, L., Wilhelmsen, C.: Visual detectibility of dents on a composite aircraft control surface: an initial study. Int. J. Appl. Aviat. Stud. **4**(2), 111–122 (2004)
14. American Society for Nondestructive Testing, Nondestructive Testing Handbook, 3rd and 4th Editions, Various publication dates (3rd edition 2001, 4th edition 2019)
15. Maldaque, X.: Infrared Methodology and Technology (Nondestructive Testing Monographs and Tracts) , 1st edn. CRC Press, Boca Raton (1994)
16. National Transportation Safety Board (NTSB): In-Flight Separation of Vertical Stabilizer American Airlines Flight 587 Airbus Industrie A300–605R, N1405, Belle Harbor, New York, 12 November 2001 Aircraft Accident Report (2001)

17. National Transportation Board Of Canada: Aviation Investigation Report A05f0047 Loss Of Rudder In Flight Air Transat Airbus A310–308 C-Gpat Miami, Florida, 90 Nm S 06, March 2005
18. Ostrom, L.T., Wilhelmsen, C., Scott, L., Whiting, E., Grover, S.: Use of three dimensional imaging technology to support visual inspection of composite materials. In: A4A NDT Symposium, Seattle, WA, September 2012 (2012)
19. Ostrom, L., Scott, R., Wilhelmsen, C.: Risk assessment framework for detecting, predicting, and mitigating aircraft material inspection. In: Final Project Methodology Report, July 2015 (2015)
20. Wilhelmsen, C., Ostrom, L.: Remote aircraft composite inspection using 3D imaging. In: Human System Interaction Conference, Portsmouth, UK, presented July 2016 (2016)
21. Avatour System, 360° Remote Presence For Inspections & Audits. https://Avatour.Co/Inspections-Audits, Accessed July 2020
22. Federal Aviation Administration, Visual Inspection For Aircraft, Date 8/14/97 AC No: 43–204 (1997)

Characterising Futuring Strategies for Biodiverse Speculative Design and Systems Design

Craig Jeffcott[1,2(✉)] and Ana Margarida Ferreira[1,2]

[1] IADE, Universidade Europeia, Lisbon, Portugal
craig@biodiverse-futures.org,
ana.margarida.ferreira@universidadeeuropeia.pt
[2] UNIDCOM/IADE, Unidade de Investigação em Design e Comunicação, Lisbon, Portugal

Abstract. In this paper, it is proposed that new tools and techniques for speculative design practice are needed to guarantee the biodiversity of global systems, and that they can be based in the non-normative strategies of futuring that exist within culturally diverse and ecologically-engaged communities. Furthermore, these tools and techniques can be used to build an ecology-centred practice of speculative design for community action. Taking a research through design approach that draws on critical theories of ecosophy, queer ecology and queer futures, the presented research, as a part of an ongoing doctoral work, aims to characterise the futuring strategies of such communities through participant and desk-based qualitative research. Preliminary reflections are that the communitarian and ecological future-making strategies described in theory are present in material and imaginative practices. Such strategies are multi-faceted and richly contextual and resonate across theoretical, practical and speculative domains. Moreover, these approaches possess common and recurring aspects of care-taking, non-linear temporality, relationality, storytelling, and playfulness.

Keywords: Speculative design · Systems design · Cultural diversity · Biodiversity · Queer ecology · Queer futures · Sustainable social change

1 Introduction

This ongoing doctoral research project aims to explore the potential of non-normative and ecological futuring strategies to underpin speculative design tools and techniques that are diverse, ecology-centred and impactful. The hope is that such strategies can form part of a more sustainable speculative design practice, and contribute to positive and sustainable social change, and biodiversity and community activism.

This paper reports on the first phase of this research project, which is to learn about non-normative futuring strategies, through participant based research with diverse and ecologically-engaged communities, and through desk-based research and strategy mapping.

This research builds towards an original framework for developing novel speculative design techniques, which will be deployed in a community speculative design

W. Karwowski et al. (Eds.): IHSED 2020, AISC 1269, pp. 277–282, 2021.
https://doi.org/10.1007/978-3-030-58282-1_44

project intended to demonstrate the potential for a new speculative design practice in taking action for more biodiverse futures.

2 Contexts

This work is situated in the context of the catastrophic destruction of the global ecosystem driven by a normative human-centred view of the future, with a United Nations report evidencing that biodiversity—a plurality of different types of organisms—is giving way to a human-centred monoculture which cannot sustain life on this planet. It is clear that action must be taken to argue for—and to achieve—alternative futures to this [1].

Speculative Design is a mode of design practice that materialises transformative futures that are an alternative to the future that is expected. As explained by Anthony Dunne and Fiona Raby in Speculative Everything, a key foundational reference for this mode of design, it is the use of design practice to ask "what if?" [2].

A promising design approach to a future-oriented biodiversity activism could possibly be found in speculative design practices that aim to materialise futures outside of the norm. However, speculative design has been criticised for lacking impact, and not being practiced in a sufficiently diverse way to produce truly transformative outcomes [3]. These concerns were initially explored by Jeffcott and Ferreira in 2019 [4].

3 Critical Lenses

As already stated, this doctoral research proposes a path towards a more diverse, ecology-centred speculative design practice by learning from existing community strategies for uncertain futures, as well as existing speculative fictions that engage with the project themes. This approach is rooted in the critical theories of ecosophy and queer theory—specifically queer ecology, and utopian queer futures.

In The Three Ecologies, Guattari argued that that the three ecological realms of the environment, society, and the subjective individual are simply one ecosystem in a state of crisis, a worldview he termed ecosophy (ecological philosophy). Strategies for responding to the cultural-ecological crisis were based around "implementing effective practices of experimentation…on a micro-social level scale" [5].

Queerness—described by Warner as a "resistance to the regimes of the normal" [6]—and queer theory allow for just such a radical reframing of the relationship between human and non-human. Morton's Queer Ecology centres on the importance of diversity of human experience as well as diversity of the beings in an ecosystem, suggesting that the queer challenge of gender and relational norms could be applied to the artificiality of the distinctions between humanity, the natural world and technology [7].

A queer lens can also provide a unique perspective on futurity and speculation. Muñoz's wrote in about how queer experiences are inherently futuring experiences. He argued that minoritarian outsiders must and do find spaces in the present where a new, more hospitable future can be imagined, enacted, and made real. By invoking and materialising futures outside the norm, this queer gesture resonates with the speculative gesture in design. This framing of queer futurity is grounded in Ernst Bloch's concrete utopia, a better future that we can imagine and work for, incrementally and together, through action in the present [8].

4 Methodological Approach

This research is methodologically situated as *research through design*: reflective, practice-based research using design activities to generate knowledge through creating artefacts and reflecting on the design process [9]. It is also situated as a *speculative* research practice: a way to move from "reacting to pre-defined research questions to devising questions and research techniques that may engender the emergence of novel and inventive responses" [10], with an emphasis on a "speculative gesture" within research, and the desirability of flexible and practice-based 'techniques' over more fixed and procedural 'methods'.

The research has four phases: to characterise these strategies, to create and validate new speculative design tools and techniques, to apply these in a community speculative design project to take action against biodiversity loss, and to evaluate their effectiveness within design practice and other fields through engagement and reflection practices.

5 Research Practice

5.1 Overview

This section of the paper gives some detail of the ongoing research activities that form the two pillars of this first phase: characterising diverse futuring strategies through exploratory qualitative participant research, and taxonomic strategy mapping from the existing literature. These two approaches are complementary, with participant research providing real world examples of theoretical or fictional approaches, and the wider literature giving new perspectives on community engagements. This research will build towards a strategic framework for the design of experimental speculative design tools and techniques.

5.2 Participant Research: Design Probes

Participant-oriented research to learn about community strategies has to date largely focused on design probes, alongside some initial and informal conversations with relevant communities and individuals; this will be complemented later with interviews, and design workshops (delayed from spring 2020 due to the coronavirus pandemic).

Design probes are a way to gather idiosyncratic and qualitative research data from research subjects while the researcher is not present, consisting of a pack of activities that can be distributed to research subjects with being collected at a later date [11].

A draft set of 11 design probe tools was created, drawing on a number of the theoretical and critical frames for the project, with a questionnaire asking how engaging, interesting and well explained each task was.

These probes were distributed to 15 reviewers who were asked to attempt any activities that interested them, returning any completed probes, and completing the questionnaire. Out of the 15 reviewers, 5 responded to the questionnaire and 3 returned some design probe responses. Questionnaire responses showed that reviewers found the probe tools were interesting, but that many of the tasks were not clearly explained, were too much of a time commitment, and may be more suited to group design workshops. One reviewer generally disliked the design probe technique itself, feeling that the technique was too "furtive", and that they'd rather engage with the topics directly.

Reflecting on these responses, the rationale of using design probes as part of the project is being re-evaluated. While it is useful method to engage participants and to generate some interesting qualitative research data, it is important to consider the cognitive effort and time commitment to complete a set of probe tools, and it is clear that well explained instructions are essential. As the research progresses, the design probes will be reframed it as a series of very quick, low cognitive-load activities that act to engage possible project participants with the themes of the project and generate some rough qualitative data, with some of the more challenging activities being reworked for collaborative design workshop settings.

5.3 Desk-Based Research: Building a Taxonomy of Futuring Strategies

The desk-based research is oriented towards characterising and mapping relevant existing strategies and techniques into a loose taxonomy; this complements the design probe activities by providing a wider context and understanding of how some of these strategies are being deployed in different domains of discourse and practice.

Within the scope of the research are: critical theoretical texts with queer, diverse and ecologically-oriented worldviews—to provide high-level and abstract strategic outlooks; community projects and strategies—to relate theoretical strategies to rich contextual real-world practices; speculative design projects and methods—to put them into the context of existing design practices; and speculative texts dealing with themes of cultural and bio-diversity—to illuminate how strategies can be engaged with through imaginative, narrative and speculative practices. These are quite distinct domains of discourse and practice with different languages, values, and ways of world-making that requires an approach that accommodates perspectives across these different domains.

An initial attempt to create a spreadsheet database of speculative design techniques quickly became an exercise of trying to categorise and hierarchise strategies, leading to the realisation that grouping the techniques into categories was reinforcing the arbitrary demarcations within and between people and nature that this doctoral project aims to challenge. Another realisation was that for any particular design intervention, piece of fiction, or community strategy, there was often a combination of different methodological and strategic approaches in play.

A more promising approach was to use a process of qualitative coding, using qualitative research software with which it is possible to 'code' a text by assigning labels to any part of it, creating a series of 'codes' that can be cross referenced within and between documents. This allows the creation of an evolving set of codes, as well as acting as a way of interrogating and finding connections between different texts. This approach enabled a mapping of strategies that was evolving and co-mingling, as resonances appeared across different research sources: broad worldviews from critical theory suggested further readings from the design that presented examples of those strategies playing out in practice alongside other strategies, which resonates with speculative fiction works in which the themes of the project are present, and which may then present their own strategies that speak to the original theoretical framings. Over time, strategies can be removed, merged, and collected, and a structure can slowly emerge.

The strategies were initially grouped under headings, such as temporal, narrative and relational strategies. Many strategies fall under multiple headings which thus make more sense as aspects that a strategy can have, rather than categories they belong to, so a more appropriate taxonomic structure will need to be devised.

6 Reflections

Reflecting on this research stage, it demonstrates that the diverse and ecological future-making strategies that are described in theory are playing out in practice. What is emerging is a set of inter-related and deeply contextual strategies and principles that are characterised by aspects of **care-taking**, **non-linear temporality**, **relationality**, **storytelling**, and **playfulness** that resonate across the domains of critical theory, community and environmental practices, and speculative fiction and design. A core strategic principle is articulated by Haraway [12] as the necessity for taking care of people, their land, and their stories, and that storytelling engenders an mutual ethical responsibility, and such themes recur across the research domains. To give some concrete examples, The Community of Arran Seabed Trust (COAST) is a community action group who set up a successful ecosystem-protecting 'No Take Zone' that legally prohibits fishing in Lamlash Bay on the Scottish island of Arran [13]. Through tourism outreach that emphasised local stories of community and biodiversity, COAST have implicated tourists with a duty of care by telling them how to recognise fishing boats that have deployed their nets, and have successfully encouraged them to contribute to the community by reporting any boats they see fishing illegally. A similar duty of care within communities is seen in the 'Queer Care' practices in Lisbon's underground queer raves, with parties such as Mina working closely with the harm-reduction charity Kosmicare, collaborating on projects such as Ravelength—a platform to provide online support for community members who feel isolated from the sociality of parties cancelled due to the cornovirus pandemic [14]. Such parties are a good example of the sorts of performative and ephemeral spaces that Munoz suggested are characteristic of queer futuring, by transiently finding better futures within a challenging present. In a final example, from speculative fiction, Le Guin's novel The Telling [15] gives an account of a queer envoy from Earth who is sent to the planet Aka and becomes implicated in the tensions between an oppressive ahistorical technostate, and the traditional adherents of 'The Telling', a collaborative sharing of information about their history, people, and land, in which the storyteller is doctor, priest, and librarian.

The text has thematic resonances with community practices, such as ethical responsibility inculcated by listening to the stories of others, caring for the people and the land through caring for stories, a non-linear temporality, and the experience of non-normative outsiders.

In conclusion, this emerging research strengthens the argument that some of the assumptions from critical theory about diverse and ecologically-based futuring strategies can be evidenced in rich and contextual practices, and that there is a resonance between such practices and speculative and imaginative engagements with these themes. It offers the promise of an emerging framework in which a more diverse, impactful and ecology-centred speculative design practice can be situated.

References

1. Brondizio, E.S., Settele, J., Díaz, S., Ngo, H.T.: Global assessment report on biodiversity and ecosystem services of the Intergovernmental Science-Policy Platform on Biodiversity and Ecosystem Services. IPBES, Bonn, Germany. (2019)
2. Dunne, A., Raby, F.: Speculative Everything: Design, Fiction, and Social Dreaming. MIT Press, Cambridge (2014)
3. Prado de O. Martins, L.: Privelage and Opression: Towards a Feminist Speculative Design. https://a-pare.de/2014/privilege-and-oppression-towards-a-feminist-speculative-design/, Accessed 09 Dec 2018
4. Jeffcott, C., Ferreira, A.M.: Can a diverse and well-disseminated approach to speculative and critical de- sign engage effectively with the complex systems and crises the Anthropocene Era? a reflective study of practice-based design research. In: Doctoral Design Conference'19: TRANSformation Proceedings of the DDC 6th Conference. Edições IADE, Lisbon, Portugal (2019)
5. Guattari, F.: The Three Ecologies, pp. 34–55. Bloomsbury Academic, London (1989)
6. Warner, M., ed: Fear of a Queer Planet: Queer Politics and Social Theory, pp. xxvi. University of Minnesota Press, Minneapolis (1993).
7. Morton, T.: Queer ecology. PMLA **125**, 273–282 (2010)
8. Muñoz, J.E.: Cruising Utopia: The Then and There of Queer Futurity. NYU Press, New York (2009)
9. Godin, D., Zahedi, M.: Aspects of Research through Design: A Literature Review, p. 14 (2014)
10. Savransky, M., Wilkie, A., Rosegarten, M.: Speculative Research : The Lure of Possible Futures, p. 113. Routledge, Abingdon (2017)
11. Gaver, B., Dunne, T., Pacenti, E.: Design: cultural probes. Interactions **6**, 21–29 (1999). https://doi.org/10.1145/291224.291235
12. Haraway, D., Endy, D.: Tools for multispecies futures. J. Des. Sci. (2019). https://doi.org/10.21428/7808da6b.05eca6f1
13. Stewart, B.D., Howarth, L.M., Wood, H., Whiteside, K., Carney, W., Crimmins, É., O'Leary, B.C., Hawkins, J.P., Roberts, C.M.: Marine conservation begins at home: how a local community and protection of a small bay sent waves of change around the UK and beyond. Front. Mar. Sci. **7**, 76 (2020). https://doi.org/10.3389/fmars.2020.00076
14. Marum, P., Nobre V, M., Cunha, M., Rafael, X., Bakr, N.: Harm Reduction and Queer Care (2020). https://youtu.be/MiEVvFkNNSE
15. Le Guin, U.K.: The Telling. Gateway (2000)

Investigating the Measurement of High Reliability Organisations for Health Care Safety

Manikam Pillay(✉), Andrew Enya, and Emmanuel B. Boateng

School of Health Sciences, The University of Newcastle, Callaghan, NSW 2308, Australia
manikam.pillay@newcastle.edu.au,
{andrew.enya, emmanuel.boateng}@uon.edu.au

Abstract. This project investigates the measurement of high reliability organisations for improving health care safety. A growing body of peer-reviewed studies continues to be published on high reliability organisations, demonstrates its recognition and importance for improving safety across high risk contexts such as health care. However, little attention has focused on developing an understanding of how it has conceptualized, and factors that have been used in its measurement. This is a significant gap which can limit its operationalization for research and practice. This paper presents a systematic review project currently underway which seeks to address this gap. After completing a systematic search and selection strategy twenty-one articles were selected for analysis. Results indicate fourteen survey instruments have been used in these studies. Seventeen different definitions of HRO were identified. Nine studies used surveys for data collection. Both independent and outcomes variables were reported, these can be used to inform an initial theoretical framework and a survey instrument.

Keywords: High reliability organisations · Collective mindfulness · Systematic review · Health care safety

1 Introduction

This paper is a work-in-progress and is part of a broader research project on advancing organisational safety through High reliability organisations (HROs) in the health care sector. Personnel working in this sector are always under pressure to provide quality and uniform services for all users, while ensuring safety is guaranteed before, after treatment, during admission, and discharge of all patients. The Australian Institute of Health and Welfare [1] has identified the improvement of safety as a key performance area that the sector needs to focus on in order to reduce preventable adverse outcome and risks to health care staff which pose a significant cost to the Australian health care system [2]. While some improvements have been observed, these have not been sustained. In part this is due to a high reliance on compliance, proceduralisation and standardization. These issues have been previously identified by the Institute of Medicine (IOM) [3] while highlighting that solving problems in the health care

W. Karwowski et al. (Eds.): IHSED 2020, AISC 1269, pp. 283–288, 2021.
https://doi.org/10.1007/978-3-030-58282-1_45

required more than procedures and standards [4]. This was because health care safety was part of a complex, socio-technical system, and the effective management of safety in such a system required embracing more advanced solutions from beyond health care sector. The IOM strongly pointed towards HROs as a framework to embrace.

HROs are able to manage and sustain a nearly-error free operation despite operating in highly complex and uncertain environments, where the consequences of errors can be catastrophic [5]. The term was first introduced in the 1980s by a group of University of California, Berkeley (UCB) researchers as an alternative to Normal Accident Theory (NAT), the basic tenet of which was that accidents were inevitable in tightly-coupled complex technological systems because such systems had become so complex and tightly coupled such that a small event could trigger a series of cascading failures in many parts of the system, leading to an eventual disaster [6]. However, the UCB researchers argued there were some organisations that were equally complex and tightly-coupled but had achieved excellent safety and production goals [7]. These authors argued that organisational accidents in such systems were preventable and could be managed effectively by relying less on structures and routines and more in processes related to 'collective mindfulness'. Traditionally, mindfulness is associated with the mental state of individual who have a heightened alertness to the environment and are able to react accordingly. In HROs, however, this is more of a cognitive mindset that is characterised by preoccupation with failure, reluctance to simplify, sensitivity to operations, commitment to resilience, and deference to expertise [5]. Conceptually, HROs have been associated with the fourth age of safety, with collective mindfulness representing as an advanced, socio-technical approach to safety management [8]. Many health care organisation in the United States have incorporated HRO guidelines into their daily operations through initiatives such as training [9], culture, behaviour and leadership [10], teamwork and communicate [11]. Recent reviews have examined its utility in, apart from healthcare, construction, major hazard facilities [12, 13], and other industries [14], resulting in a rich body of knowledge around its conceptualisation. However, limited attention has been devoted to developing a comprehensive understanding of how it has been defined, the characteristics and instruments that have been used for measuring collective mindfulness. This paper seeks to address this gap and propose a framework that can be used to benchmark and improve safety in the Australian health care sector.

2 Method

A systematic review was used for this research, and the specific methods utilized are described in Pillay, Enya and Boateng [15]. The Preferred Reporting Items for Systematic Reviews and Meta-Analyses (PRISMA) [16] adapted from a previous review [12] were used, with articles searched in five electronic databases (CINAHL, EMBASE, MEDLINE, PsycINFO and SCOPUS). Papers published in English between January 2003 and September 2018 which focussed on HRO in health care and used an instrument were included in the final selection. The quality of the final set of articles were evaluated based a ten-item checklist adapted from Gillman and Pillay [17] and Enya, Pillay and Dempsey [12]. Key information relating to research aims, study design, characteristics measured, and instruments were exported into an Excel spreadsheet.

3 Results

The results of the search and final selection based on the PRISMA flowchart is illustrated in Fig. 1. To summarize, the five electronic databases generated 1424 articles from which 630 duplicates were removed. Title and abstract screening for the remaining 794 led to the exclusion of 712 articles, with 82 eligible for full-text screening. A further 61 were excluded resulting in a final list of 21 articles for synthesis.

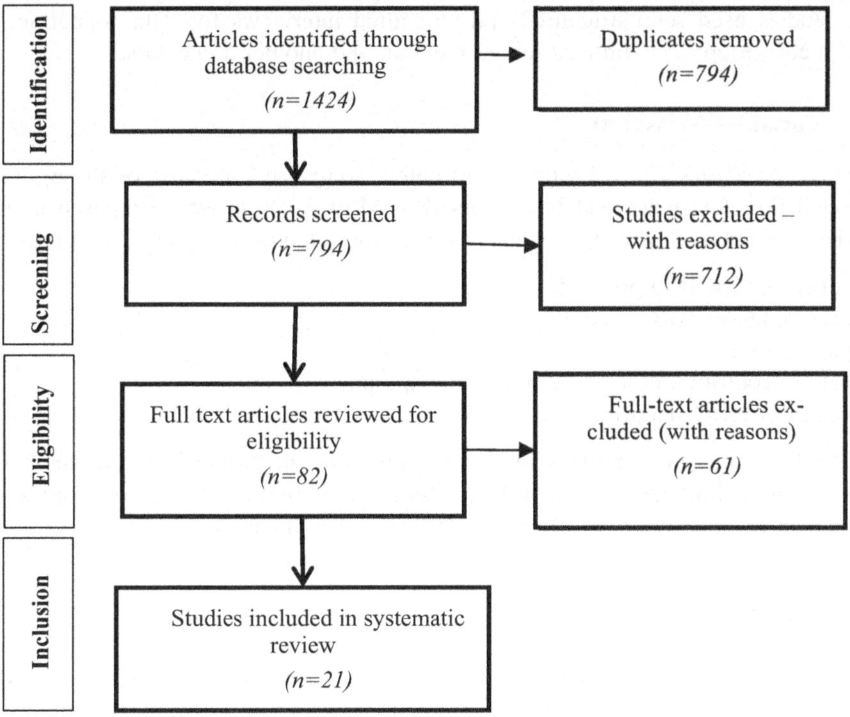

Fig. 1. PRISMA flowchart of results

3.1 General Characteristics of Studies Selected

Quantitative studies of HROs did not receive much attention until 2004, approximately two decades after these were first conceptualized by UBC researchers. The target population investigated included a mix of top management, middle management and front-line practitioners, which covers the socio-technical system of health care. The research settings were mostly hospitals and a few medical academic departments. The sample size for participating medical organisations ranged from 4–200 hospitals and medical facilities, and 15–1685 in general healthcare. Four theoretical frameworks or models were used in these studies.

3.2 Definitions of HROs

Seventeen different definitions of HRO were identified, with most basing it on the works of key proponents of HROs [18–20]. These centered around the management of risks and risk during catastrophes, complexity and hazardous contexts.

3.3 Research Instruments Used for Measuring Collective Mindfulness

Nine studies used surveys for data collection, with the instruments comprised of 9–55 items. The survey instruments used Likert response scales ranging from 3–7-points. Five studies used semi-structured and structured interviews for data collection, two used focus groups, one utilised mixed method, and another a checklist.

3.4 Variables Measured

The five principles for collective mindfulness were used in most of these studies, although in a few these had been expanded. Most of these were proposed as independent variables. Seven studies also reported on outcome variables, including:

- adherence to HRO principles
- knowledge of HRO model
- patient safety
- patient contribution to mindful organizing, and
- stages of HRO maturity

While an in-depth analysis of these is currently in progress, the research team believes these findings can be used to inform a theoretical framework and survey instrument for measuring HRO in the Australian health care sector.

4 Conclusion

This research is work-in-progress so any conclusive statements regarding how HROs have been defined and measured in the literature cannot be made at this stage. However, the preliminary findings suggest that questionnaire surveys are the most common tools used for investigating, so this can be used to conduct benchmarking studies in the field. Some key characteristics for measuring HROs through expanded principles of collective mindfulness have been identified, together with examples of outcome variables they are expected to affect. These findings provide a way forward for developing an appropriate theoretical framework and a supporting survey instrument for conductive evaluations and benchmarking studies. Future work will report on the development, pilot testing and validation of such surveys.

Acknowledgements. This research was supported through an internal University of Newcastle School of Health Science Strategic Pilot Grant Awarded to the first author in 2018.

References

1. Australian Institute of Health and Welfare: Safety and quality in Australian hospitals. Australian Institute of Health and Welfare (2016)
2. Australian Commission on Safety and Quality in Healthcare: The state of patient safety and quality in Australian Hospitals 2019. Australian Commission on Safety and Quality in Healthcare (2019)
3. Institute of Medicine (IOM): To Err is Human: Building a Safer Health System. The National Academies Press (2000)
4. Affare, S.C., Tolk, J., Cantu, J.: Analyzing healthcare tools practitioners use to implement HRO theory. In: IIE Annual Conference and Expo 2015, pp. 493–498 (2015)
5. Hopkins, A.: Learning From High Reliability Organisations. CCH Australia Limited, Sydney (2009)
6. Perrow, C.: Organizing to reduce the vulnerabilities of complexity. J. Contingencies Crisis Manage. **7**, 15–155 (1999)
7. Weick, K.E., Obstfeld, D.: Organizing for high reliability: process of collective mindfulness. In: Boin, A. (ed.) Crisis Management, vol. III, pp. 31–65. JAI Press, Los Angeles (2008)
8. Pillay, M.: Accident causation, prevention and safety management: a review of the state-of-the-art. Procedia Manuf. **3**, 1838–1845 (2015)
9. Riley, W., Lownik, E., Parrotta, C., Miller, K., Davis, S.: Creating high reliability teams in healthcare through in situ simulation training. Administrative Sciences **1**, 14–31 (2011)
10. Frankel, A.S., Leonard, M.W., Denham, C.R.: Fair and just culture, team behaviour, and leadership engagement: the tools to achieve high reliability. Health Res. Educ. Trust **41**, 1690–1709 (2006)
11. Yip, L., Farmer, B.: High reliability organizations - medication safety. J. Med. Toxicol. **11**, 257–261 (2015)
12. Enya, A., Pillay, M., Dempsey, S.: A systematic review on High reliability organisational theory as a safety management strategy in construction. Safety **4**, 6 (2018)
13. Lekka, C.: High reliability organisations: a review of the literature. Health and Safety Executive, London (2011)
14. Cantu, J., Tolk, J.N., Fritt, S., Gharehyakeh, A.: High reliability organization (HRO) systematic literature review: discovery of culture as a foundational hallmark. J. Contingencies Crisis Manage. (2020)
15. Pillay, M., Enya, A., Boateng, E.B.: High reliability organisations and collective mindfulness for improving healthcare safety management: a scoping review protocol of factors, measures and instruments. Int. J. Occup. Environ. Saf. **3**, 8–13 (2019)
16. Liberati, A., Altman, D.G., Tetzlaff, J., Mulrow, C.D., Gotzsche, P.C., Ioannidis, J.P.A., Clarke, M., Devereaux, P.J., Kleijnen, J., Mohler, D.: The PRISMA statement for reporting systematic reviews and meta-analyses of studies that evaluate health care interventions: explanation and elaboration. PIOS Medicine **6**, W1–W30 (2009)
17. Gillman, M., Pillay, M.: An integrative literature review: what are the barriers that stop organisations from learning the lessons highlighted in serious incident investigations? In: AHFE 2017 International Conference on Safety Management and Human Factors, 17–21 July, pp. 627–634. Springer (2017)

18. Roberts, K.H.: Some characteristics of high-reliability organizations. Organization Science **1**, 160–177 (1990)
19. Schulman, P.R.: The analysis of high reliability organizations: a comparative framework. In: Roberts, K.H. (ed.) New Challenges to Understanding Organisations, pp. 33–53 (1993)
20. Weick, K., Sutcliffe, K.: Managing the Unexpected: Resilient Performance in an Age of Uncertainty. Wiley, New York (2007)

Investigating Resilience Engineering Through Safe Work Method Statements in Residential Construction

Manikam Pillay[1(✉)] and Michael Tuck[2]

[1] School of Health Sciences, The University of Newcastle, Callaghan, NSW 2308, Australia
manikam.pillay@newcastle.edu.au
[2] School of Engineering, Information Technology and Physical Sciences, Federation University, Ballarat, VIC 3353, Australia
m.tuck@federation.edu.au

Abstract. This paper reports on an investigation of resilience engineering as a construction safety strategy through safe work method statements. Safe work method statements are a regulated strategy for controlling health and safety risks in Australia. However, there is little research published on safe work method statements, so the specific role they play in construction are unknown. Previous studies have suggested similar approaches, such as safety rules and/procedures were not followed to the latter but adapted to suit the local context of work and environment, creating gaps between the two. Such gaps are also important in Resilience engineering, a new paradigm for safety management. While RE is attention in construction safety, the specific links between safe work method statements and resilience engineering have not been investigated, so it is not clear whether safe work method statements enhance or hinder resilience engineering. This study seeks to address this through a case study in a residential construction project.

Keywords: Resilience engineering · Safe work method statements · Construction safety · Case study

1 Introduction

This paper is a part of a broader research project on advancing organisational safety through Resilience engineering (RE) in the Australian construction sector. Like most countries, construction continues to be singled out for its poor safety performance in Australia. The industry currently employs 9% of the workforce but is responsible for more than 11% of the workers compensation costs and 16% worker fatalities [1]. The industry also ranked third highest for both the number of fatalities and workers compensation claims for serious injuries in the last five years. What is concerning is that workers continue to be affected by the same type of health and safety hazards they were exposed to a decade ago. These include, for example, falls from heights, being hit by falling objects, and impacted by vehicles [1]. This is more pronounced in residential construction where work environment changes constantly, work crews are relatively

W. Karwowski et al. (Eds.): IHSED 2020, AISC 1269, pp. 289–294, 2021.
https://doi.org/10.1007/978-3-030-58282-1_46

smaller and comprised of young workers or migrants, on-site safety advisors are rare and safety innovations lag behind commercial construction projects [2]. Over two and half decades ago it was observed that the manner in which construction safety was managed had not changed [3]. Recent studies suggest this continues to be the case, with the industry continuing to rely on contemporary strategies to manage construction safety risks. As authors such as Wachter and Yorio [4] note, such strategies become institutionalised through policies, plans and procedures; and generally not able to be adjusted to inevitable changes in work, the environment in which such work is conducted, or any emergent risks that may be encountered.

One contemporary approach includes safe work method statements (SWMS) that was introduced in Australia under the harmonized construction safety regulations [5]. However, there is a paucity of research on SWMS, so the role they play in construction safety are largely unknown [6]. What is known is that similar approaches, such as safety rules and/procedures were not followed to the latter but adapted to suit the local context of work and environment, creating gaps between the two [7]. Contemporary safety management generally treats these gaps as violations. Advanced approaches such as Resilience engineering (RE), however, treat such adaptations as human variability and part of normal human performance, and a necessary requirement for achieving safe performance by recognising, adapting and absorbing variations and changes [8]. Accordingly the gap between work-as-imagined (WAI) and work as done (WAD) is an important facet of RE [9, 10], which is gaining some attention in construction safety [11]. The specific links between SWMS and RE, however, have not been investigated [6, 12]. This study reports on the findings of a study which sought to address this gap on a residential housing project. It specifically focuses on the WAI aspect of SWMS from the perspective of managers.

2 Theoretical Framework

Fostering RE involves developing an understanding of how work practices evolve in normal work settings, as opposed to how it was imagined (or prescribed) [9]. Nathanael and Marmaras [13] argued that the transformation of prescriptions to work practices involved a series of four interconnected loops which could be decomposed at two broad levels. This is illustrated in the modified prescriptions-repetitions-distinctions-descriptions (MPRDD) model in Fig. 1.

The top level included the prescriptions loop which represented an organization's goals and intentions as expressed in its policies, standards, procedures and/or work instructions, and communicated downwards as assigned responsibilities, specific objectives norms, standard operating procedures, task and work descriptions [13]. In Fig. 1 this is conceptualized as work-as-imagined (WAI).

The bottom level is about the actual practice, which evolved through a series of double loops, each encapsulated inside the other and unfolding at different times [13]. The repetitions (R) loop involves the development of 'safe work habits' via re-enactment and reinforcement during normal operations; while the distinctions (D) loop their situated challenging following a breakdown where members distinguished new ways of acting [13]. The authors suggested some of these also triggered reflection-in-action

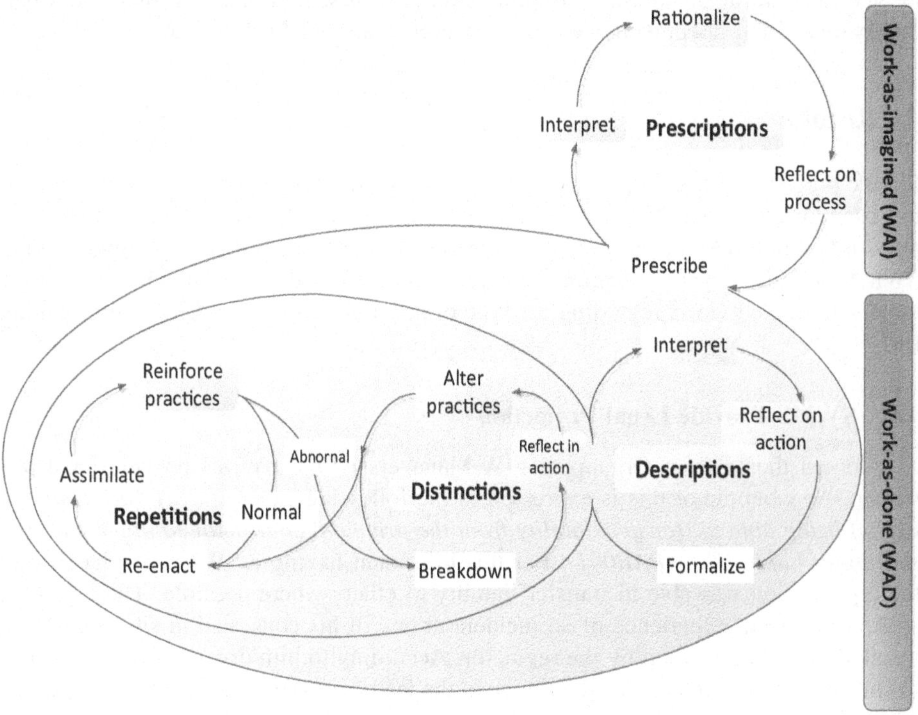

Fig. 1. Modified prescriptions-repetitions-distinctions-descriptions model (M-PRDD)

in the groups, leading to modified practices which were subsequently immersed into the prescriptions cycle. The descriptions (D) loop provided an opportunity for detached reflection in instances where workgroups acted not only in the present but also placed their identity outside of their evolving work experience and reflected upon it. This reflection-on-action was generally disconnected from experience and allowed groups to generate their own interpretations which were formalised through dialogue and discussions [13]. Successful adaptation depended on an organization's ability to provide a stable foundation for their interpretation and formalisation. In Fig. 1 this is conceptualised as WAD.

The MPRDD model acted as the theoretical framework for collecting and analysing data on the prescription and practice of SWMS.

3 Methodology

A case study was used for this project, a comprehensive research strategy where data is collected over a period through a combination of methods in order to illuminate the phenomena being investigated [14]. The research setting, which is identified in this paper as Organisation A, included a medium-density residential construction project located in Victoria, Australia. A triangulation [15] of semi-structured interviews, field observations, and documents were used to collect and analyze the data.

4 Results

For the purposes of this paper, the results of interviews with managers is reported, with the pseudonyms PAR### used to indicate participant number. These represent the prescription of SWMS according. The five themes that arose at this level suggested that SWMS provided legal protection, involved a process, had a role in safety. However, there were mixed views regarding the type of construction work SWMS were required for.

4.1 SWMS Provide Legal Protection

A common theme from Organisation A Managers is that SWMS provide legal protection. An example of this is expressed in the following excerpt: *"...I find them as a way of being able to transfer liability from the principal contractor to other parties... shifting of liability"... (PAR001).* For this informant having a SWMS in place meant the organisation was able to transfer liability to others where possible. The informant spoke of a recent experience of an incident at one of his construction sites which had resulted in an intervention by the regulator. According to him organisation A was able to shift the liability for the incident back to the lead contractor, in this case a concreter: *"It had basically gone from a builder's point of view, from I'd done the paperwork, and then the liability went to the concreter for taking unnecessary works outside his scope. So now any of the liability that would have come to use from the Regulator has now gone straight across to the concreter..." (PAR001).*

4.2 SWMS Involve a Process for Driving Efficiency

A second theme from Organisation A managers is that SWMS establish a process; an example of this is expressed by the following excerpt: *"Well, I think they are quite good because they set the process...." (PAR003).* The informant elaborated that *"... instead of wily-nily turning up to a job and just going for it, but if there's a set process that guys can work through.... That it's actually a quicker way to do it by these devised processes.* The second part suggested this process was about achieving the work faster i.e. driving efficiencies by following a set process.

4.3 Work Contexts for SWMS

A third theme related to the types of construction work that warranted a SWMS. There were two different views in this regard. The first was that these were required for some types of construction, as ex-pressed in the following: *"Well, obviously we've got regulatory requirements to make sure we've got it for all high risk works"... (PAR001).* However, another suggested these were required for nearly all types of construction:

"So you'd almost say that our generic Safe Work Method Statements that we have would apply to 95% of the works that we do."... (PAR0003).

4.4 SWMS Have a Role in Safety

A fourth themes related to SWMS and safety, generally expressed in the following way: "*Obviously it's to ensure the safety of the guys on site as well. That's a primary function of it from that point of view...*" *(PAR004).* Another provided some context around this *"...it's to keep people safe and to prevent them from completing activities in the incorrect manner that could possibly increase their risk of personal injury"... (PAR004).* In this the suggestion is safety can be achieved through working correctly i.e. procedure following.

5 Discussion

This study is part of a larger study aimed at investigating the links between SWMS and RE. A M-PRDD model was developed and used as the theoretical framework to investigate the prescription of SWMS in a residential construction project. These preliminary findings from semi-structured interviews with managers suggested SWMS provided legal protection, involved a process and aimed at achieving efficiency by following a set process. These finding are, to some extent, at odds with a previous findings from regulators [6], which suggested they were a safe system of work, a live strategy for controlling risks, a cognitive artefact, and a tool for social interactions. There were two different views regarding the type of construction work for which SWMS were required. The first is that these are required for some work, while the second suggesting required for nearly all work. The view that they were required for some types of construction is consistent with previous views of the regulators; while the view they are required for most types of construction resonates with those of industry association [6]. The view that SWSM contributed to safety by rule following also was different to previous findings from regulators, who suggested that their use as a safe system of work and social interactions were more important.

Future work will report on the findings from semi-structured interviews with supervisors and workers, observations on the use of SWMS used in a range of construction activities, and analysis of documents.

References

1. Safe Work Australia: Priority industry snapshot: Construction. Safe Work Australia (2018)
2. Kaskutas, V., Dale, A.M., Nolan, J., Patterson, D., Lipscomb, H.J., Evanoff, B.: Fall hazard control observed on residential construction sites. Am. J. Ind. Med. **52**, 491–499 (2009)
3. Brown, P.E.: Total integration of safety professional into project management. In: 1st International Conference of CIB Working Commission, vol. 99, pp. 137–144. International Council for Research in Building and Construction (1996)

4. Wachter, J.K., Yorio, P.L.: A system of safety management practices and worker engagement for reducing and preventing accidents: an empirical and theoretical investigation. Accid. Anal. Prev. **68**, 117–130 (2014)
5. Pillay, M.: Harmonisation of construction health and safety laws in Australia. In: CIBWO 99 Safety and Health in Construction, pp. 97–109. CIB (2013)
6. Pillay, M.: Resilience engineering, gaps and prescription of safe work method statements part 1: the view of organisational outsiders. In: Advances in Safety Management and Human Factors, pp. 261–272. Springer International Publishing, Switzerland (2016)
7. Dekker, S.W.A.: Failure to adapt or adaptations that fail: contrasting models on procedures and safety. Appl. Ergon. **34**, 233–238 (2003)
8. Woods, D.D., Hollnagel, E.: Prologue: resilience engineering concepts. In: Hollnagel, E., Woods, D.D., Leveson, N. (eds.) Resilience Engineering: Concepts and Precepts, pp. 1–6. Ashgate Publishing, Aldershot (2006)
9. Nemeth, C.P.: Resilience engineering: the birth of a notion. In: Hollnagel, E., Nemeth, C.P., Dekker, S. (eds.) Resilience Engineering Perspectives, Volume 1: Remaining Sensitive to the Possibility of Failure, pp. 3–9. Ashgate Publishing, Aldershot (2008)
10. Dekker, S.W.A.: Resilience engineering: chronicling the emergence of confused consensus. In: Hollnagel, E., Woods, D.D., Leveson, N. (eds.) Resilience Engineering Concepts and Precepts, pp. 77–92. Ashgate Publishing Ltd., Aldershot (2006)
11. Hollnagel, E.: Resilience engineering and the built environment. Build. Res. Inf. **42**, 221–228 (2014)
12. Pillay, M., Borys, D.: Towards an exploration of safe work method statements as a safety management strategy in construction: background and theory. In: Seventh International Conference on Construction in the 21st Century (CITC-VII) (2013)
13. Nathanael, D., Marmaras, N.: On the development of work practices: a constructivist model. Theor. Issues Ergon. Sci. **9**, 359–382 (2008)
14. Yin, R.K.: The case study as a serious research strategy. Knowl. Creation Diffus. Utilization **3**, 97–114 (1981)
15. Jack, E.P., Raturi, A.S.: Lessons learned from methodological triangulation in management research. Manag. Res. News **29**, 345–357 (2006)

Characterizing High-Speed Serial Transceivers for a Multi-processor Parallel Architecture

Mireya Zapata[1(✉)], Bernardo Vallejo-Mancero[1,2], and Liliana Topon-Visarrea[1]

[1] Research Center of Mechatronics and Interactive Systems - MIST, Universidad Tecnológica Indoamérica, Machala y Sabanilla, Quito, Ecuador
{mireyazapata, blancatopon}@uti.edu.ec
[2] Instituto Superior Tecnológico Central Técnico, Av. Isaac Albéniz E4-15 y El Morlán, Quito, Ecuador
bvallejo@istct.edu.ec

Abstract. Emulation of large scale neural networks is a growing research field that tries to understand how the brain works. Different approaches based on hardware and software have been developed for this purpose. However, in this paper, we focus on dedicated parallel hardware implemented with FPGA. In this context, brain connectivity is one of the biggest challenges to overcome for neuromorphic circuits. To establish an efficient communication link in multi-FPGA architectures, high-speed serial transceivers GTX are an excellent alternative. Through hardware tests with Kintex 7 and Zynq ZC706 platforms, we compare eye pattern, and BER results, in order to obtain the optimal line rate to establish communication between both boards. The maximum transmission speed achieved without signal degradation was 5 Gpbs.

Keywords: GTX transceivers · Aurora 8b/10b · Parallel hardware · FPGA · Zynq

1 Introduction

The growing interest in the development of neuromorphic applications with high biological realism has boosted the search for solutions through hardware and software. Software-based systems have the advantage of supporting a high degree of programmability. However, given its inherently sequential processing, the increase in delay is proportional to the size of the network, affecting runtime and drifting far from real-time. Simulators such as BRIAN [1], NEURON [2] are some of the most used for applications that require few neurons. In turn, hardware developments characterized by parallel processing can offer execution times between 2 and 3 orders of magnitude faster than their counterpart in software for high-scale networks. The proposed architectures are in the analog and digital field. For the present case study, we focus on FPGA-based digital architectures. This technology allows parallelism, high flexibility, and programmability at a low cost of implementation. Besides, they have IP cores to speed up the description of the hardware. Architectures like NeuroFlow [3] and Panni

W. Karwowski et al. (Eds.): IHSED 2020, AISC 1269, pp. 295–300, 2021.
https://doi.org/10.1007/978-3-030-58282-1_47

[4] based on this technology are capable of supporting networks of 98304 and 1440 neurons per core, respectively, staying within real-time limits.

In this context, neural connectivity is one of the determining factors that has prevented hardware from reaching biological scales. It is due to the number of synaptic connections required to connect each neuron. There are an estimated one hundred billion neurons in the brain of an adult, and each neuron typically has 1,000 to 10,000 synapses [5].

Despite its significant advance, electronic technology presents a fan in and fan out well below what is required to achieve a biological scale. Thus, dedicated connections to each neuron are unfeasible. As an alternative, digital systems take advantage of multiplexing techniques to improve the transmission capacity per unit of time, with a latency penalty that is proportional to the number of neurons. The bandwidth of the communication channel is one of the factors that limit the growth of a neural network. To transmit the events generated by a large-scale network, a channel with high throughput is required, which is restricted by the limitations of the technology.

The AER (Address Event Representation) asynchronous protocol proposed by Mahowald and Silvolotty [6] is one of the most used to model a large number of spiking-type neuronal interconnections encoding the direction of the presynaptic neuron. This is an alternative one-way point-to-point communication of spike events between neural arrays. An event corresponds to the code belonging to the firing neuron. Among its limitations is the information loss caused by collisions when several neurons fire simultaneously on a multiplexed channel. Variations of the original protocol, such as the one proposed in [7] handle synchronous pipeline communication with the advantage of being free of collisions. It improves scalability and transmission efficiency when using high-speed serial connections implemented with GTX transceivers available on high-end FPGA like Kintex 7 and Zynq-7 ZC706. In the proposed manuscript, a performance study of different transmission rates is proposed, as well as the quantification of the bit error rate (BER) between the Zynq ZC706 and Kintex 7 platforms by obtaining eye diagrams.

2 Multiprocessor Architecture

The development of hardware neuromorphic applications on embedded systems requires efficient systems to implement its configuration and learning process, especially when addressing large-scale systems. However, due to a large amount of information that must be processed and sent to each dedicated processing unit responsible for computing neural and synaptic algorithms, these tasks are often performed off-line. In this sense, communication from and to the outside to carry out configuration, monitoring, and learning tasks must be managed efficiently, so that a bottleneck does not occur and ends up slowing down the system in general.

The neuromorphic hardware on which this proposal is developed corresponds to a scalable architecture that emulates spiking neural networks based on FPGA [8]. In order to increase its scalability this architecture allows multi-chip communication in a ring topology between Zynq ZC706 and Kintex 7 FPGA platforms. During its execution cycle, after the neural processing is done the spike distribution takes place. During this phase the communication is carried out on the basis of the Xilinx Aurora protocol, which allows point-to-point communication between nodes using high-speed serial lines, supported by GTX transceivers. In this context, to ensure the establishment of a secure and efficient communication link while maintaining the real-time architecture operation, the integrity of the data on the serial link must be guaranteed at the maximum transmission rate, and a minimal bit error rate.

2.1 GTX Transceivers

GTX transceivers are embedded on Xilinx Kintex 7 FPGA and Zynq-7 ZC706 board. GTX are configurable hard-tiles with common components like Phase Locked Loop (PLL) and reset logic. These have the advantage of maximizing the data flow, reaching speeds from 500 Mbps up to 12.5 Gbps for the GTX family. They have an efficient consumption of power and resources [9]. Its function is to collect parallel data generated to an fclk1, serialize it to an fclk2 to be transmitted in a bitstream over a serial link to the receiving side where it is deserialized for being treated again as parallel data to fclk1. The transceivers are located in Quads equivalent to a group of 4 transceivers. In the case of the ZC706 [10], there are 16 GTX Quad that distinguishes each transceiver by their X and Y coordinates. The Quad requires a differential clock reference, independent of the clock that controls the rest of the programmable logic. This signal can be obtained from Si5324 jitter attenuator programmable chip.

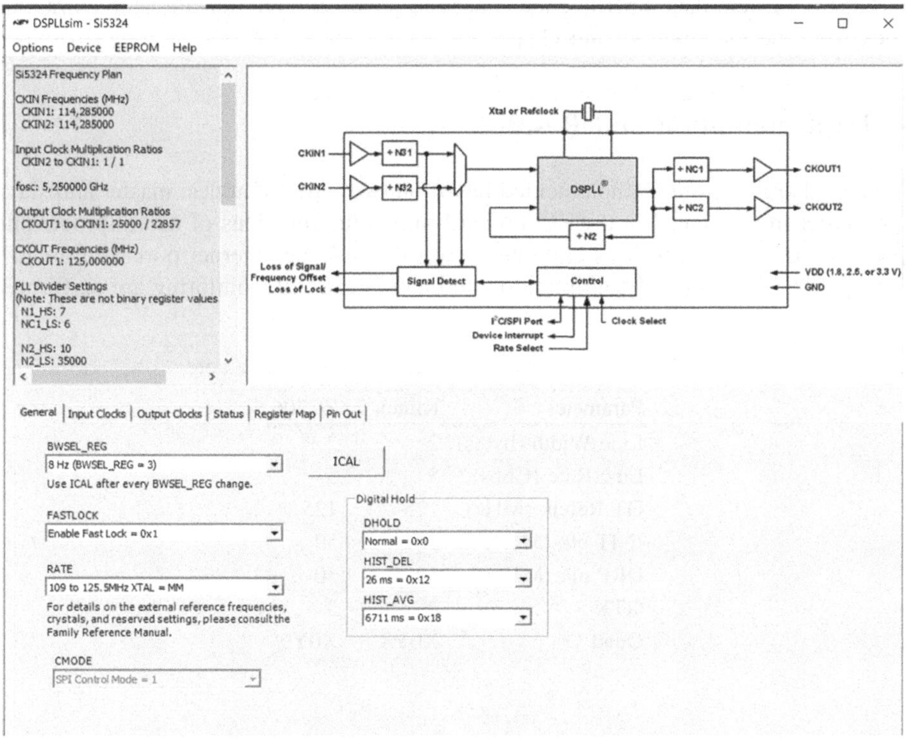

Fig. 1. DSPLLsim programming interface

Si5324 is a high precision circuit used to generate clock signals with ultra-low delay jitter. It is a configurable 36-pin integrated circuit via SPI or I2C communication. This clock circuit can generate any frequency within the range of 2 kHz to 945 MHz and select frequencies at 1.4 GHz from an input frequency of 2 kHz at 710 MHz. It can provide any frequency combination across its operating range.

Zynq devices, which include an ARM microcontroller-based processing system (PS) can program the Si5324 controller through the I2C interface. In order to generate the required frequency value, the controller has to be initialized by the PS at boot time by writing the controller configuration registers used for modifying the frequency.

DSPLLsim (see Fig. 1) is a useful software tool used to determine the value that each register should have (143 in total). The application allows that according to the desired frequency and other specifications, to generate a map of register with the corresponding values. The Zynq PS is programmed through an application created using the Xilinx SDK in C language, enabling I2C communication and including the registers' configuration values of the Si5324 controller.

2.2 Aurora Core

Xilinx Aurora 8b/10b IP, it is a serial lightweight protocol used as point-to-point communication which includes error detection and recovery data. The bit stream contains the protocol encapsulated to differentiate between data/control words and start/end of current data. 8b/10b code is used to obtain enough transitions to extract the clock from the incoming stream [11].

3 Implementation and Results

The neural architecture is implemented based on two types of nodes: master and slave. The master implemented in the ZC706 card fulfills the functions of neural processing, spike transmission, as well as communication through the ethernet port with an API. Through this, functions of configuration and neural activity monitoring are performed.

Table 1. Configuration parameters to set Aurora core

Parameter	Kintex 7	ZC706
Line Width (bytes)	2	2
Line Rate (Gbps)	5	5
GT Refclk (MHz)	125	125
INIT clk (MHz)	50	50
DRP clk (MHz)	50	50
GTX	2	2
Quad	X0Y8	X0Y9

The slave implemented on Kintex 7 cards, performs neural processing and spike transmission to the rest of the network nodes. Communication between master and slave is done based on GTX transceivers connected to SMA terminals. To determine the maximum transmission rate and quantify the error rate (BER) between both platforms, the Xilinx core iBERT IP was used to stablish the functionality and get information about the transceiver channels status and its capability. Tests were performed at 3.25, 5, and 6 Gbps obtaining the eye diagram in each case. The highest transmission rate at which the channel does not degrade was 5 Gbps (see Fig. 2) with a BER of 1.57E-12. For the 6 Gbps case, a completely closed eye diagram was obtained, which suggests a high probability of transmission errors occurring.

Once the line rate was obtained, the Aurora protocol was implemented as a communication protocol for data serialization and deserialization. Table 1 shows the core configuration parameters for both platforms.

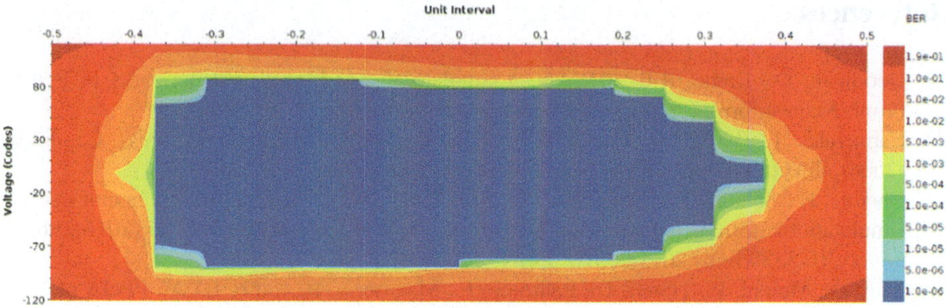

Fig. 2. Eye diagram for ZC706 MGT_X0Y9 5Gbps.

Figure 3 shows hardware result in board ZC706 from Aurora Protocol. It can be seen that the channel is up and transmission of data is performed without error at 5 Gbps rate. To physically connect the two training cards to their respective GTX channels, coaxial cable was used. A pair of SMA cables were necessary for the cross-connection between the transmitter and receiver since the signals are differential for each transceiver.

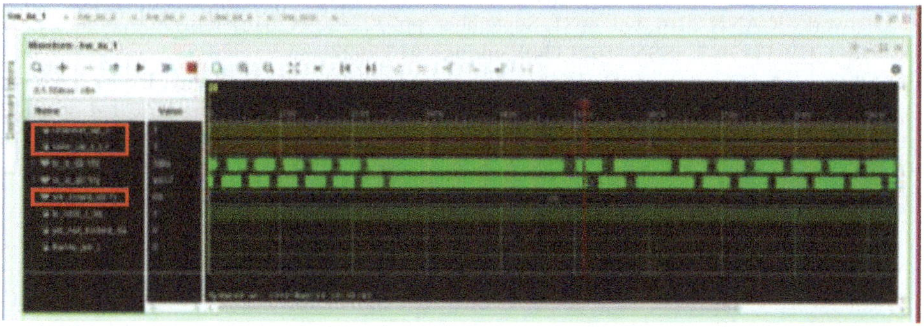

Fig. 3. ILA core, hardware implementation result with ZC706 board for Aurora Protocol.

4 Conclusions

The limitations in connectivity and bandwidth imposed by the traffic generated by spike events in high-scale networks straiten the scalability of the network and even more so its ability to keep its operation within real-time limits. In hardware neuromorphic architectures, the use of high-speed serial transceivers combined with highlight protocols such as Aurora allows an alternative to solve the bottleneck in data transmission, given the low overhead introduced by control packages.

Tests carried out on hardware with Kintex, and ZC706 cards determined that the maximum line rate achieved with GTX transceivers through SMA connectors and using shielded coaxial cable for the physical interconnection, with an acceptable BER is 5Gbps. This data constitutes a reference point for future applications in any field.

References

1. Goodman, D., Brette, R.B.: BRIAN: a simulator for spiking neural networks in Python. Front. Neuroinform. (2008). https://doi.org/10.3389/neuro.11.005.2008
2. Carnevale, N., Hines, M.: The NEURON Book. Cambridge University Press, Cambridge (2006). direcci´on: http://www.neuron.yale.edu/neuron/docs
3. Cheung, K., Schultz, S.R., Luk, W.: NeuroFlow: a general purpose spiking neural network simulation platform using customizable processors. Front. Neurosci. **9**, 1–15 (2016). https://doi.org/10.3389/fnins.2015.00516. ISSN: 1662453X
4. Pani, D., Meloni, P., Tuveri, G., Palumbo, F., Massobrio, P., Raffo, L.: An FPGA platform for real-time simulation of spiking neuronal networks. Front. Neurosci. **11**, February 2017. https://doi.org/10.3389/fnins.2017.00090, ISSN: 1662-453X
5. Cattell, R., Parker, A.: Challenges for brain emulation: why is building a brain so difficult? Natural Intelligence, Feb. de 2012
6. Mahowald, M.: VLSI Analogs of Neuronal Visual Processing: What does the retina know about a Synthesis of Form and Function. Tesis doct., California Institute of Technology y Pasadena (1992)
7. Dorta, T., Zapata, M., Madrenas, J.: AER-SRT: scalable spike distribution by means of synchronous serial ring topology address event representation. Neurocomputing **171**, 1684–1690 (2016). https://doi.org/10.1016/j.neucom.2015.07.080
8. Zapata, M., Madrenas, J.: Synfire chain emulation by means of flexible SNN modeling on a simd multicore architecture", English. In: 25th International Conference on Artificial Neural Networks ICANN 2016, vol. 8681, September, pp. 222–229 (2016). https://doi.org/10.1007/978-3-319-11179-7, ISSN: 16113349, eprint: 1412.7927
9. Athavale, A., Christensen, C.: High-speed serial I/O made simple. Xilinx, Inc., 1.0 edition, April 2005. http://www.xilinx.com/publications/archives/books/serialio.pdf. Cited on pages 9, 12, and 13
10. Design, Z.: Zynq-7000 All Programmable SoC, vol. 585, pp. 1–1836 (2014)
11. Tang, Z.: High speed IO using Xilinx Aurora. Electr. Eng. **54**(1), 13–34 (2010)

Ergonomics of Firefighting Protective Clothing: A Short Review on Fit and Sizing Issues

Anna S. P. Moraes[1(✉)], Miguel A. F. Carvalho[1], Rachel S. Boldt[1], Fernando B. N. Ferreira[1], Fernando M. Duarte[2], Susan P. Ashdown[3], and Linsey Griffin[4]

[1] Department of Textile Engineering, University of Minho, Campus Azurem, Guimaraes, Portugal
sophiapiacenza@gmail.com
[2] Department of Polymer Engineering, University of Minho, Campus Azurem, Guimaraes, Portugal
[3] Department of Textiles and Apparel, Cornell University, Ithaca, USA
[4] Department of Design, Housing, and Apparel, University of Minnesota, Minneapolis, USA

Abstract. Firefighting is considered one of the most challenging professions. Personal protective equipment plays a crucial role in protecting firefighters against numerous occupational hazards. Despite the advances in materials technology, current firefighting personal protective clothing is considered heavy, bulk, stiff, and thick. Many studies evaluated firefighters' perceptions of current equipment. Fit and sizing are frequently reported as issues in firefighting protective clothing. This paper provides a literature review with the objective of offering a better understanding of how fit and sizing of firefighting protective clothing have been addressed by researchers. A review of the main issues faced by firefighters as well as a synthesis of suggestions for improved design and better purchasing as reported in the literature are provided.

Keywords: Comfort · Firefighter · Performance · Personal protective equipment · Protection

1 Introduction

Firefighting is considered one of the most dangerous, arduous, and physically demanding occupations [1–3]. In their daily duties, firefighters face numerous occupational hazards that can result in injuries and illnesses.

W. Karwowski et al. (Eds.): IHSED 2020, AISC 1269, pp. 301–307, 2021.
https://doi.org/10.1007/978-3-030-58282-1_48

Firefighters are required to wear different personal protective equipment according to the type of mission and environmental conditions[1]. When fighting structural fires, for example, firefighters wear protective clothing (e.g. coverall or turnout coat and trousers) and other accessory items (e.g. helmet, hood, face mask, self-contained breathing apparatus, gloves, footwear, and harness).

Ergonomics of protective equipment is a multifaceted subject involving comfort, protection, and performance. In the case of firefighting PPC, Ergonomics is being addressed from many perspectives: thermal protection, effects of load carriage, freedom of movement, functional balance, visibility and conspicuity, and so on.

1.1 Remaining Issues in Firefighter Personal Protective Clothing

Firefighter PPC has evolved over the years. Great efforts have been applied to the investigation of new material, textiles, and finishing processes. Nonetheless, many obstacles remain despite functional improvements.

Studies have pointed out that higher thermal protection obtained due to the advances in materials technology has greatly decreased burn injuries yet inevitably increased the weight and bulkiness of firefighters' gear [4]. Current firefighter protective clothing is considered heavy, cumbersome, stiff, and thick. Nonetheless, fit and sizing remain critical challenges when designing firefighting protective clothing so as to enhance firefighters' comfort and performance.

This paper provides a literature review with the objective of offering a better understanding of how fit and sizing of firefighters' protective clothing have been addressed by researchers. A review of fit and sizing issues faced by firefighters when wearing turnout gear reported in current literature as well as a synthesis of suggestions from a range of authors for an improved design and better purchasing and selection is presented. Nonetheless, a short description of the main characteristics of personal protective clothing for structural and proximity firefighting[2] is presented as follows.

1.2 Characteristics of Structural Firefighter Personal Protective Clothing

Firefighting PPC is primarily designed to protect firefighters' upper and lower torso, neck, legs, and arms and is currently available in different styles and materials. However, firefighting PPC can be individualized only to a certain extent, through choosing different styles and types of pockets, trim, snaps, rivets, and closures [1].

[1] The personal protective equipment worn in structural and proximity firefighting, for example, is different from the equipment worn in wildland firefighting. A definition of structural and proximity firefighting and wildland firefighting is available in the *NFPA 1971 - Standard on Protective Ensembles for Structural Fire Fighting and Proximity Fire Fighting* and in the *NFPA 1977 - Standard on Protective Clothing and Equipment for Wildland Fire Fighting*, respectively.

[2] In this paper, we focus on Structural and Proximity Firefighting Protective Clothing, and more precisely, on coat and trousers, which will be referred as PPC from now on.

Commercially available firefighting PPC should meet design requirements and performance requirements set by standard organizations[3]. Firefighting PPC must comprise three layers: an outer shell, a moisture barrier, and a thermal barrier. An additional internal comfort layer may also be included [5, 6].

Turnout coat length usually goes down to hips and must overlap the pants to provide adequate protection. The coat may have a straight hem or a tailed back. A high collar to protect the neck region is required. Sleeves must be long and have a protective wristlet or another interface component to cover and provide protection to the wrist area. The wristlet may also contain a loop strap to be attached to the thumb to prevent the sleeves from going up and consequent skin exposure.

Turnout trousers usually have a high waist to protect the abdomen region. Suspenders are attached to turnout trousers in order to stand up against the weight added to the pockets or when trousers become wet. Reinforcement pads may be included in the knee area. Additionally, garment shall have fluorescent and retroreflective trim permanently attached to the outer shell to meet visibility requirements.

Concerning sizing systems, the NFPA 1971 standard provides a table with size ranges for both men and women as well as the sizing increments for 4 dimensions: chest, sleeve, waist, and trousers inseam [7].

2 Methods

Research papers addressing fit and sizing of firefighters PPC were selected for the present review. An initial search was made in the Scopus subscribed database. The search criteria included the following keywords and operators: ('clothing' OR 'gear' OR 'ensemble' OR 'suit') AND ('firefighter' OR 'firefighting') AND ('fit' OR 'fitting') AND ('size' OR 'sizing'), appearing in the article title, abstract, or keywords. Secondly, references cited within all relevant retrieved papers were examined for finding additional papers. Additional searches in open-access databases were randomly performed aiming to find further relevant references.

3 Results and Discussion

Current literature in firefighting PPC has been addressing fit and sizing of firefighting PPC from different perspectives. A fair amount of research studied the impacts of wearing heavy and bulky turnout gear in firefighters' mobility and movement restriction [2, 3, 8, 9]. Likewise, the relevance of PPC fit and sizing in protection is being demonstrated, which is causally related to the volume of air gaps in between the clothing and the wearer [10, 11].

Several studies addressed gender differences, pointing out that female firefighters commonly wear PPC designed for men. Regarding this aspect, authors highlight that such clothing is incorrectly sized and ill-fitting for the female body sizes and proportions

[3] Such as the U.S. National Fire Protection Association (NFPA), European Standards Committees (EN), and the International Organization for Standardization (ISO).

[12–14]. Furthermore, fit and sizing of PPC still appear as common complaints among firefighters, impacting their comfort and performance [3–6, 8, 13–16].

3.1 Structural Firefighter Personal Protective Clothing Fit Issues

Firefighter gear must fit properly to perform at its highest level [1]. The fit of PPC is directly related to the freedom of movement and comfort of firefighters and it is especially critical when performing movements. In this regard, results of the Park & Hahn [15] study show that the level of satisfaction of the participants with the fit of PPC was lower evaluated in active body movements than in simple walking. More recently, similar results were found by McQuerry [8].

Concerning fit issues in specific areas of the turnout coat, the chest region received the lowest scores by participants of the Park & Hahn [15] study. Gender differences were significant, evidencing this as one of the main areas of discomfort for female firefighters [15]. As reported in Lee et al. [16], discomfort in chest area for female firefighters is increased by the chest straps of the self-contained breathing apparatus.

A poor fit of the coat sleeves was also mentioned by many authors, including the sleeve length, the armhole, and the sleeve fit. Issues were mainly reported by female participants of the Park & Hahn [15] study, and included length of sleeves (too long), cuff width (too wide) and upper arm area where extra fabric and lining can cause folds. Male participants requested the opposite: longer sleeves and looser arm area [15]. Participants of the Lee et al. [16] study commented on the need for larger armholes in order to enhance the range of motion in shoulder area. The coat length was reported as being either too long [15] or too short [15, 16]. Participants of the Boorady et al. [4] study expressed a preference for coats with a tail in the back.

Concerning specific areas of the turnout trousers, the crotch, thigh, and knee areas were the most critical in terms of fit. The crotch was the lowest rated among all areas of the turnout trousers evaluated in the Park and Hahn [15] study. Low crotch trousers were mentioned in the Boorady et al. [4] and McQuerry [8] studies. Huang et al. [6] comment the thigh area was the most restricted part while wearing PPC, followed by the knee area. Park & Langseth-Schmidt [13] highlighted that although the female participants wore largely looser-fitting turnout trousers than the male participants, women experienced a tighter fit at the hips and crotch.

Furthermore, studies focused on the impacts of fit on thermal protection of firefighting PPC. In this regard, Nawaz and Troynikov [11] demonstrated how the female body shape affects the air gap volume in firefighting protective coats, particularly under the chest/bust, around the stomach, and in the back and hips.

3.2 Structural Firefighter Personal Protective Clothing Sizing Issues

Sizing issues were also mentioned in a fair amount of research. Authors have highlighted the fact that some occupational groups may have different body measurements from the general population. Results of the Hsiao [17], Boorady [1], and Laing et al. [18] studies found that firefighters exemplify this assumption. A proper understanding of the anthropometrics of the firefighter population is needed in order to create a sizing scheme for this target market [1].

Moreover, authors warned that sizing systems applied to female equipment may not be appropriate. McQuerry et al. [12] mention that sizing requirements for female turnout suits are a simple downscaling of male suit designs, proportions, and sizes. The authors [12] explain how this rationale can be observed in the NFPA 1971 standard.

Additionally, some manufacturers may not offer a full range of sizes [1]. Nonetheless, authors also pointed out that there is no real consistency between the same clothing items from different manufacturers [19], which is also highlighted in the NFPA 1971 standard [7]. The fact that some manufacturers of the U.S. market offer a wide choice of sizes by chest and waist circumferences but are limited in their offering of sleeve and trousers inseam lengths without custom ordering was also mentioned [1]. This may be related to the fit issues reported in other studies.

3.3 Recommendations and Suggestions

Recommendations and suggestions to overcome some of the aforementioned issues can help to improve the design as well as the purchasing process of firefighting PPC.

The development of custom-fitted turnout gear is suggested as an effective way to address fit issues [5, 15]. However, the commercial feasibility of specific tailored firefighting PPC is still a challenge for manufacturers.

Alternatively, the development of improved sizing systems considering the firefighters' anthropometrics was recommended [1]. Regarding this aspect, the relevance of updating body dimensions in dynamic postures was emphasized [1, 5].

In addition, size adjustment systems for minor adjustments, particularly for girth measurements, were advised by Boorady et al. [4]. Park et al. [9] suggested similar solutions for reducing bulkiness of trousers legs in the ankle area. Many authors [8, 11–14] have adverted to the importance of addressing fit and size of PPC considering female anthropometric differences.

Nonetheless, purchasing recommendations were also provided. Boorady [1] proposes that procurement agents should be familiar with the sizes of their crew. Since manufacturer's patterns may vary, it is suggested that measurement for sizing should be done by a trained professional to ensure a proper fit [7].

4 Final Considerations

Protective clothing is necessary to protect firefighters against the numerous occupational hazards encountered in their daily duties. However, firefighters' protection is a challenging task as complex interactions among the wearer, the protective equipment, and the environment can be observed. All these interactions must be understood to reach the goals of Ergonomics, i.e. optimize human well-being and the overall system performance. This literature review demonstrates how fit and sizing can be challenging when selecting equipment for fire brigades. These issues ought to be addressed by researchers and manufacturers when designing firefighting personal protective clothing.

Acknowledgments. We would like to acknowledge the 2C2T-Centre for Textile Science and Technology of the University of Minho. This work is financed by FEDER funds through the Competitive Factors Operational Program (COMPETE) POCI-01-0145-FEDER-007136, by national funds through the FCT-Portuguese Foundation for Science and Technology under the project UID/CTM/000264, by Fundo de Apoio às Vítimas dos Incêndios de Pedrógão, and by ICC/Lavoro.

References

1. Boorady, L.M.: Bunker gear for fire fighters: does it fit today's fire fighters? J. Text. Apparel Technol. Manag. **9**, 1–15 (2015)
2. Coca, A., Williams, W.J., Roberge, R.J., Powell, J.B.: Effects of fire fighter protective ensembles on mobility and performance. Appl. Ergon. **41**, 636–641 (2010). https://doi.org/10.1016/j.apergo.2010.01.001
3. Coca, A., Roberge, R., Shepherd, A., Powell, J.B., Stull, J.O., Williams, W.J.: Ergonomic comparison of a chem/bio prototype firefighter ensemble and a standard ensemble. Eur. J. Appl. Physiol. **104**, 351–359 (2008). https://doi.org/10.1007/s00421-007-0644-z
4. Boorady, L.M., Barker, J., Lee, Y.A., Lin, S.H., Cho, E., Ashdown, S.P.: Exploration of firefighter turnout gear Part 1: identifying male firefighter user needs. J. Text. Apparel Technol. Manag. **8**, 1–13 (2013)
5. Wang, S., Park, J., Wang, Y.: Cross-cultural comparison of firefighters' perception of mobility and occupational injury risks associated with personal protective equipment. Int. J. Occup. Saf. Ergon., 1–9 (2019). https://doi.org/10.1080/10803548.2019.1607027
6. Huang, D., Yang, H., Qi, Z., Xu, L., Cheng, X., Li, L., Zhang, H.: Questionnaire on firefighters' protective clothing in china. Fire Technol. **48**, 255–268 (2012). https://doi.org/10.1007/s10694-011-0214-0
7. National Fire Protection Association: NFPA 1971 - Standard on Protective Ensembles for Structural Fire Fighting and Proximity Fire Fighting (2018). https://www.nfpa.org/codes-and-standards/all-codes-and-standards/list-of-codes-and-standards/detail?code=1971
8. McQuerry, M.: Effect of structural turnout suit fit on female versus male firefighter range of motion. Appl. Ergon. **82**, 102974 (2020). https://doi.org/10.1016/j.apergo.2019.102974
9. Park, H., Trejo, H., Miles, M., Bauer, A., Kim, S., Stull, J.: Impact of firefighter gear on lower body range of motion. Int. J. Cloth. Sci. Technol. **27**, 315–334 (2015). https://doi.org/10.1108/IJCST-01-2014-0011
10. Mandal, S., Lu, Y., Wang, F., Song, G.: Characterization of thermal protective clothing under hot water and pressurized steam exposure. AATCC J. Res. **1**, 7–16 (2014). https://doi.org/10.14504/ajr.1.5.2
11. Nawaz, N., Troynikov, O.: Firefighters' protective jackets: Fit to female form and its effects on attributes relevant to thermal comfort. J. Occup. Environ. Hyg. **15**, 792–802 (2018). https://doi.org/10.1080/15459624.2018.1506587
12. McQuerry, M., Kwon, C., Johnson, H.: A critical review of female firefighter protective clothing and equipment workplace challenges. Res. J. Text. Appar. **23**, 94–110 (2019). https://doi.org/10.1108/RJTA-01-2019-0002
13. Park, J., Langseth-Schmidt, K.: Anthropometric fit evaluation of firefighters' uniform pants: a sex comparison. Int. J. Ind. Ergon. **56**, 1–8 (2016). https://doi.org/10.1016/j.ergon.2016.08.011
14. Boorady, L.M., Barker, J., Lee, Y.A., Lin, S.H., Cho, E., Ashdown, S.P.: Exploration of firefighter bunker gear part 2: assessing the needs of the female firefighter. J. Text. Apparel Technol. Manag. **8**, 1–12 (2013)

15. Park, H., Hahn, K.H.Y.: Perception of firefighters turnout ensemble and level of satisfaction by body movement. Int. J. Fash. Des. Technol. Educ. **7**, 85–95 (2014). https://doi.org/10.1080/17543266.2014.889763
16. Lee, J.-Y., Park, J., Park, H., Coca, A., Kim, J.H., Taylor, N.A.S., Son, S.-Y., Tochihara, Y.: What do firefighters desire from the next generation of personal protective equipment? Outcomes from an international survey. Ind. Health **53**, 434–444 (2015). https://doi.org/10.2486/indhealth.2015-0033
17. Hsiao, H., Whitestone, J., Kau, T.Y., Whisler, R., Routley, J.G., Wilbur, M.: Sizing firefighters: method and implications. Hum. Factors **56**, 873–910 (2014). https://doi.org/10.1177/0018720813516359
18. Laing, R.M., Holland, E.J., Wilson, C.A., Niven, B.E.: Development of sizing systems for protective clothing for the adult male. Ergonomics **42**, 1249–1257 (1999). https://doi.org/10.1080/001401399184929
19. Stull, J.O., Stull, G.G.: When it comes to PPE, size matters (2012). https://www.firerescue1.com/fire-products/apparel/boots/articles/when-it-comes-to-ppe-size-matters-gYpKXO4l6BGvm23J/

Strategic Priorities for Socio-economic Development of Ukraine in Comparison with the Republic of Poland

Olena Pavlova[1], Kostiantyn Pavlov[1], Oksana Novosad[1], Inna Irtyshcheva[2], Nazariy Popadynets[3(✉)], Iryna Hryhoruk[4], Nataliia Gelich[1], Alla Suriak[1], Oksana Makara[1], Olha Zhuk[4], Yevheniya Boiko[2], and Iryna Kramarenko[5]

[1] Lesya Ukrainka Eastern European, National University, Volya Avenue, 13, Lutsk 43025, Ukraine
strishenez@gmail.com
[2] Admiral Makarov National University of Shipbuilding, Heroyiv Ukrayiny Avenue, 9, Mykolayiv 54025, Ukraine
innauamd@gmail.com
[3] SI "Institute of Regional Research named after M. I. Dolishniy, of the NAS of Ukraine", Kozelnytska, Street 4, Lviv 79026, Ukraine
popadynets.n@gmail.com
[4] Vasyl Stefanyk Precarpathian National University, Shevchenko, Street 57, Ivano-Frankivsk 76018, Ukraine
ira.hryhoruk@gmail.com
[5] Mykolaiv Interregional Institute of Human Development of the University "Ukraine", Viyskova, Street 22, Mykolayiv 54000, Ukraine
ukraina_09@ukr.net

Abstract. The article deals with the general social and economic features of the formation of the state of Ukraine comparing with similar indicators characterizing the Republic of Poland. The indicators of the dynamics of a gross domestic product; generalized indicators of economic efficiency; and the geographical structure of indicators of foreign trade relations of Ukraine and Poland have been analyzed. Studying the issues of intensification of Ukrainian-Polish cooperation, as well as the study of different socio-economic indicators of these countries in all dimensions, is one of the most pressing scientific problems from establishing effective mechanisms for bringing Ukraine closer to the European Union, as well as forming the appropriate basis and tools for mutual relations.

Keywords: Socio-economic development · Gross domestic product · Economic efficiency · Foreign trade · Comparative analysis

1 Introduction

The problems of the Ukrainian-Polish political and economic cooperation are studied in various aspects by many domestic and foreign scholars. In particular, features and perspectives of economic and social development were analyzed in [1–3]. At the same

W. Karwowski et al. (Eds.): IHSED 2020, AISC 1269, pp. 308–314, 2021.
https://doi.org/10.1007/978-3-030-58282-1_49

time, insufficient attention was paid to the study of social and economic indicators of Ukraine through the prism of the achievements of the Republic of Poland, which led to our decision to research the chosen topic.

The purpose of the study is to carry out an analytical review of the current state and prospects of further socio-economic development of Ukraine in comparison with the Republic of Poland to determine strategic priorities for Ukraine.

Studying the issues of intensification of Ukrainian-Polish cooperation, as well as the study of different socio-economic indicators of these countries in all dimensions, is a very important scientific topic. Research in this field enables establishing effective mechanisms for bringing Ukraine closer to the European Union, as well as forming the appropriate basis and tools for mutual relations with neighboring countries that are members of the European Union.

The Republic of Poland is the most consistent and effective partner of Ukraine in its European integration aspirations among EU countries, which share borders with Ukraine, historical and cultural traditions, as well as related geo-economic interests. From ensuring the evolutionary development of economic relations between Ukraine and neighboring Poland, the most revealing is the identification of the main problems and socio-economic features of their cooperation, which has been formed over a long period. Poland is one of Ukraine's main strategic partners in supporting efforts to move closer to the EU and its possible future membership status [4–8].

2 Materials and Methods

Compose comprehensive and systematic approaches to understanding the identities of two countries were applied. The basic principles of economic theory, national and regional economics, psychology, sociology, geography are the base of research. A special place belongs to the complex scientific and special methods, the application of which allowed to analyze thoroughly and in-depth these issues. Also, we used the following methods: scientific abstraction, historical method, comparative analysis, theoretical and logical; dialectical, induction and deduction, empirical-historical, statistical method, methods of analysis, and economic modeling.

3 Results and Discussion

Ukraine has been undertaking radical reforms in many leading industries, including the banking system, the financial sector, the legal field, and so on. At the same time, the process is rather slow. On the other hand, the Republic of Poland has made its difficult journey from a state a former satellite of the Union of Soviet Socialist Republics to Modern Poland - a progressive state that has been a member of the European Union for thirty years. As a result, the Republic of Poland is far ahead of Ukraine in terms of development [9].

The reform processes that took place in the Republic of Poland were quite painfully reflected on the social level among the population. But the Republic of Poland has diversified its economic processes to reorient them from markets in the post-Soviet countries to Europe [10, 11]. These were the same processes that Independent Ukraine has been experiencing for the last years [12–14].

After the so-called transition period, which took place in the market processes of the Republic of Poland, the main economic indicators tended to increase. They slowed down in one period, grew up in another. Since the emergence of the present-day Poland, the value of its gross domestic indicator has been characterized only by growth (average value) (Fig. 1), even at a time of a major financial crisis in 2008.

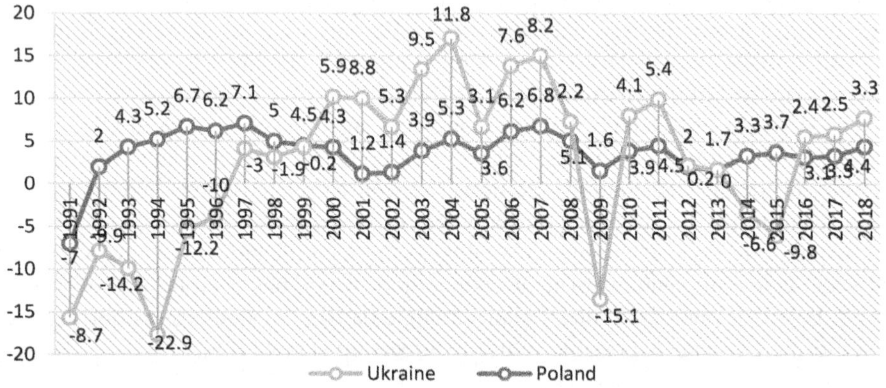

Fig. 1. Dynamics of Gross Domestic Product of Ukraine and the Republic of Poland for the Period 1991–2018, (in%, to the previous year), (2014–2018 without taking into account the temporarily occupied territories in Donetsk, Luhansk, and Crimea). Source: [15–17].

It is also important to note that the population of the Republic of Poland is 3.5 million lower than in Ukraine and its GDP is more than four times higher (Table 1).

Table 1. Economic performance indicators of Ukraine and Poland, as of 01.01.2019

Indicator	Value	
	Ukraine	Poland
Gross domestic product, billion USD	130,8	549,5
Population, thousand people	42153,2	38637,9

Source: [15–17]

In general, the present-day Republic of Poland began its development at about the same time as Independent Ukraine and with almost identical situational economic positions. As of today, it outperforms our Ukraine by more than three times in terms of gross domestic product per person (Fig. 2).

In general, even though the number of socio-economic indicators of Ukraine has been increasing since 2015 to the present, to catch up with the values of the social and economic development of our Western partner, the Republic of Poland, Ukraine will

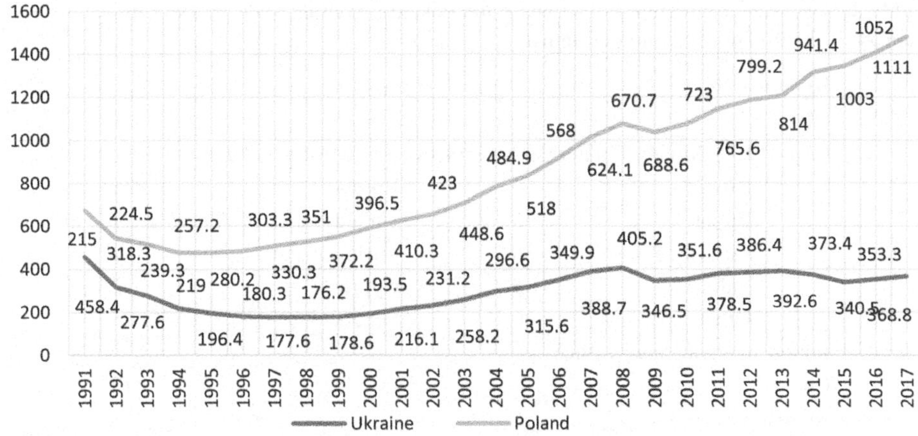

Fig. 2. Dynamics of gross domestic product by purchasing power parity per person, (US $, January 1, 2018). Source: [15–18].

need at least half a century. Based on the value that we managed to get as a result of our calculations we have the following systematic indicators.

Provided that the GDP growth rate of Ukraine is at the level of 2.5% in 2017, Ukraine will need 48-50 years. However, if the value of growth is at the level - 3.5% (the indicator of Ukraine in 2018), it will take 35–37 years to achieve the "Polish" value. In turn, provided that Ukraine's GDP growth rate is reached at the level of 5%, it will take 25 years; provided 10% - 12–14 years. In other words, to catch up with the huge detachment of the Republic of Poland's indicators from the realities that are present in Ukraine today, there is a need to increase the socio-economic values of the latter by - 240% [15, 16].

As we have already noted, the standard of living of the citizens of the Republic of Poland is almost three times higher than the similar value in Ukraine and is less than 12 thousand US dollars per capita. It should be noted that the level of this indicator is low compared to other countries of the European Union. For example, Portugal - Thousand US dollars United States, Germany - $ 36 thousand United States, Norway - $ 69 thousand US per capita [18].

When it comes to indicators of international cooperation, trade partnerships in different fields and industries also have a significant difference. While the foreign trade policy of Ukraine reflected the concept of multi-vector, Poland chose the EU and the implementation of energy-saving measures. As a consequence, the path of the Republic of Poland to the integration with the European Union has been overcome in about a decade. The aggregate indicators of foreign trade relations are shown in Table 2.

The potential competitive advantages, which in the "right direction" of the government can also be attributed: the ability to provide a lower level of tax rates for the capital raised; a simpler mechanism for regulating investment and business processes; low level of bureaucracy; the faster process of implementation of state procedures.

Table 2. Geographic structure of Ukraine's and Poland's external trade relations indicators as of January 1, 2019 (Top five ranking items)

Ukraine, %					
Export					
Goods	EU (Mostly Poland)	37	Services	EU (Mostly Poland)	30
	Russia	10		Russia	31
	Egypt	6		United States of America	7
	Turkey	6		Turkmenistan	4
	India	5		Cyprus	2
	Others	36		Others	25
Import of goods and services					
China					15
Russia					13
Belarus					6
United States of America					5
Turkey					3
Others					58
Republic of Poland,%					
Export, import of goods and services					
Export	Germany	27	Imports	Germany	23
	Great Britain	6		China	12
	Czech Republic	6		Russia	7
	France	6		Italy	5
	Italy	5		France	4
	Others	50		Others	49

Source: [15–21]

4 Conclusions

In Ukraine it is common to compare our country with the Republic of Poland. Of course, today such a comparison will not benefit Ukraine. The fact that the reform processes taking place in the country today have not yet been completed and ten years and the initiative of the people implementing the reforms is not as powerful as that of the Euro-activists in our western neighbor. At the same time, reforms in Ukraine are often sabotaged by pro-Russian politicians who are constantly working against the "Europeanization" of Ukraine.

We can draw the following conclusions from the experience of the Republic of Poland. Although in the early 1990s, the starting points of the economic conditions of the two states were identical, there was a significant difference in the heads and awareness of the population itself. After all, the length of roads or a number of factories do not matter, unless the population of the state has a full understanding of what changes and measures for their implementation will lead the state to social and economic success.

It should also be remembered that since 1991 and later, the population of the Republic of Poland has been difficult enough and may not have had any idea of Russian memes about victory and betrayal, but the government's policy was understood to be a total betrayal. The citizens of Poland were well aware of the need for rather unpleasant and painful reforms and were able to bring about responsible reformers. And, as a consequence, after a long time of implementation of the changes, the Republic of Poland from a loyal state to the Soviet Union turned into a state with a sufficient standard of living, high economic and social indicators, which is a good example for Ukraine.

References

1. Andrushkiv, B.: Ukraine and Poland - economic aspects of cooperation. Econ. Ukraine **11**, 82–85 (2005)
2. Borshchevsky, V.V.: Ukrainian-Polish Economic Cooperation in the Conditions of European Integration: Monograph. Avers, Lviv (2007)
3. Nykyruy, L., Yakubiv, V., Wisz, G., Hryhoruk, I., Zapukhlyak, Zh., Yavorskyi, R.: Renewable energy in Ukraine-Poland region: comparison, critical analysis, and opportunities. In: Al Qubeissi, M. (ed.) Renewable Energy - Resources, Challenges and Applications (2020)
4. Strilchuk, R.S., Pavlova, O.M., Pavlov, K.V.: Current Problems of Analytical Economy: Monograph. Lutsk: SPD Gadiak Zhanna, printing house "Volynpoligraf" (2019)
5. Pavlov, K., Pavlova, O., Kupchak, V.: Integral indicators based on competitiveness capacity characteristics of regional real estate markets of Ukraine. J. Competitiveness **11**(3), 87–108 (2019)
6. Yakubiv, V., Panukhnyk, O., Shults, S., Maksymiv, Y., Hryhoruk, I., Popadynets, N., Bilyk, R., Fedotova, Y., Bilyk, I.: Application of economic and legal instruments at the stage of transition to bioeconomy. In: Ahram, T. (ed.) AHFE 2019. AISC, vol. 965, pp. 656–666. Springer, Cham (2020). https://doi.org/10.1007/978-3-030-20454-9_64
7. Vasyltsiv, T., Irtyshcheva, I., Lupak, R., Popadynets, N., Shyshkova, Y., Boiko, Y., Ishchenko, O.: Economy's innovative technological competitiveness: decomposition, methodics of analysis and priorities of public policy. Manag. Sci. Lett. **10**(13), 3173–3182 (2020)
8. Kramarenko, I., Irtyshcheva, I., Shults, S., Stegney, M., Blishchuk, K., Boiko, Y., Popadynets, N., Samotiy, N., Rogatina, L., Hryshyna, N.: The model of economic system management for the Black Sea region of Ukraine in the sustainable development context. Accounting **6**(4), 387–394 (2020)
9. Poland's lost property. The Financial Times. https://www.ft.com/content/be98d9aa-360d-11df-aa43-00144feabdc0 (2019)
10. Main documents of the Mutual Economic Assistance Council. Moscow (1981)
11. CMEA is a new type of international economic relations. Kyiv (1988)
12. Kryvets, N.V.: The Council of Economic Mutual Assistance. Vol. 9. Encyclopedia of the History of Ukraine: in 10 volumes. Institute of History of Ukraine NAS of Ukraine. Naukova Dumka, Kyiv (2012)

13. Irtyshcheva, I., Stehnei, M., Popadynets, N., Danylo, S., Rogatina, L., Bogatyrev, K., Boiko, Y., Hryshyna, N., Ishchenko, O., Voit, O.: Business process management in the food industry under the conditions of economic transformations. Manag. Sci. Lett. **10**(14), 3243–3252 (2020)
14. Pylypiv, N., Piatnychuk, I., Halachenko, O., Maksymiv, Y., Popadynets, N.: Balanced scorecard for implementing united territorial communities' social responsibility. Probl. Perspect. Manag. **18**(2), 128–139 (2020)
15. World Bank Group (2020). https://www.worldbank.org
16. State Statistics Service of Ukraine (2020). http://www.ukrstat.gov.ua
17. Statistics Poland. Statistics Service of Poland (2019). https://stat.gov.pl/en/
18. The World Bank Group. Inflation, GDP deflator (2017). https://data.worldbank.org/indicator/NY.GDP.DEFL.KD.ZG?end=2017&locations=PL-UA&start=1996
19. Carpathian Explorer. Ukrainian exports. What? Where? How? Ukraine sells and makes money (2019). https://carpatobserver.eu/український-експорт-що-куди-як-украї/
20. The Observatory of Economic Complexity (2019). https://oec.world/ru/
21. Economic Globalization - Classification of Countries. The Global Economy (2019). https://ru.theglobaleconomy.com/rankings/kof_econ_glob/

Human Cyber-Physical Systems Interactions

Optimizing the Human Psyche in a Pandemic

Sherry Palamara(✉)

Sherry A. Palamara, Psy.D., P.A., 407 Lincoln Road,
Miami Beach, FL 33139, USA
shpalamara@aol.com

Abstract. While the human psyche is not a machine, during said Pandemic, undeniable considerations for human nature is pertinent to all studies. Tesla [1], emphasizes human as "mass moved by a force". Perceiving Mass as thought then moves force to either deplete or break through friction. James [2] identifies habits as "fundamental properties of matter"; plasticity "structure weak enough to yield an influence, but strong enough not to yield all at once". "All at once" is this pandemic influence. Bringing Maslow's Motivational Hierarchy [3] from optimizing potential to basic sustenance and safety needs. The very foundation that enhances self-actualization progression while climbing Social Belongingness and Self-Esteem hierarchy. Creative redesign may optimize a paradigm shift from Pandemic influence via efficient-solution-optimizing Breakthrough, efficient-non-optimizing Heuristic, non-efficient- optimizing Algorithm and non-efficient-optimizing Brainstorming. If removing friction accelerates momentum, habitual routines maintain constancy for new accelerated mass and humans grow mentally stronger. Deepening common-sense-reasoning, psychosocial-etiquette, and conscientious-awareness is discussed.

Keywords: Optimize · Resilience · Human · Psychology · Pandemic

1 Introduction

The following presents ideas prepared to receive commentary and inspire discussion about optimizing the human psyche during these current times. According to the World Health Organization, Pandemic is defined as "The world-wide spread of a new disease" [4]. "New" suggests uncertainty for which scientific examination is anticipated as a method to deepen understanding. "Disease", although not a factor in and of itself to optimize the human psyche, for now, is excused from elaboration and can be identified on a continuum of impaired functioning.

Maslow's Motivational Hierarchy [3] is a well known psychological reference in the pre-pandemic era since the ideology was introduced by Abraham Maslow in 1943. Maslow emphasizes setting goals that aggregate proficiency. The Hierarchy comprises basic fundamental needs of Biology [3], such as Carbon Dioxide-Oxygen exchanges [5], food and water; needs of Safety, such as availability of personal and professional resources; as well as midpoint needs of Social Belongingness, such as meaningful relationships [3] akin to in-group inclusion [6] and social comparison [7]; Self-Esteem, such as prestigious accomplishments [3] as well as supporting the utility of personal

W. Karwowski et al. (Eds.): IHSED 2020, AISC 1269, pp. 317–324, 2021.
https://doi.org/10.1007/978-3-030-58282-1_50

Strengths [8] and counterfactual thinking [9]. At the pinnacle, is optimization of Self Actualization, such as the realization of full potentiality [3].

Pre-pandemic psychologist William James [2] identifies substantial influences as needing to first yield impressions in sensory and blood pathways. The deepening of existing pathways, referred to as habits, determine innate instincts instilled since birth. James conjectures that attempts made to define habits seem to naturally generate interest in the "fundamental properties of matter". James further submits that Nature's Law encompasses "immutable habits" of disparate elementary matter which proceeds in their interactions, i.e., "actions and reactions upon each other" [pp. 104, 2]. However, while elementary particles of matter are not designed to change, compound mass of matter can change. According to James, this is due to the variability of compound structure that modifies instinctual habits by external forces and internal tension. James is essentially referring to the ability to maintain integrity of a body while plasticity reconfigures new internal or external shape of the compound structure; a changed habit of cohesion of sorts, similar to a lock functioning better after usage for a period of time. If the Habit be arbitrary as instructed by Wittgenstein's illustration of an arbitrary polygon [10], then the essential properties may be undeniably unique and set estranged from familiar or possibly pre-pandemic states of habits. Well established pre-pandemic habits are likely challenged by the current pandemic's macrocosmic forces; yielding reflexive mental and physical discharges until forming new habits suitable to survive a better quality of life in the post-pandemic era.

To reinstate appropriated resolution of the human psyche with regards to the pandemic, a model of problem solving and decision making is submitted for consideration. There are four categories to examine: 1) Brainstorming, 2) Algorithms, 3) Heuristics and 4) Breakthroughs. As professed by Einstein, "imagination" is designated as having greater importance than "knowledge" [11]. Brainstorming, has its roots in creative imagination as a useful application to build group participation without criticism or judgement [12]. Algorithms optimize ideas and quantity of optional perspectives to resolve problems [13]. Heuristics foster perfunctory decision selection based on minimal cerebral effort and information [14–16]. If Breakthroughs [16] inspire significant scientific discovery, conceptual synthesis of knowledge as well as substantial motivation to advance ideas and produce forward action, then Breakthroughs could set the precedence for an anticipated post-pandemic foundation of better quality human habits. Also, Einstein's belief in "intuitions and inspirations" [11] could be considered important factors that may lead to achieving Breakthrough outcomes.

Predominant focus for living in this pandemic, is now the supply of sufficient oxygen. Oxygen is not able to be reserved in the body yet this resource is ceaselessly necessary to sustain life. Changes in the availability of oxygen can substantially effect cell function whereby attempts are then made to reinstate balance [17]; perhaps, with incentive, a type of reconfigured rebalance which enhances the benefit of physical and mental properties for an improved quality of life in the post-pandemic era, may be feasible.

Pre-pandemic understanding of human advancement once thought to be, for the most part, building blocks of progress now remain in an undetermined state of friction with apparent ambiguity about the anticipated post-pandemic period. Nevertheless, if a new foundation of ample supports and accessible resources perpetuates mental and physical momentum while removing unwanted friction, this is certain to be advantageous for the collective future of the human species.

2 Paper Presentation

The current pandemic influences psychological distress and interminable behaviors. In an effort to explore the enigmatic nature of the human response to this pandemic, the initial analysis examines awareness of sensations that drive the mind and body to take action until discontinued by transformative interaction. The utility of formulae seems a suitable constant to parallel psychological understanding of variant human resilience and is, in part, yet not solely relied upon in the focus of this presentation [18, 19].

Initial inspection ponders mass, force and matter in relation to humans. To begin, Tesla's [1] view emphasizes the human factor to be "mass moved by a force".

Consider first, the Force of the 2020 pandemic initially activating movement of mass. If this be a matter of human thought, initiating cognitive momentum activated by pandemic Force creates a chain reaction in which thought is a catalyst to accelerate physical movement.

$$m \cdot a = F. \tag{1}$$

Where m, is human mass, a, is the acceleration of the mass and F, is the Force it takes to move the mass; then to mobilize the sum of human mass in a world-wide Pandemic requires a proportional amount of Force.

Pre-pandemic habits are expected to fractionalize as unprecedented successive Pandemic Forces inspire the drive to develop resilient habits that proficiently recalibrate challenges, indefinitely.

$$m' \cdot a' = F'. \tag{2}$$

Whereby reconfigured mass, m', times acceleration, a', equals Force, F'. Subsequent chain reaction is anticipated after initiating momentum. New patterns of resilience are foreseen to counter compounding Force and counteract unwanted friction.

$$\frac{1}{2}mv^2 = KE. \tag{3}$$

Half the human's mass is, m, times the mass's velocity squared which is equal to Kinetic Energy; KE. If Potential Energy from the pre-pandemic period converts to Kinetic Energy during the Pandemic period, then velocity of Kinetic Energy increases at the time of the Pandemic period.

What will change in the post-pandemic era? If thoughts are a catalyst to action, then what is the Pandemic's effect on the thinking process? Do viewpoints such as macrocosmic and microcosmic, conscious and unconscious, upward and downward counterfactual thinking, and upgrade and downgrade social comparison, subsume Kinetic movement and directionality or are they Kinetic precursors with Potential Energy?

$$m \cdot g \cdot h = P. \tag{4}$$

Whereby mass, m, times the acceleration of gravity, g, times height equals Potential Energy, P. If Potential Energy is a matter of thinking, then presume pre-pandemic level of Potential Energy determines directionality of Potential and Kinetic Energy during pandemic and post-pandemic periods.

Will a post-pandemic era result in amass Force sufficient enough to sustain exponential Acceleration of human faculties such as those of mind and body? To what extent will the collective human species function as a singularity?

Is pre-pandemic Psychology sufficient to sustain an advancement in understanding the human psyche during this pandemic? Has psychology reconfigured methods to detect the impact of psychological modifiers on the newly emerging psyche? What psychological modifiers support the newly emerging psyche? Consider the following: As the build up of friction from the Force of the pandemic increases, the practice of pre-pandemic habits decrease and inevitably sets into motion a new state of psychological resilience.

The pandemic of today imposes friction on the human psyche's pre-pandemic habits. Pre-pandemic Psychologist William James [2] identifies habit as an aspect of the "fundamental properties of matter". Paradoxically, although the fundamental properties of pre-pandemic matter seem to be losing resilience, emerging from the pandemic is gaining a new psychological resilience; a plasticity, according to James [2, pp. 105] with "structure weak enough to yield an influence but strong enough not to yield all at once". The pandemic "yields an influence all at once," and challenges the resilience of pre-pandemic habits. As the pandemic incites a new human psyche, psychology's pre-pandemic paradigm is Forced to align with the current human psyche. How do human's relieve themselves from pre-pandemic habits and align with the newly maturing present day human psyche when uncertainty is at the forefront of decision making? How can the Force of the pandemic inspire new states of psychological resilience? What are the value-loaded pre-pandemic psychological concepts that replenish human needs during the pandemic period? What psychological tools will integrate a higher quality of human capability?

A pre-pandemic psychological perspective that seems to retain value in this pandemic state thus far, is Maslow's Motivational Hierarchy [3]. Self-actualization is the highest attainable need in Maslow's Hierarchy. The lesser tiers constitute basic fundamental needs of food, water and safety. On the Self-Actualization climb to realize one's optimal level of functioning, is the need for Social Belongingness and Self-Esteem. The immediate response to the pandemic activates a drive to acquire a satiated level of non-perishable items, like toilet paper, and sanctum of protective barriers, like hand sanitizer and the N95 mask; habits [20] that probably reveal themselves at a functionality consistent with Maslow's survival needs on the hierarchy.

As the pandemic brings to light coinciding needs with survival decisions, a method for efficient decision making and guaranteed solution outcomes may offer support during unknown human responses to the pandemic. Such as, a solution optimizing Breakthrough, efficient and non-optimizing Heuristic, non-efficient and optimizing Algorithm and all inspiring yet neither efficient nor optimizing Brainstorming (Refer to Fig. 1).

Little effort is put forth in Heuristics as the approach relies on readily available information [14]. As indicated, 1) Heuristics identify efficient means of finding an answer by implementing a decision-making and problem-solving strategy without guarantee of correct outcomes. On the other hand, 2) Algorithms are designed to generate alternative solutions to problems [13] by a less efficient problem-solving strategy but guarantee solution outcomes to problems. Additionally, 3) Brainstorming is an uninhibited and spontaneous creative process to generate ideas [12]; usually by a group without immediate critical judgement about concepts with potential value. Intuitive instinct may be a risky approach to decision making [15], but the combination could lead to a breakthrough discovery. A 4) Breakthrough is efficient and guarantee solution outcomes. Breakthroughs suggest a significant leap forward [16]. Their outcome has the potential to integrate knowledge, develop concepts and propel major scientific discoveries.

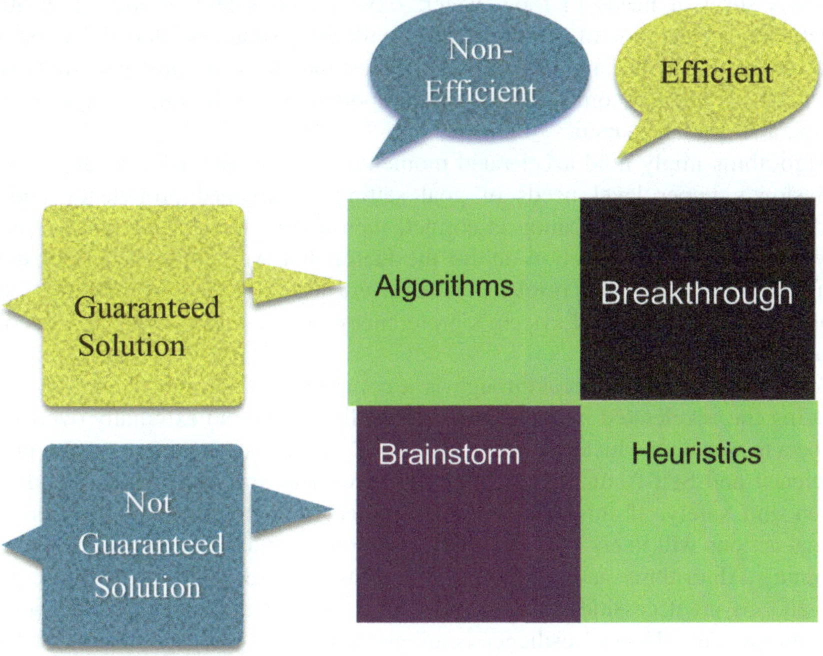

Fig. 1. Problem-Solving Decision Making Model: 1) Heuristics: Efficient, No Guarantee, 2) Algorithm: Guarantee, Not Efficient, 3) Brainstorming: Not Efficient, No Guarantee, 4) Brainstorming: Efficient, Guarantee.

With regard to remarks on problem solving and decision making, Heuristics plus Algorithms equal Breakthroughs.

$$H + A\imath = B. \quad (5)$$

Where H, equals Heuristics, Aı, equals Algorithms and B, equals Breakthroughs.

It stands to reason that if Heuristics plus Algorithms equal Breakthroughs, then Heuristics minus Algorithms is equal to Brainstorming.

$$H - A\imath = B\imath. \quad (6)$$

Where H, equals Heuristics, Aı, equals Algorithms and Bı, equals Brainstorming.

Heuristics assumes a greater number than Algorithms during Brainstorming sessions. Decision-making via Brainstorming is more likely to derive at a sense of finality with unfiltered ideas rather than precise Algorithms. Even though many ideas may transpire from a Brainstorming session, it is possible that zero ideas are chosen to solve a pertinent problem. Brainstorming is more a matter of taking a break to refresh the mind without pressure to solve issues within time constraints.

Heuristics imply fluid accelerated momentum that is externally overt; analogous to Maslow's survival needs of food, water, oxygen and safety. If this implication is correct, then acceleration of external actions will likely suppress internal friction due to the factor that Heuristics and Algorithms are presumed to be negatively correlated. The overt necessity then becomes the focus of importance and the internal system's preferences, a secondary pursuit.

Algorithms imply fluid accelerated momentum that is internally covert, analogous to Maslow's upper level needs of goal setting, aggregated proficiency and Self-Actualization. If this implication is correct, then acceleration of internal actions will likely suppress external friction due to the factor that Algorithms and Heuristics are presumed to be negatively correlated. The covert necessity then becomes the focus of importance and the external system's preferences, such as basic hygiene, a secondary pursuit.

While proclaiming a Breakthrough is a compound of Heuristics and Algorithms, activating the accelerated momentum of internally covert and externally overt Forces, may be a means with which to integrate Maslow's emphasis on goal setting, aggregated proficiency and Self-Actualizing tendencies as well as survival needs of food, water, oxygen and safety. If this implication is correct than acceleration of internal and external actions will likely have no suppression on internal and external friction due to combining Algorithms and Heuristics to constitute a Breakthrough. Subsequently, overt and covert necessities as well as external and internal preferences become co-occurring pursuits. Hence, resilience is anticipated in an aggregate position of internal and external frictionless momentum. Furthermore, Brainstorming is an additive to unique preferences and creative applications with possible relevant value-load that may ensure a better quality Breakthrough outcome.

In conclusion, the top priorities in the current pandemic is health and safety. Inversely, this comprises the lower tiers of Maslow's motivational hierarchy. While wearing a mask is an important part of Personal Protective Equipment (PPE), comparatively, it may deplete the efficacy of expelling Carbon Dioxide (CO2). Considerations to replenish an organic surplus of Oxygen (O2) after wearing a mask, especially for long periods of time, seem practical to include in PPE protocol.

Unobstructed movement indicates absent friction and could imply exponential acceleration until stopped or slowed by forces or choices. Sufficient Force of the pandemic, initiates instinctive decisions and intuitive senses to acquire survival and functional supplies. Also, sustaining fluid momentum of thought and physicality during the present pandemic, is likely to avail quality life goals. With the future in mind, if the integrated psyche that maintains fluid momentum is not swayed by counterfactual thought, this synthesized psyche may become the quintessential foundation in the post-pandemic period.

To lessen pre-pandemic friction, the fluid momentum generated by the Force from the pandemic can serve as a catalyst to align with the fluid momentum of the new human psyche. To maintain continuous momentum in a post-pandemic period, releasing pre-pandemic hindrances and social comparisons while aligning with pandemic functionality, may inspire ongoing Breakthroughs and optimize high quality accelerated propulsion habits; both mentally and physically. If removing friction from human mass accelerates momentum, then functional routines are suspected to maintain a state of constancy for the accelerated newly reconfigured habits. If this is true, then the human species grows mentally stronger with the acceleration of thinking more conceptually and with the pragmatic practice of habitual routines.

Heuristics combined with Algorithms are considered equal to Breakthroughs. On the other hand, Brainstorming is an additive to Heuristics, Algorithms and Breakthroughs. Additionally, Brainstorming may also be a creative approach that merges instinctive and intuitive aspects of Maslow's lower and upper tiers which may synthesize to form choice actions of intuitive-instinct during these Pandemic times.

Furthermore, deepening higher cortical functioning and conceptual magnitude of Common Sense Reasoning, Psychosocial Etiquette, and Conscientious Awareness may be worth analyzing as an additive to strengthen the new internal representational mentality of the human species as more Breakthroughs are anticipated to transpire. In essence, relying on an internal cognitive reference may serve as a protective structural barrier between related strengths and unrelated or non-preferred cognitions. For instance, applying the VIA Survey of Character Strengths to improve awareness of positive perspective may be a readily available means with which to strengthen a positive mental frame of reference [21]. The VIA Strengths need only be identified by noticing rather than stringently practicing them.

Lastly, applying Directionality, Velocity and Time to the exploration of improving behavioral habits, may clarify mechanisms with which to further optimize the human psyche in the current pandemic. Pertinent to this ideology is to spotlight if amassed bodily and vehicular momentum in relation to the Earth's rotation has impacted human functions. How does this change, if any, impact life on Earth?

Eventually, it may be inevitable to shift from pre-pandemic to post-pandemic behavioral habits. Hence, a Breakthrough model that leads to enduring resiliency, may be a sensible roadmap to follow now and in the future. As the strong Forces of the

Pandemic persistently redirect humans, the human species as a whole is likely to be challenged with growing genuinely stronger in their adaptive ability while awaiting the arrival of the post-pandemic era.

References

1. Tesla, N.: The problem of increasing human energy. Cen. Il. Mo. Mag. 183–226 (1900)
2. James, W.: Habit. In: The Principles of Psychology, vol. I., pp. 104–127. Dover Publications Inc., New York (1890)
3. Maslow, A.: A theory of human motivation. Psychol. Rev. **50**, 370–396 (1943)
4. World Health Organization.: What is a Pandemic? (2010). http://www.who.int/csr/disease/swineflu/frequentlyasked_questions/pandemic/en/
5. Vadasz, I., Hubmayr, R.D., Nin, N., Sporn, P.H., Sznajder, J.I.: Hypercapnia: a nonpermissive environment for the lung. Am. J. Respir. Cell Mol. Biol. **46**(4), 417–421 (2012)
6. Valcke, B., Van Hiel, A., Van Assche, J.: The need for inclusion: the relations between relational and collective inclusion needs and psychological well-and ill-being. Eur. J. Soc. Psychol. **50**(3), 579–596 (2019)
7. Wood, J.V.: Theory and research concerning social comparisons of personal attributes. Am. Psychol. Assoc. **106**(2), 231–248 (1989)
8. Toback, R.L., Graham-Bermann, S.A., Patel, P.D.: Outcome of a character strengths-based intervention on self-esteem and self-efficacy of psychiatrically hospitalized youths. Psychiatr. Serv. **67**, 574–577 (2016)
9. Roese, N.J.: Counterfactual thinking. Am. Psychol. Assoc. **121**(1), 131–148 (1997)
10. Wittgenstein, L.: Remarks on the Foundations of Mathematics. MIT Press, Caimbridge (1994)
11. Viereck, G.S.: What life means to Einstein. Sat. Eve. Post **26**, 17–117 (1929)
12. Osborn, A.: Applied Imagination: Principles and Procedures of Creative Problem-Solving, 3rd edn. Charles Scribner's Sons, New York (1953)
13. Imanirad, R., Yang, X.-S., Yeomans, J.S.: A co-evolutionary, nature-inspired algorithm for the concurrent generation of alternatives. GSTF. J. Comput. **2**(3), 101–106 (2012)
14. Chaiken, S.: Heuristic versus systemic information processing and the use of source versus message cues in persuasion. J. Pers. Soc. Psychol. **39**(5), 752–766 (1980)
15. MacGregor, D.G.: Affect, risk and decision making. Health Psychol. **24**(4, supl.), 535–540 (2005)
16. VandenBos, G.R.: APA Dictionary of Psychology. American Psychological Association, Washington, D.C. (2007)
17. Snyder, C.M., Chandel, N.S.: Mitochondrial regulation of cell survival and death during low-oxygen conditions. Antioxid. Redox Signal. **11**(11), 2673–2683 (2009)
18. Newton, I.: Newton's Principia: The Mathematical Principles of Natural Philosophy, 1st edn (A. Mott Trans.). Daniel Adee, New York (1846)
19. Einstein, A.: Relativity: The Special and the General Theory, 15th Revised edn. Three Rivers Press, New York (1961)
20. McSweeney, F.K.: Dynamic changes in reinforcer effectiveness: satiation and habituation have different implications for theory and practice. Behav. Anal. **27**, 171–188 (2004)
21. Seligman, M.E.P.: Authentic Happiness: Using the New Positive Psychology to Realize Your Potential for Lasting Fulfillment. Free Press, New York (2002)

Towards Intelligent Pick and Place Assembly of Individualized Products Using Reinforcement Learning

Caterina Neef[1(✉)], Dario Luipers[1], Jan Bollenbacher[2], Christian Gebel[1], and Anja Richert[1]

[1] Cologne Cobots Lab, TH Köln - University of Applied Sciences, Betzdorfer Str. 2, 50679 Cologne, Germany
{caterina.neef, dario.luipers, anja.richert}@th-koeln.de

[2] Institute of Telecommunications Technology, TH Köln - University of Applied Sciences, Betzdorfer Str. 2, 50679 Cologne, Germany

Abstract. Individualized manufacturing is becoming an important approach to fulfill increasingly diverse consumer expectations. While there are various solutions for the manufacturing process, such as additive manufacturing, the subsequent automated assembly remains a challenging task. As an approach to this problem, we aim to teach a collaborative robot to successfully perform pick and place tasks by implementing reinforcement learning. For the assembly of an individualized product in a constantly changing manufacturing environment, the simulated geometric and dynamic parameters will be varied. Using reinforcement learning algorithms capable of meta-learning, the tasks will first be trained in simulation, and then performed in a real-world environment where new factors are introduced that were not simulated in training to confirm the robustness of the algorithms. A concept comprised of selected machine learning algorithms, hardware components as well as further research questions to realize the outlined production scenario are the results of the presented work.

Keywords: Machine learning · Reinforcement learning · Meta-learning · Individualized manufacturing · Collaborative robotics

1 Introduction

For decades, robots have been used to automate tasks in the industry sector. Conventional industrial robots are taught to perform one task at a time, are competent at executing this single task and can perform thousands of repetitions accurately. While the automation of such tasks has led to an increase in efficiency and a decrease in manufacturing costs for mass production, it is less applicable for the individualized consumer expectations of today's economy. Globalization, digitalization and the resulting growth of markets have led to an increasing number of product variants and shorter product life cycles [1]. Customers now demand highly individualized products that are designed specifically for them. This change is observable in a wide range of industrial fields [2]. An example is the health-care sector, where personalized medicine

W. Karwowski et al. (Eds.): IHSED 2020, AISC 1269, pp. 325–331, 2021.
https://doi.org/10.1007/978-3-030-58282-1_51

is becoming increasingly important, and additive manufacturing is being used for the production of biomaterials, implants and prosthetics [3].

To enable such individualized manufacturing, traditional programming of machines with repetitive tasks is no longer applicable [2]. Robots with inherent collaborative properties, however, are well suited for constantly changing tasks as they are mobile, highly flexible and adaptable, and can quickly be taught new tasks, even by unskilled workers. We therefore propose to use reinforcement learning (RL) algorithms capable of meta-learning (ML) combined with visual and tactile sensor data to enable collaborative robots to accomplish highly individualized pick and place tasks.

2 State of the Art: Machine Learning and Robotics

Recently, RL has achieved great success in a wide range of different tasks and complex games. The implementation of RL and ML seems promising to enable a robot to perform a pick and place task for unknown objectives and destinations. In RL, an agent interacts with the environment and receives its state. Based on this state, the agent takes an action and receives a new state and reward for the chosen action. Each RL algorithm is designed specifically for a certain task in terms of its architecture and training, and the RL agent needs to be trained from scratch for each task.

ML is an approach to overcome this shortcoming by designing an algorithm in such a way that the agent learns how to learn from a broad distribution of similar tasks. Similar to human learning, an ML agent can apply knowledge it has gained from previously solved corresponding tasks to learn a new task with only a small amount of data.

Recent ML algorithms suitable for RL can be divided into two categories depending on their architecture and optimization goals:

1. Model-based ML approaches [4, 5] generalize to a wide range of learning scenarios, seeking to recognize the task identity from a few data samples and adapting to the tasks by adjusting a model's state (e.g. long short-term memory [LSTM] internal states)
2. Model-Agnostic Meta-Learning (MAML) [6] seeks an initialization of model parameters so that a small number of gradient updates will lead to fast learning on a new task, offering flexibility in the choice of models

Another relevant RL approach is introduced by OpenAI, who have trained a robotic hand to solve a Rubik's Cube despite external perturbations [7].

The main points of this approach are:

- An actor-critic consisting of an artificial neural network (ANN) equipped with LSTM cells to install internal memory
- Automatic domain randomization (ADR) to generate diverse environments with randomized physics and dynamics (e.g. weight and size of the manipulated object)

This results in a system with high robustness and high success rates in the transfer from simulation to testing in the real-world environment. Due to the combination of internal memory and ADR this approach also shows signs of emerging ML.

In RL, it is necessary to provide the learning agent with an extrinsic reward signal. This enables the agent to determine if the actions applied to the environment have a positive effect in the long run. Extrinsic reward signals are called sparse if the reward for a certain action is temporally disentangled from the reward, e.g. only a positive reward is given after every successful task. To tackle this problem of sparse extrinsic rewards, we can divide the approaches in literature into two classes. First, by changing the reward function, e.g. using curiosity-driven exploration [8], which introduces an intrinsic reward function. Second, hierarchical RL methods which try to divide the main task into a sequence of sub-goals can be used. While the main goal is to successfully perform the task, the agent first learns to find a policy for the sub-goals. One popular candidate for this are FeUdal Networks [9] in which the agent is split into two parts. The manager learns to formulate goals and the worker is intrinsically rewarded to follow the goal. A similar approach is Hierarchical Actor-Critic [10] in which the agent learns to set sub-goals to reach the main goal.

3 Concept for Manufacturing Scenario

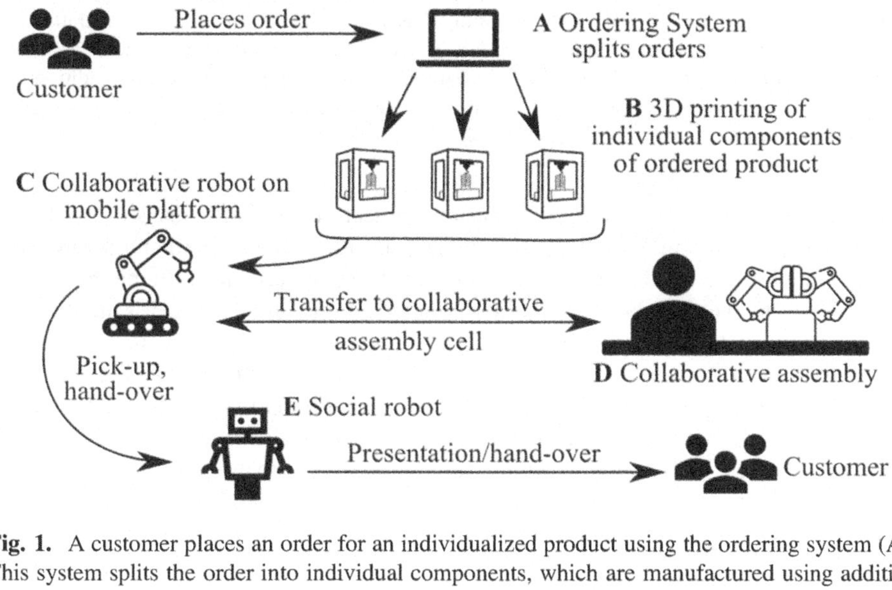

Fig. 1. A customer places an order for an individualized product using the ordering system (A). This system splits the order into individual components, which are manufactured using additive manufacturing (B). Once the manufacturing process is completed, the collaborative robotic arm Panda (Franka Emika, Germany), mounted on an automated guided vehicle (C), removes the components from the 3D printers and transports them to the collaborative assembly cell (D). This cell consists of the collaborative robot YuMi (ABB, Switzerland) and a human worker collaboratively assembling the individual components into the final product. Once the assembly process is completed, the mobile robotic arm transports the finished product and hands it over to a social robot (E), which presents and hands over the product to the customer.

The development of intelligent pick and place tasks for the assembly of products using RL is an integral part of the manufacturing scenario we are setting up in the Cologne Cobots Lab, shown in Fig. 1. It combines individualized production using additive manufacturing, autonomous mobile systems that transport components, and collaborative and social robotics.

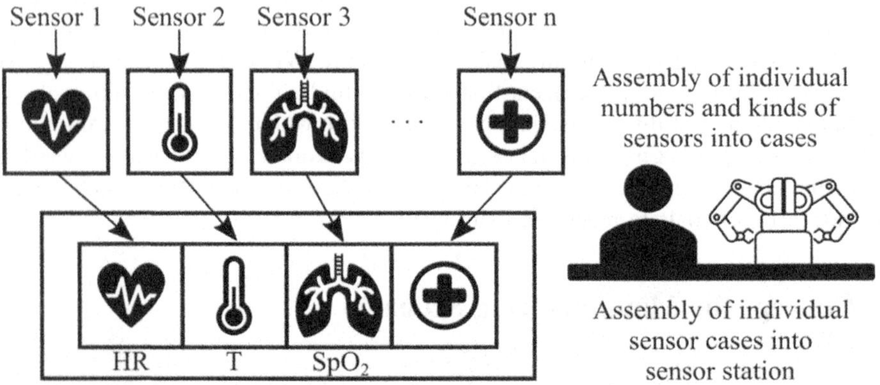

Fig. 2. The goal of the assembly task is to pick individual health care sensors, (e.g. sensors by Maxim Integrated (USA): electrocardiogram (EKG, MAX30003), body temperature (T, MAX30205), blood oxygen saturation (SpO_2), MAX30102), place them into their cases, then place the sensor cases in a health care sensor station for the monitoring of health conditions. As the sensor combination changes with each user, reinforcement learning will be used to successfully train an agent to perform each individual assembly task.

As described in Sect. 2, we aim to perform object manipulation tasks using RL in a real-world scenario, in which a specific and useful product is manufactured and assembled. Our goal is to assemble individualized sensor cases for different health care sensors (e.g. to measure body temperature, heart rate or blood oxygen saturation), as shown in Fig. 2. These cases will be used in our research concerned with social robots conducting health assessments [11]. The assembly of individualized products is desirable, as different users with different health conditions require different kinds of information. The long-term goal regarding these sensors is to create individualized wearable devices with different kinds and numbers of health care sensors. These parameters offer a promising approach to a hybrid job shop scheduling or action planning system in which human and robot actions are combined in an optimized way.

4 Approach: Hardware and Machine Learning

In our manufacturing scenario, we will be using various hardware and software/machine learning components, described in the following.

The individual sensor cases will be manufactured using fused deposition modeling (FDM)/fused filament fabrication (FFM). For the assembly, we will use the YuMi collaborative robot (ABB, Switzerland), which has two arms with 7 degrees of freedom (DoF) each, a payload of 0.5 kg and a pose repeatability of 0.02 mm. To receive additional feedback during object manipulation tasks, its gripper will be outfitted with tactile sensors. This will improve the efficiency and robustness of the grasping task. The sensor provides feedback regarding the grip quality confidence and enables slip detection of the grasped object, which can then be counteracted by improving the applied force of the gripper on the object. For the object detection, we will use a 3D scanner to create heightmaps of the objects. Several area scan cameras will be implemented for vision-based information from various angles and to determine the orientation of the objects.

In order to successfully teach the robot to perform the pick and place task, the trained machine learning algorithm needs to recognize which produced element belongs to the corresponding case. To accomplish this, the tools presented in Sect. 2 will be implemented and combined. By implementing ADR [7], the agent will be trained to realize a robust system with a high success rate in the transfer from simulation to the real world. Additionally, due to the implementation of ML and solutions for sparse reward, the learning time of the agent will be decreased. The main consideration for this approach is:

- How can we apply (a combination of) machine learning algorithms to generalize pick and place assembly tasks (i.e. various weights, sizes, geometries, quantities) for individualized products?

The Meta-Learners by Finn, Duan, and Mishra [4–6, 12] have been successfully taught to perform simple RL tasks, such as locomotion tasks. In Meta-World [13], these algorithms have been tested for a variety of advanced robotic tasks. However, the results for the presented problems have a maximum rate of successfully solved tasks of 50%. This does not seem high enough to confidently and robustly perform the presented pick and place operations. We therefore aim to reach a higher success rate with the following concept:

1. Generate and define the described collaborative tasks in the simulated environment using Mujoco (Roboti LLC, USA)
2. Find a suitable configuration and combination of ML algorithms which will solve the pick and place task
3. Implement and fine tune the trained algorithm in the real-world production scenario, applying the concept of transfer-learning

5 Conclusion and Outlook

In this paper, we propose an approach to successfully perform intelligent pick and place tasks for the assembly of individualized products using RL algorithms capable of ML.

The developed demonstrator will then be used to answer the following research questions, which are both of technical and socio-technical nature:

- Which algorithms have which impact on the robustness of the system? How can we assure that the robustness reached in simulation can be transferred to the real-world environment, e.g. using ADR?
- How can we implement a dynamic work space for the robot when working collaboratively with a human? How can a human be integrated into the collaborative assembly process in a way that is both sensible and effective?

In the future, we plan on fully implementing the developed assembly process into our manufacturing scenario described in Sect. 3. The manufacturing scenario includes the transportation of individual parts using automated guided vehicles and presenting the final product to the customer. A further goal is to study the collaborative assembly process between humans and robots. This is the focus of another research project in our lab, which aims to achieve adaptive human-robot collaboration through the implementation of sensors to detect the user's status (e.g. focus, stress). The combined results of these projects will contribute to an optimal collaborative working process.

References

1. Brettel, M., Friederichsen, N., Keller, M., Rosenberg, M.: How virtualization, decentralization and network building change the manufacturing landscape: an Industry 4.0 perspective. Int. J. Mech. Ind. Sci. Eng. **8**, 37–44 (2014)
2. Outón, J.L., Villaverde, I., Herrero, H., Esnaola, U., Sierra, B.: Innovative mobile manipulator solution for modern flexible manufacturing processes. Sensors **19**, 5414 (2019). https://doi.org/10.3390/s19245414
3. Zadpoor, A.A., Malda, J.: Additive manufacturing of biomaterials, tissues, and organs. Ann. Biomed. Eng. **45**, 1–11 (2017). https://doi.org/10.1007/s10439-016-1719-y
4. Duan, Y., Schulman, J., Chen, X., Bartlett, P.L., Sutskever, I., Abbeel, P.: RL^2: fast reinforcement learning via slow reinforcement learning. arXiv:1611.02779 [cs, stat] (2016)
5. Mishra, N., Rohaninejad, M., Chen, X., Abbeel, P.: A simple neural attentive meta-learner. arXiv:1707.03141 [cs, stat] (2018)
6. Finn, C., Abbeel, P., Levine, S.: Model-agnostic meta-learning for fast adaptation of deep networks. In: Proceedings of the 34th International Conference on Machine Learning-Volume 70, pp. 1126–1135 (2017). http://www.JMLR.org
7. OpenAI, Akkaya, I., Andrychowicz, M., Chociej, M., Litwin, M., McGrew, B., Petron, A., Paino, A., Plappert, M., Powell, G., Ribas, R., Schneider, J., Tezak, N., Tworek, J., Welinder, P., Weng, L., Yuan, Q., Zaremba, W., Zhang, L.: Solving rubik's cube with a robot hand. arXiv:1910.07113 [cs, stat] (2019)
8. Pathak, D., Agrawal, P., Efros, A.A., Darrell, T.: Curiosity-driven exploration by self-supervised prediction. In: Proceedings of the IEEE Conference on Computer Vision and Pattern Recognition Workshops, pp. 16–17 (2017)
9. Vezhnevets, A.S., Osindero, S., Schaul, T., Heess, N., Jaderberg, M., Silver, D., Kavukcuoglu, K.: Feudal networks for hierarchical reinforcement learning. In: Proceedings of the 34th International Conference on Machine Learning-Volume 70, pp. 3540–3549 (2017). http://www.JMLR.org
10. Levy, A., Konidaris, G., Platt, R., Saenko, K.: Learning multi-level hierarchies with hindsight. arXiv:1712.00948 [cs] (2019)

11. Richert, A., Schiffmann, M., Yuan, C.: A nursing robot for social interactions and health assessment. In: International Conference on Applied Human Factors and Ergonomics, pp. 83–91. Springer (2019)
12. Finn, C., Xu, K., Levine, S.: Probabilistic model-agnostic meta-learning. In: Bengio, S., Wallach, H., Larochelle, H., Grauman, K., Cesa-Bianchi, N., Garnett, R. (eds.) Advances in Neural Information Processing Systems 31, pp. 9516–9527. Curran Associates, Inc. (2018)
13. Yu, T., Quillen, D., He, Z., Julian, R., Hausman, K., Finn, C., Levine, S.: Meta-world: a benchmark and evaluation for multi-task and meta reinforcement learning. arXiv:1910.10897 [cs, stat] (2019)

A Probabilistic Model of Taking-Over Control from Semi-autonomous Vehicles

Hermann Kaindl[1(✉)] and Gunther Paul[2]

[1] TU Wien, Vienna, Austria
hermann.kaindl@tuwien.ac.at
[2] James Cook University, Townsville, QLD, Australia
gunther.paul@jcu.edu.au

Abstract. In automated driving scenarios with semi-autonomous vehicles of today and the near future, a request for taking-over control may at any time be communicated to a human driver. Empirical studies exist of, e.g., the effects of different modalities used in this human-machine interaction, and of drowsiness of drivers involved, depending on manual or automated driving. However, there is no model available yet of how the decision-making time in the course of taking-over control from the automated vehicle depends on the driver's drowsiness. Hence, we present a probabilistic model of the decision-making time as influenced by certain properties characterizing the driver's drowsiness according to previous work.

Keywords: Human Factors and Ergonomics (HFE) · Probabilistic models · Automated Driving · Human-System Integration (HSI)

1 Introduction

Only vehicles of Level 5 according to the Automated Driving (AD) Roadmap ERTRAC 17 [1] will not need human interference, and this will not be achievable anytime soon. Human Factors and Ergonomics (HFE) studies suggest that there are several problems involved especially with vehicles of Level 3, when humans are supposed to take over control within a short time span in order to handle safety-critical situations that the machine cannot [4, 11]. In an earlier probabilistic study [14], we already showed how bad it can be for humans and their safety having to take over control all of a sudden from a semi-autonomous vehicle in a situation that requires emergency braking [3].

There are situations that are less dangerous but still require a human driver to take over control, if and when the machine requests it in a situation like the following. When driving on a freeway, the automated system may spot something it cannot handle, like a construction site. If the human driver will not take over control within a certain short time span, a serious accident may occur at Level 3, or the vehicle will have to safely stop at Level 4. The former situation is clearly safety-critical, while the latter is still unpleasant.

Generally, in any situation where a semi-autonomous vehicle requests the human driver to take over control, the decision-making time of this human matters. We model

W. Karwowski et al. (Eds.): IHSED 2020, AISC 1269, pp. 332–337, 2021.
https://doi.org/10.1007/978-3-030-58282-1_52

it probabilistically and we are interested in how its probability distribution can be determined based on certain properties related to drowsiness, like the driving time or the age of the driver. Much as in our previous study of emergency braking [14], the basic approach to modeling human-in-the-loop (HITL) is the one of [12]. This new study is actually complementary to our previous one, since it determines the specification of the very probability distribution modeling the human decision-making time that is a major part of the probabilistic approach for determining the probability of an accident.

The remainder of this paper is organized in the following manner. First, we sketch some background material and discuss related work, in order to make this paper self-contained. Then we present our new theoretical model, a deterministic approach first and a probabilistic approach based on it. Finally, we draw a few conclusions and sketch future work on our way towards probabilistic analysis of human-system integration (HSI) in automated driving.

2 Background and Related Work

2.1 Empirical Studies on Human-Machine Interaction (HMI)

Empirical studies like [11] try to find out how unimodal or multimodal signals may support control transitions in semi-autonomous vehicles, i.e., they focus on the *modalities* of human-machine interaction. For our current paper, studies like [4] are much more relevant, since they investigate inherent properties of the situation and of the human driver.

In particular, the driving mode (manual/automated), driving time, and the driver's age (young/old) have been shown to have significant effects on drowsiness in [4]. The participants got more drowsy in automated than in manual driving, with the younger subjects achieving higher levels of drowsiness. Even though there was no clear correlation shown between subjective drowsiness and reaction time in this work, we assume that, in general, a dependency between such properties and decision-making time exists, which we try to capture in our probabilistic model.

2.2 Modeling and Simulation of Vehicle-Driver Integration Issues in HFE

These findings in [4] are consistent with HFE theory. A comprehensive overview of the foundations of such a system can be found in [7]. A system dynamic (i.e., control) model, as commonly used in systems engineering might be applied to the problem, with the addition of a dead-time element to a PID human controller, representing the reaction time of the driver. However, the reaction time, is unknown, and will vary significantly. Without going into too much detail, it is theoretically well established that such a system that includes a dead-time element is intrinsically unstable, a finding that we will see confirmed in experimental studies. It is sufficient here to conclude from such a model that, while a vehicle-driver control system may become unstable if the D element is poorly configured (e.g., the driver-controller overreacts), the system is much more likely to become unstable (i.e., the vehicle will crash) if the controller involves a large enough dead-time element.

The model approach typically used in HFE for this problem is a cognitive performance model [8], which considers human cognitive capacity allocation rather than the dynamics of the system. The logic behind this model selection and approach is based on Hollnagel and Woods [2], who argued that human-machine systems should be designed and analyzed at a cognitive level. Rasmussen's decision-making and information-processing model is structured in three levels, a skill-based level, rule-based level and knowledge-based level (for this reason, the model is also called SRK model). From sensory input (i.e., the instruction of the AD system to take over, by voice and/or visual and/or tactile), feature information may lead to an automated sensory-motor reaction as the fastest possibility. However, such a pattern must be 'imprinted' into the skill base of an operator. In the case of taking-over control from an AD system, this is highly unlikely, if not impossible. On the second and higher level of rule-based behavior, the feature information must be recognized (i.e., it is processed as a sign), associated with a task, connected with a learned rule, and will then lead to allocation of sensory-motor patterns. A behavioral response at this rule-based level will take significantly longer than a response at the skill-based level. We expect an AD system-driver taking-over response to normally occur at this level.

In this paper, driver drowsiness comes into play, because both recognition of a signal and association with a task can be assumed affected by drowsiness. At the highest level of knowledge-based behavior, feature information from the AD system would be processed as a symbol, which requires identification before a goal-driven decision may kick-off a planning process, which will eventually allocate learned rules and response patterns. Behavior at this cognitive level takes the most processing effort and time, and if several exacerbating conditions coincide (e.g., no or poor training, little experience, little awareness of the AD system, drowsiness, etc.), AD taking-over control may effectively occur at this level, causing a very significant dead-time.

2.3 Our Own Previous Approach Towards Probabilistic Analysis

Probabilistic analytical modeling (PAM) for human-in-the loop (HIL) systems [12] could effectively complement computational *simulations* (see, e.g., [6, 10]) or *model checking* (see, e.g., [5, 9]). Based on PAM, we presented and analyzed an example, where the automated system all of a sudden hands over to the driver in the face of an obstacle [14]. In this study, the decision-making time has been modeled as a random variable distributed in accordance with the Rayleigh law. Further random variables modeled deceleration time and distances probabilistically as well. This allowed deriving a formula for the probability that an accident occurs.

3 Theoretical Model

We are specifically interested in a theoretical model of the relationship between drowsiness of drivers and their decision-making times. Hence, we define a function that maps the former to the latter. We define a deterministic function for this mapping first, and then derive a probabilistic approach based on it.

3.1 Deterministic Approach

Drowsiness is a complex phenomenon that we cannot fully cover here. Still, we can characterize it as inspired by [4], where the relationship between certain properties and (subjectively measured) drowsiness was studied. While the driving mode was one of the independent variables in this study, we are only interested here in automated driving. Hence, we exclude the driving mode from our model.

Let us define a property space for characterizing drowsiness as a tuple (D, A), where

1. D is the real-valued time *duration* of automated driving.
2. A is the integer-valued *age* of the driver.

T is the decision-making *time* for given drowsiness as characterized through these properties. This leads us to the following scheme of a mapping function:

For given $d \in D$ and $a \in A$, the function f: $D \times A \rightarrow T$ maps (d, a) to the decision-making time t.

3.2 Probabilistic Approach

It would be unrealistic to assume that this mapping is deterministic. Hence, we propose a probabilistic approach for defining a function that maps drowsiness of the driver to the decision-making time.

Much as in [14], we assume that the decision-making time T is a *random variable* distributed in accordance with the Rayleigh law. The decision-making time t is non-negative, and it is usually not very long. In such a situation, small time values are more likely than high time values, and, because of that, their probability density distribution functions should be heavily skewed in the direction of short time values.

The Rayleigh distribution $f(t;\sigma) = \frac{t}{\sigma^2}\exp\left(-\frac{t^2}{2\sigma^2}\right)$ possesses all these important properties, where $\sigma > 0$ is the *scale parameter* of the distribution. For an illustration, see Fig. 1. The key idea behind our probabilistic approach taken in this paper is to map from the property space for characterizing drowsiness to this scale parameter σ of the Rayleigh distribution:

For given $d \in D$ and $a \in A$, the function f_p: $D \times A \rightarrow \Re^+$ maps (d, a) to the scale parameter σ of the Rayleigh distribution of the random variable T, which specifies the decision-making time.

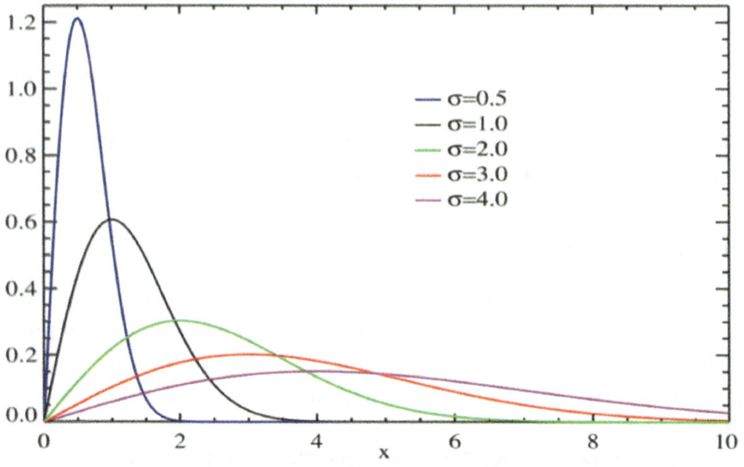

Fig. 1. Rayleigh probability density function.

This function f_p (d, a) will have to be fully specified yet, of course, but our model already defines how drowsiness characterized by certain properties can be mapped to a probability distribution. With regard to the human driver, only the age is included here yet. Based on previous work [12], constituents of the *human capacity factor* (HCF) may be included as additional properties, for example.

Once the scale parameter of the Rayleigh distribution is determined, the probability of an accident can be calculated in situations where the automated vehicle requests taking-over control for emergency braking [14]. It may also be used for other situations with requests for taking-over control, once they will have been modeled probabilistically.

4 Conclusion and Future Work

In this paper, we propose a probabilistic model of a driver's decision-making time after a request for taking-over control from a semi-autonomous vehicle, depending on certain properties related to drowsiness. Inspired by [4], we included here as properties the time duration of automated driving, and the age of the driver. Other properties may be added, such as constituents of the *human capacity factor* (HCF), see, e.g., [12]. More precisely, we proposed defining a function that maps such properties to the scale parameter σ of the Rayleigh distribution of the random variable T, which specifies the decision-making time. Based on the resulting specification of this probability distribution, the probability of an accident can be calculated in situations where a probabilistic model like the one in [14] is available.

Future work will have to fully specify this mapping function. For real-world use, an approach for that feasible today could be *machine learning* while actually driving and collecting masses of data for that purpose.

References

1. ERTRAC 17, Automated Driving Roadmap, Version 7.0, May 2017
2. Hollnagel, E., Woods, D.D.: Cognitive systems engineering: new wine in new bottles. Int. J. Man Mach. Stud. **18**, 583–600 (1983)
3. Kudarauskas, N.: Analysis of emergency braking of a vehicle. Transport **22**(3), 154–159 (2007)
4. Kundinger, T., Riener, A., Sofra, N., Weigl, K.: Driver drowsiness in automated and manual driving: insights from a test track study. In: Proceedings of the 25th International Conference on Intelligent User Interfaces (IUI' 2020), pp. 369–379. ACM (2020)
5. Luckender, C., Kaindl, H.: Systematic top-down design of cyber-physical models with integrated validation and formal verification. In: 40th International Conference on Software Engineering: Companion. ACM/IEEE (2018)
6. Luckender, C., Rathmair, M., Kaindl, H.: Investigating and coordinating safety-critical feature interactions in automotive systems using simulation. In: 50th Hawaii International Conference on System Sciences, pp. 6151–6160 (2017)
7. Paul, G.: Modeling and simulation of human systems. In: Salvendy, G., Karwowski, W. (eds.) Handbook of Human Factors and Ergonomics, 5th edn. John Wiley & Sons, New York (2020)
8. Rasmussen, J.: Skills, rules, and knowledge; signals, signs and symbols, and other distinctions in human performance models. IEEE Trans. Syst. Man Cybern. **13**(3), 257–266 (1983)
9. Rathmair, M., Luckender, C., Kaindl, H.: Minimalist qualitative models for model checking cyber-physical feature coordination. In: 23rd Asia-Pacific Software Engineering Conference (APSEC' 2016). IEEE Press (2016)
10. Rathmair, M., Luckender, C., Kaindl, H., Radojicic, C.: Semi-symbolic simulation and analysis of deviation propagation of feature coordination in cyber-physical systems, best paper award. In: 51st Hawaii International Conference on System Sciences (HICSS 2018), pp. 5655–5664 (2018)
11. Salminen, K., Farooq, A., Rantala, J., Surakka, V., Raisamo, R.: Unimodal and multimodal signals to support control transitions in semiautonomous vehicles. In: Proceedings of the 11th International Conference on Automotive User Interfaces and Interactive Vehicular Applications (AutomotiveUI' 2019), pp. 308–318. ACM (2019)
12. Suhir, E.: Human-in-the-Loop: Probabilistic Modeling of an Aerospace Mission Outcome. CRC Press, Boca Raton, London, New York (2018)
13. Suhir, E.: Assessment of the required human capacity factor (HCF) using flight simulator as an appropriate accelerated test vehicle. IJHFMS **6**(1), 71–74 (2019)
14. Suhir E., Paul G., Kaindl H.: Towards probabilistic analysis of human-system integration in automated driving. In: Ahram T., Karwowski W., Vergnano A., Leali F., Taiar R. (eds) Intelligent Human Systems Integration (IHSI' 2020). Advances in Intelligent Systems and Computing, vol. 1131, pp. 9–14. Springer, Cham (2020)

Selecting the Right Tool for the Task: A Hard-Soft Cake Eating Experiment with a Spoon and Fork

Sondre Bjarkum[1], Maximilian Stiegler Chadoyan[1], Fausto Orsi Medola[2], and Frode Eika Sandnes[1,3(✉)]

[1] Oslo Metropolitan University, 0130 Oslo, Norway
{s305075, s334005, frodes}@oslomet.no
[2] Sao Paulo State University (UNESP), Bauru, Brazil
fausto.medola@unesp.br
[3] Kristiania University College, 0153 Oslo, Norway

Abstract. It is generally accepted that one need to select the correct tool for a given task. The same holds for eating utensils. A cake eating experiment was designed to determine if the visual appearance of cakes with different associated hardness would affect the choice of eating utensil. A total of 25 participants was recruited. The results did not reveal any effect of cake type, or presentation layout on the choice of eating utensil. It therefore seems that other factors, such as, for instance, customs and/or habits, are more influential in the choice of cake eating tool choice.

Keywords: Eating utensil · Design · Fork spoon · Preference · Hand dominance

1 Introduction

The preferences and mechanisms controlling the selection of tools in various situations has been extensively studied, including at industrial level [1] and tool level [2]. Generally, tools are linked to certain problems, training, habits, customs and culture. In order to decide about the most appropriate tool for a task an individual must first identify the nature of the problem. Then, an individual is likely to choose a tool that he or she is trained to use, or according the individuals' habits.

Culture also plays a part. In Chinese Heritage Cultures chopsticks are mainly used for most eating, with the occasional accompanying by a spoon, while in Western cultures knives, forks and spoons are used to achieve the same goals. Some individuals are trained in eating utensil etiquette from childhood, such as holding the knife and the fork in the "correct" way and in the "correct" hands. Others, on the other hand, just use whatever seems more practical and natural with less regard for etiquette.

Figure 1 illustrates one classic table setting which may be used on formal occasions so that a meal can be consumed according to a set protocol and custom. In this convention, utensils are used from the outside and inwards for the starter and main course, while the utensils on the top are used for the desserts, being it pudding or cake.

W. Karwowski et al. (Eds.): IHSED 2020, AISC 1269, pp. 338–343, 2021.
https://doi.org/10.1007/978-3-030-58282-1_53

Notice that the dessert utensils often comprises both a cake fork and a dessert spoon. There seems to be no generally agreed convention for what is the "correct" way to eat cake. A search on this topic on discussion forums reveals that most people prefer forks for eating cake as they do not have to "chase the last piece around the plate with the spoon", and the cake fork often have a knife-like edge that can be used for cutting. Other argue that the spoon is useful for sorbets, puddings and other soft desserts, and sometimes hard cakes are served on the same dessert platter as a sorbet, pudding or mousse. Others claim that desserts should be formally eaten with a spoon, with a fork as an optional extra utensil.

Fig. 1. A classic table setting with spoon and fork for desserts.

This study set out to study whether the hardness associations triggered by the visual appearance of a food affected the participants' choice of eating utensils. Cake was chosen as the food category as most people like cake and it would simplify recruiting participants. Cake and desserts are usually eaten with a spoon or a fork, and these two categories of utensils were therefore used in the experiment. It was predicted that with soft cake the participants would choose a spoon as the spoon will be more efficient in preserving the shape of the cake, while it was predicted that a fork would be preferred when eating a hard cake as more force is needed to break the firm cake into pieces.

The rest of this paper is organized as follows. Section 2 present related work, Sect. 3 outlines the methodology, while the results are presented in Sect. 4 and discussed in Sect. 5. Conclusions are presented in Sect. 6.

2 Related Works

The literature on eating utensils have focused on the design of utensils in general [3], to facilitate children [4] and individuals with reduced motor function such as Parkinson's disease [5–7] and arthrogryposis multiplex congenital [8].

Eating with utensils have also been used to study hand preference among children [9]. Eating utensils have also been studied from a more technological perspective, such as eating utensils with sensors that can detect eating problems among children [10, 11]. Moreover, interactive eating utensils have been used as a platform for developing fine motor skills [12] and to slow down eating speeds [13].

3 Method

3.1 Experimental Design

A within groups controlled experimental design was chosen with one independent variable, namely visual food hardness with two levels, hard or firm (brownie) and soft (cream cake). The dependent variable was utensil preference. This variable was a dichotomous variable with values fork and spoon. Utensil placement (left-right) was also used as a random variable.

3.2 Participants

A total of 25 participants was recruited for the experiment among the students in the first authors' University, of which 12 were female and 13 were male. Of these, 6 participants were left-hand dominant, while 19 participants were right-hand dominant.

3.3 Materials

Two sets of cakes were baked, namely hard backed brownies and three-layer cream cake. Both cakes were assumed to be familiar to participants both in taste and consistency. Brownies are usually hard and sticky while cream cakes are usually soft, and it was assumed that the participants would expect the hardness of the cake type according to its visual appearance before eating.

3.4 Procedure

The participants were tested individually in isolation. First, each participant was brought into a room and placed in front of a desk with one cake type on a plate with a metal spoon and knife on a napkin next to the cake. Next, each participant was informed about the experiment and were asked about any potential allergies. Then they were asked to choose one of the utensils and eat the cake. When finished, they were asked to turn around while the other cake type and the utensils in opposite order were placed in front of the participants. They were then asked to turn back and asked to select one of the utensils and eat the cake. The order of the cakes was randomized, and so was the initial left and right placements of the knife and the fork to minimize the risk of bias. During each eating trial the participants' choice of eating utensil was recorded. Spearman correlations confirmed that there was no correlation between the presentation side of utensil and the participants' choice of utensil, hence no effect of the random variable utensil position.

3.5 Analysis

Statistical analyses were performed with JASP version 0.11.0.0 [14].

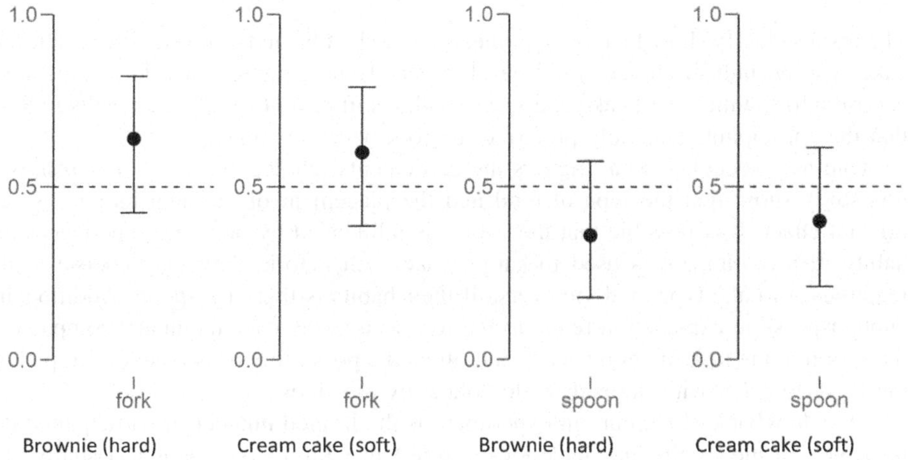

Fig. 2. Proportions of fork and spoon choices for hard and soft cake with 95% confidence intervals.

4 Results

Figure 2 shows the results of the experiment. According to our predictions 100% of cake eaters will choose a fork to eat a hard cake while 100% will choose a spoon to eat a soft cake. With our cohort of students ($N = 25$), a Binomial test revealed that the proportion of students who prefer forks while eating hard cakes was significantly less (64.0%, $p < .001$) and those using spoons significantly more (36.0y6%, $p < .001$) than expected. Moreover, the proportion of students who prefer spoons while eating soft cakes was significantly less (40.0%, $p < .001$) and those using forks was significantly more (60.0%, $p < .001$) than expected. A hypothesis based on 50/50 distribution of choices cannot be rejected. A contingency table analysis further confirms that there was no significant difference in choice of utensil for the two cake types ($\chi^2(1) = 1.418$, $p = .234$).

There appear to be some differences between right dominant participants and left dominant participants, although the number of left dominant participants is too small to be the basis for any conclusions. Among right hand dominant participants, 72.2% prefer the fork with the hard cake, while only 55.6% prefer fork with the soft cake. For left hand dominant participants only 40% preferred fork with the hard cake and 60% preferred fork with the soft cake. There is a moderate non-significant correlation between handedness and the utility choice for hard cake ($r_s(25) = 0.359$, $p = .078$). A contingency table analysis further confirms that there is a significant difference in choice of utensil for the two cake types for right-hand dominant participants ($\chi^2(1) = 3.997$, $p = .046$).

5 Discussion

The results clearly show that the hypothesis had to be rejected as no significant effect of cake type on utensil choice could be observed. Hence, participants do not distinctly choose a fork with a hard cake and spoon with a soft cake. Overall, the results indicate that the participants generally preferred the fork over the spoon.

One may speculate what triggers the participants' choice of utensil. The results of this study show that the type of food and the placement of the utensils had no significant effect. It is possible that the choice is influenced by previous experiences and habits. If a participant is used to eating cake with a fork, they may choose a fork regardless of cake type, and vice versa if their habits is that of a spoon. Additionally, another possible explanation refers to the fork as a more versatile utensil compared to the spoon in the task of eating cake, as it offers the possibility of two actions to pick up the food: to prick with the ends or to collect by the sides.

One drawback of the current experiment is the limited number of participants. It is possible that the results may have looked different with a larger sample. Usually, it is relatively challenging to recruit participant to voluntarily participate in controlled experiments. This was not the case in this study. In fact, once rumor got around about the experiment, several students contacted the authors with a request to participate. The prospect of getting free cake to participate in a short and easy experiment served as an effective incentive. The practicality of having enough material (cake) to scale up the experiment is the main obstacle in this experiment.

The forks used in the experiment were slightly larger than the spoons. It may be that this has caused some bias if participants preferred to use the largest utensil.

6 Conclusions

An experiment was conducted to assess if the type cake with distinct hardness characteristics would affect the choice of eating utensil. The results did not show that the type of cake affected the choice of eating utensil. About 60% of the participants preferred to use a fork and the remaining participants chose to use a spoon. Further studies exploring different types of food might bring information that could elucidate this question.

References

1. Yurdakul, M.: AHP as a strategic decision-making tool to justify machine tool selection. J. Mater. Process. Technol. **146**(3), 365–376 (2004)
2. Oh, S., Radwin, R.G.: Pistol grip power tool handle and trigger size effects on grip exertions and operator preference. Hum. Factors **35**(3), 551–569 (1993)

3. Laohakangvalvit, T., Achalakul, T., Ohkura, M.: A proposal of model of Kawaii feelings for spoon designs. In: International Conference on Human-Computer Interaction, pp. 687–699. Springer, Cham (2017)
4. Liu, B.S., Chen, Y.L.: Evaluation of dining utensil design for children. In: Ergonomics in Asia: Development, Opportunities and Challenges: Proceedings of the 2nd East Asian Ergonomics Federation Symposium (EAEFS 2011), CRC Press (2012)
5. Ma, H.I., Hwang, W.J., Tsai, P.L., Hsu, Y.W.: The effect of eating utensil weight on functional arm movement in people with Parkinson's disease: a controlled clinical trial. Clin. Rehabil. **23**(12), 1086–1092 (2009)
6. Doughty, T.A., Bankus, N.: Mechanical modeling and design for reduction of parkinsonian hand tremor. In: ASME 2010 International Mechanical Engineering Congress and Exposition, pp. 521–527. American Society of Mechanical Engineers Digital Collection (2010)
7. Doughty, T., Heintz, J., Ishii, M.: Reducing parkinsonian hand tremor with a novel dynamic eating utensil. In: ASME 2013 International Mechanical Engineering Congress and Exposition. American Society of Mechanical Engineers Digital Collection (2013)
8. Lieberman, B., Alqumeran, M., Scharf, J., Damcott, M., Foulds, R.: Design of an assistive eating utensil for an individual with arthrogryposis multiplex congenital. In: Proceedings of the 2010 IEEE 36th Annual Northeast Bioengineering Conference. IEEE (2010)
9. Gordon, K.: A study of hand and eye preference. Child Dev. **2**(4), 321–324 (1931)
10. Kadomura, A., Li, C.Y., Chen, Y.C., Tsukada, K., Siio, I., Chu, H.H.: Sensing fork: eating behavior detection utensil and mobile persuasive game. In: CHI' 2013 Extended Abstracts on Human Factors in Computing Systems, pp. 1551–1556. ACM (2013)
11. Zuckerman, O., Gal, T., Keren-Capelovitch, T., Karsovsky, T., Gal-Oz, A., Weiss, P.L.T.: DataSpoon: overcoming design challenges in tangible and embedded assistive technologies. In: Proceedings of the TEI' 2016: Tenth International Conference on Tangible, Embedded, and Embodied Interaction, pp. 30–37. ACM (2016)
12. Chia, F.Y., Saakes, D.: Interactive training chopsticks to improve fine motor skills. In: Proceedings of the 11th Conference on Advances in Computer Entertainment Technology (2014)
13. Hermsen, S., Hermans, R.C.: Take It Slow: can feedback from a smart fork reduce eating speed?. In: Frontiers in Public Health Conference Proceedings (2016)
14. JASP Team: JASP (Version 0.11.1)[Computer software]. (2019)

Exploring Relationship Between User Satisfaction and Impacts of Digital Competence Certification System in Schools

Igor Balaban(✉) and Aleksandra Sobodić

Faculty of Organization and Informatics, University of Zagreb, Pavlinska 2, 42000 Varaždin, Croatia
{igor.balaban, aleksandra.sobodic}@foi.unizg.hr

Abstract. This research developed a measurement instrument to assess user satisfaction and impact of digital competence system in primary and secondary schools. The testing platform was a prototype of a digital competence certification system that has been developed and piloted during 2019 in primary and secondary schools in six European countries involving more than 800 teachers and 6.000 students. In the process of evaluation, 298 teachers' responses were analysed, and the measurement instrument was evaluated for validity and reliability. Results revealed a strong positive relationship between user satisfaction and impacts the DC system has on teachers and their work. Such a relationship is expected to engage teachers to use the system as well as their students to acquire new competences. Moreover, it is concluded that satisfaction contributes to motivation, user experience and performance within the system.

Keywords: User satisfaction · Net impacts · Digital competence certification

1 Introduction

Today, a young person needs to acquire the set of digital skills most commonly represented as a digital competence (DC) to be able to enter the labour market without the risk of exclusion. The literature review has shown that the best solution for students to acquire digital competence is to integrate it into the formal educational curriculum [1, 2]. This is further supported by [3–6] who suggest that education and assessment of DC should be started from the earliest age of students and promoted throughout the curriculum. That way, schools could timely identify the lack of a specific DC and intervene with a certain plan of development.

However, very few studies have been reported to deal with the assessment of DC at any level of education, especially in primary and secondary education. A three-year longitudinal study [7] concluded that digital skills do not develop equally over the years of education, e.g. creating information skill has been developing most slowly. An interesting fact is that numerous papers are investigating the competences of teachers [5, 8, 9] and their adoption of technology [10] in education since they are considered to be the driving force behind the development of students' DC on all levels of education [11].

Therefore, it is necessary to study the ways in which teachers are using DC certification systems and to indicate the benefits of such systems for teachers. With this

W. Karwowski et al. (Eds.): IHSED 2020, AISC 1269, pp. 344–349, 2021.
https://doi.org/10.1007/978-3-030-58282-1_54

respect, the main aim of this paper is to assess teachers' satisfaction with the CRISS DC platform and to identify and assess the impact such a system has on teachers' work.

2 Research Aims

Since recent research findings (see for example [11] and [12]) suggest that teachers are the main drivers for the incorporation of DC assessment and certification into curricula, we focus our research on teachers' perception and define the main research questions:

1. In what ways an online digital competence certification system in primary and secondary schools impacts teachers' work?
2. What is the relationship between teachers' satisfaction with DC certification system and its impact on the work of teachers?

In order to answer the proposed questions, the research aims are:

1. To develop and validate a survey instrument that assesses the teachers' satisfaction and impacts of DC certification system.
2. To examine the relationships between elements of teachers' satisfaction and impacts of DC certification system.

3 Research Context

Our research context was based on the online CRISS platform for DC acquisition, evaluation and certification that was developed in order to pilot DC evaluation in primary and secondary schools in Europe. It is based on the CRISS Digital Competence framework (hereinafter referred to as CRISS DC Framework) [13] that decomposes digital competence into five areas and twelve sub-competences. Each sub-competence is composed of a set of performance criteria (PC) that translate the sub-competences into more specific elements of what a student should be able to demonstrate. Teachers are responsible to plan the learning, to provide feedback and to evaluate activities and tasks that relate to an individual sub-competence. The activities and tasks are retrieved by the CRISS repository and teachers can apply them with or without further adaptations. The students should conduct the activities by performing one or more tasks and generate pieces of evidence to prove the acquisition of a specific sub-competence.

The assessment of digital (sub-)competence is performed through the CRISS platform with two types of interventions: human and technological. Human interventions are carried out by teachers and students using tools like Rubrics, Check Lists, Scales, etc., that are automatically generated by the CRISS platform and customized by teachers. The technological intervention is executed by the CRISS platform automatically which is set to track the students while working in their assigned activities and to collect relevant information i.e. the indicators of the evaluation of the pieces of evidence. The CRISS platform has been piloted in six European countries (Spain, Sweden,

Croatia, Greece, Romania and Italy) for several months with a targeted population of students aged between 9 and 16 years.

4 Measurement Instrument Development

We used two constructs from the DeLone & McLean Model [14] revised in 2016 to assess and identify the most relevant factors of teachers' satisfaction and impacts of DC certification system. The first construct, User satisfaction measures users' level of satisfaction with reports, platform, and support services. The second construct, Net impacts measures the extent to which the CRISS platform contributes to the success of teachers. The instrument development phase was conducted by following the recommendations from [15–17]. We started with the operationalization of research constructs based on the existing measures and modified it with a set of new target-specific measures. Content validity was ensured, besides using an extensive literature review, by using focus groups that involved experts in the field of pedagogy, e-learning, assessment, and teaching methodology. The final measurement instrument (see Table 1) was translated into all target languages and administered to the teachers using the LimeSurvey.

Table 1. Measurement instrument after content validity

ITEM
User satisfaction
US1 I feel comfortable using the CRISS platform
US2 I find the CRISS platform useful for additional assessment of my students
US3 I think it is worthwhile to use the CRISS platform
US4 I feel confident using the CRISS platform
US5 I am satisfied with the CRISS platform possibilities
Net impacts
NI1 The CRISS platform helps me to improve the engagement of my students
NI2 The CRISS platform enables me to provide clear evaluation criteria to my students
NI3 I am able to provide better feedback to my students through the CRISS platform
NI4 I am able to provide timely feedback to my students
NI5 The CRISS platform extends my capacity for assessment
NI6 The CRISS platform saves me time by supporting my teaching activities (planning process, guiding students, assigning tasks, monitoring students' activities, etc.)
NI7 The CRISS platform allows me to track the progress of my students much better than I could do without CRISS platform
NI8 I am able to detect underperforming students more quickly than I would do it without CRISS platform
NI9 The CRISS platform helps me to make more suitable decisions to enable students' progress
NI10 The CRISS platform enables me to propose tasks that allow students to be creative in solving them (ingenious, original)
NI11 The CRISS platform enables me to track my students' reasoning when solving the tasks

Answers on 1–5 point Likert-type scale (1 – Strongly disagree; 2 – Disagree; 3 – Uncertain; 4 – Agree; 5 – Strongly agree; NA – not applicable).

5 Sample

The assessment of the platform has been done between May and September 2019, four months after the CRISS platform has been released to selected primary and secondary schools. A dataset with 298 useful responses was further processed into descriptive statistics using R [18]. The demographic structure of respondents, whose answers were taken into account, was the following: Greece (19%), Spain (31%), Italy (19%), Romania (8%), Sweden (3%) and Croatia (19%). Sixty-eight per cent (68%) of the online instrument's respondents were women, 68% of them were aged between 30 and 49 years. The majority of teachers work in secondary schools (77%) with over 11 years of teaching experience (64%).

6 Data Analysis and Results

The measurement instrument was assessed with the variance-based SEM using SmartPLS to test the relationships between the indicators and constructs, and the paths between constructs. The measurement instrument was also examined for the internal consistency (reliability), convergent and discriminant validity. After dropping out items that did not converge into their prospective constructs (factor loadings below 0.60) or have cross-loaded highly on both constructs, reflective measurement models indicated satisfactory reliability and validity.

The internal consistency of each sub-scale was tested using Cronbach's alpha, which was calculated as 0.90 for User satisfaction and 0.94 for Net impacts. In addition to the previous measure, composite reliability was also calculated for User Satisfaction (0.93) and Net Impacts (0.95). Both structures showed highly reliable. The convergent validity is confirmed by measuring AVE for US (0.72) and NI (0.63). All indicators loaded over 0.60 into their posited construct (User Satisfaction or Net Impacts) [16]. Discriminant validity showed to be present since HTMT value is significantly different from one [19]. Regarding the structural model, no problem related to collinearity was detected because all VIF values were below the threshold of 5. The R^2 value of endogenous construct NI (0.70) is considered substantial. Satisfaction ($\beta = 0.84$, $p = 0.00 < 0.01$) was a significant predictor of the Net impacts. The result suggested that the overall value of the system could be improved by focusing on users' satisfaction.

The mean values for user satisfaction and net impacts were 3.17 (SD = 1.11) and 3.25 (SD = 1.05), respectively. As expected, the median value for both question categories was 4.00 which indicates mostly positive opinion towards the system. Teachers expressed their concerns regarding the comfortability (US1) while using the system (M = 2.88, SD = 1.12). Furthermore, teachers to some extent doubt the system saves them time (NI8) by supporting teaching activities (M = 2.78, SD = 1.15). On the other hand, they consider the system very useful as an additional assessment of students' work (item US2; M = 3.30, SD = 1.13) and their digital competence (item NI2; M = 3.70, SD = 0.94).

7 Conclusion

This research addressed the problem of measuring and investigating the relationship between user satisfaction and net impacts of cloud-based infrastructure for acquisition, evaluation and certification of digital competence in primary and secondary education focusing on teachers' perspective. In addressing the research questions, this research has validated the measurement instrument and revealed the elements of a DC certification system that impact teachers work (RQ1). By using the developed instrument, it was shown that user satisfaction is indeed a significant predictor of net impacts of DC certification system (RQ2).

According to our knowledge, CRISS platform is the first attempt to create a comprehensive, cloud-based solution for digital competence acquisition, evaluation and certification in Europe, and to pilot such solution within a formal curriculum of primary and secondary schools in six European countries. From a practical point of view, using the measurement instrument, schools can evaluate their own systems for certification of digital competences and find out, for example, whether those systems need improvement or whether the school needs to improve the quality of services for its teachers. It should be encouraging for those who plan to implement such systems or for those who plan to develop them.

References

1. Tudor, S.L.: The open resources and their influences on the formation of specific competencies for the teaching profession. In: Proceedings of the 10th International Conference on Electronics, Computers and Artificial Intelligence, ECAI 2018. IEEE (2018). https://doi.org/10.1109/ECAI.2018.8679010
2. Varela, C., Rebollar, C., García, O., Bravo, E., Bilbao, J.: Skills in computational thinking of engineering students of the first school year. Heliyon. **5** (2019). https://doi.org/10.1016/j.heliyon.2019.e02820
3. Siddiq, F., Hatlevik, O.E., Olsen, R.V., Throndsen, I., Scherer, R.: Taking a future perspective by learning from the past - a systematic review of assessment instruments that aim to measure primary and secondary school students' ICT literacy. Educ. Res. Rev. **19**, 58–84 (2016). https://doi.org/10.1016/j.edurev.2016.05.002
4. Siddiq, F., Gochyyev, P., Wilson, M.: Learning in Digital Networks – ICT literacy: a novel assessment of students' 21st century skills. Comput. Educ. **109**, 11–37 (2017). https://doi.org/10.1016/j.compedu.2017.01.014
5. Casillas Martín, S., Cabezas González, M., García Peñalvo, F.J.: Digital competence of early childhood education teachers: attitude, knowledge and use of ICT. Eur. J. Teach. Educ. (2019). https://doi.org/10.1080/02619768.2019.1681393
6. Zabotkina, V., Korovkina, M., Sudakova, O.: Competence-based approach to a module design for the master degree programme in translation: challenge of tuning russia tempus project. Tuning J. High. Educ. **7**, 67 (2019). https://doi.org/10.18543/tjhe-7(1)-2019pp67-92
7. Lazonder, A.W., Walraven, A., Gijlers, H., Janssen, N.: Longitudinal assessment of digital literacy in children: findings from a large Dutch single-school study. Comput. Educ. **143**, (2020). https://doi.org/10.1016/j.compedu.2019.103681

8. Hinojo-Lucena, F.-J., Aznar-Diaz, I., Caceres-Reche, M.-P., Trujillo-Torres, J.-M., Romero-Rodriguez, J.-M.: Factors influencing the development of digital competence in teachers: analysis of the teaching staff of permanent education centres. IEEE Access. **7** (2019). https://doi.org/10.1109/ACCESS.2019.2957438
9. McGarr, O., Ó Gallchóir, C.: Examining supervising field instructors' reporting and assessment of technology use by pre-service teachers on school placement. Comput. Educ. **146** (2020). https://doi.org/10.1016/j.compedu.2019.103753
10. Scherer, R., Siddiq, F., Tondeur, J.: All the same or different? revisiting measures of teachers' technology acceptance. Comput. Educ. **143** (2020). https://doi.org/10.1016/j.compedu.2019.103656
11. Cordero, D., Mory, A.: Education in system engineering: digital competence. In: 2019 IEEE 6th International on Industrial Engineering and Applications ICIEA 2019, pp. 677–681 (2019). https://doi.org/10.1109/IEA.2019.8715223
12. Scherer, R., Siddiq, F., Tondeur, J.: The technology acceptance model (TAM): a meta-analytic structural equation modeling approach to explaining teachers' adoption of digital technology in education. Comput. Educ. **128**, 13–35 (2019). https://doi.org/10.1016/j.compedu.2018.09.009
13. Guárdia, L., Maina, M., Juliá, A.: Digital competence assessment system: supporting teachers with the CRISS platform. In: Central European Conference on Information and Intelligent Systems, Varazdin, pp. 77–82 (2017)
14. DeLone, W.H., McLean, E.R.: Information systems success measurement. In: Foundations and Trends® in Information Systems in Information Systems (2016)
15. Vogt, D.S., King, D.W., King, L.A.: Focus groups in psychological assessment: Enhancing content validity by consulting members of the target population. Psychol. Assess. **16**, 231–243 (2004). https://doi.org/10.1037/1040-3590.16.3.231
16. Straub, D., Gefen, D.: Validation guidelines for IS positivist research. Commun. Assoc. Inf. Syst. **13**, 380–427 (2004)
17. Malgady, R.G., Rogler, L.H., Cortés, D.E.: Cultural expression of psychiatric symptoms: idioms of anger among Puerto Ricans. Psychol. Assess. **8** (1996). https://doi.org/10.1037/1040-3590.8.3.265
18. R Core Team: R: A Language and Environment for Statistical Computing (2017). https://www.r-project.org/
19. Hair, J.F., Hult, G.T.M., Ringle, C.M., Sarstedt, M.: A Primer on Partial Least Squares Structural Equation Modeling (PLS-SEM). Sage Publications, Thousand Oaks (2017)

Using Augmented Reality and Step by Step Verification in Industrial Quality Control

João Alves(✉), Bernardo Marques, Paulo Dias, and Beatriz Sousa Santos

DETI - Department of Electronics, Telecommunications and Informatics, IEETA - Institute of Electronics and Informatics Engineering of Aveiro, University of Aveiro, Aveiro, Portugal

Abstract. Quality control procedures are extremely important among industrial applications. Generally, these tasks include many repetitive tasks that require manual intervention. Given their complexity, quality control tests are often detailed in video recordings, paper instructions, photos or diagrams to guide workers throughout the process. Augmented Reality (AR) has been making significant progress in the last decades, becoming mature enough to be used in industrial scenarios. While some AR systems have been proposed to support quality control procedures, most of them only present information to workers but do not track or validate the process in real-time being used only to guide it. Another limitation of existing systems is the generation of virtual instructions used by AR systems to guide the operator. In this work, we propose an AR-based tool to guide users by overlaying information in a video stream while performing real-time validation during the execution of quality control procedures. The main objective is to provide dynamic support and decrease the mental workload needed to complete the procedure as well as the number of errors, facilitating the procedure execution by untrained workers. Besides this, the tool allows to create virtual content that can be used to generate step-by-step instructions automatically based on human demonstrations. By making the virtual instruction creation effortlessly it is possible to eliminate the user's need for memorizing new instructions with each change of the product lines. While presenting task relevant information the system uses computer vision techniques to keep track of the procedure stage, verifying its completion and switching automatically to the next step without requiring any interaction from the user. A comparison between the time taken to perform the procedure with and without validation was made. The results show that the validation process would confer the process a significant efficiency boost, while avoiding possible human errors.

Keywords: Augmented Reality · Computer vision · Quality control procedure · User guidance · Action validation · Authoring · Industrial efficiency

1 Introduction

Conventional assembly processes often resort to instructions available on paper or in digital format (photos, videos or diagrams) to guide users across different types of industrial procedures. Typically, users are required to map these instructions to actions

W. Karwowski et al. (Eds.): IHSED 2020, AISC 1269, pp. 350–355, 2021.
https://doi.org/10.1007/978-3-030-58282-1_55

to be performed on real objects, without any feedback or additional help [1]. While some processes are automated, a significant number of assembly operations still require manual intervention due to their complexity. In this context, the use of information aid systems using Augmented Reality (AR) might increase significantly task efficiency by keeping the worker focused on the task and not dividing his/her attention between the tasks and the instructions [2]. AR makes possible to display digital contextual information [3, 4] overlaid on top of the real-world, being potentially useful for quality control processes with step-by-step instructions, 3D illustrations, or other relevant data [5]. Specifically, by providing 3D relevant information AR tools can provide a guide to help users navigate through unfamiliar or complex use cases [6].

However, ready-for-market AR tools are still rarely used and as consequence its benefits are not demonstrated often [7]. One obstacle is the generation of virtual instructions that are generally a tedious and time-consuming process [8]. Another obstacle is the open-loop nature of most AR system that only present information without any awareness about the procedure current state of the assembly sequence [9].

In this paper, we address both problems in a real-world scenario, aiming to improve the efficiency of the industrial process. Our work, which is part of an ongoing project with partners from the industrial sector, aims to leverage AR methods in industrial shop floors to enrich the way instructions are presented and ease repetitive tasks associated with quality control. The paper describes the AR tool developed to guide quality control procedures, proposes the creation of virtual instructions based on human demonstration, and presents a real-time error detection algorithm to validate the assembly process.

The remaining of this paper is structured as follows: Sect. 2 describes the AR tool developed to guide a worker and check his/her actions in a quality control procedure. In Sect. 3, we discuss the impact of our method in a real industrial scenario. Finally, in Sect. 4 we draw some remarks and ideas for future work.

2 AR with Verification in a Quality Control Procedure

This section describes a real-world problem, inspired by an industrial shop floor scenario based on a quality control procedure. First, we describe the method used to aid workers in their labour through AR. Afterwards, we present the verification method employed to determine if the worker is executing the procedure steps correctly. To close the loop, we show how the validation procedure can be used to produce virtual instructions that will be overlaid on top of the assembly video stream.

2.1 Scenario

The industrial quality control task that motivates our work requires a worker to check if the distance intervals of an automotive part at specific positions are within a pre-established interval. This procedure is needed to ensure that the assembled piece is within the final client requirements. Not fulfilling the quality check leads to the disposal of the part involved in the process.

The procedure is performed with a wireless comparator (measurement device) handled by the employee. Measurements are sent to an external computer and displayed in a monitor above the quality control cell. Figure 1 illustrates one step of the process: the measurement device is in one of the predefined positions. After each of the nine measures, the worker needs to move away from the cell to verify the measurements in the display (see Fig. 2). This procedure is sequential and in each step the worker must position the comparator in a specific location of the piece and issue an order to trigger the measurement.

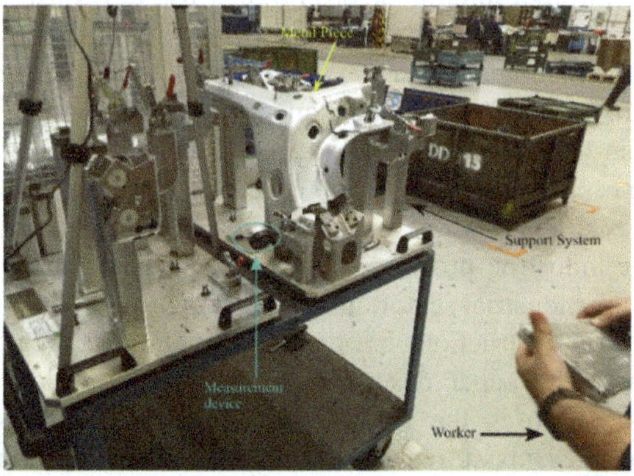

Fig. 1. Quality control process. A gauge ensures the correct placement of the automotive piece under evaluation. The measurement device determines the distance intervals of an automotive part at specific positions during the nine steps measurement process.

2.2 User Action Verification During Procedure

One problem identified in the procedure depicted above was the time required to trigger the measurement since the worker had to move away from the support system to look at the monitor to validate the measurement on the display before moving to the next action. This also requires the interaction with a wireless keyboard during the process (see Fig. 1).

With this in mind, we developed a computer vision process to verify the correct localization of the comparator. The system can trigger automatically the measurement when the device is correctly positioned advancing to the next stage showing the collaborator the next location to be measured.

The correct information about the measurement device placement is critical to trigger the measurement and enable a correct control of transitions between assembly stages. To obtain it, an algorithm based on a template matching approach to compare two 3D point clouds produced by the same perspective is used validating if the measurement device is correctly positioned. The point cloud is processed to extract only the objects that are not present in the initial template acquired before the start of the procedure.

With this in mind, we developed a computer vision process to verify the correct placement of the comparator. The system can trigger automatically the measurement when the device is correctly positioned advancing to the next stage showing the worker what is the next location to be measured.

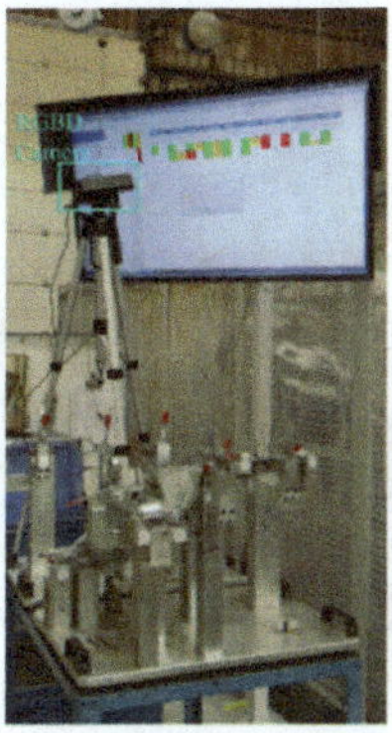

Fig. 2. Verification setup with a fixed depth camera held by a tripod, looking down to the support platform where the quality control procedure is executed

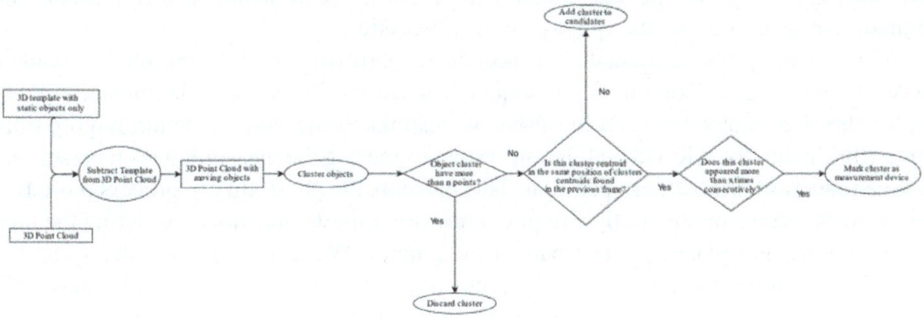

Fig. 3. Algorithm workflow aimed to perform the verification process

Figure 3 presents the workflow of the verification process, which starts by capturing a point cloud using the camera (mounted in such a way as to detect the support system 2) and send it to a computer for processing. Subsequently, dedicated software running on the computer filters the objects outside a pre-established working area and segments the pieces which were not present in the initial point cloud template, to extract the associated clusters of points. After this step, the number of points of each cluster is checked and the ones below a certain threshold are discarded. The algorithm considers that the measurement device is in a specific position if a cluster is in the same position for more than 2 s. To perform this verification procedure, we set up a RGBD camera (ORBBEC ASTRA) held by a tripod, looking down to the support platform (Fig. 2).

The validation mechanism presented above can be also used to leverage the creation of virtual content based on a demonstration. For example, in Fig. 4, a worker placed the comparator in a certain location and the system creates automatically the green arrow indicating the device position. This process can be used to easily create step-by-step instructions for this specific use case or a similar one.

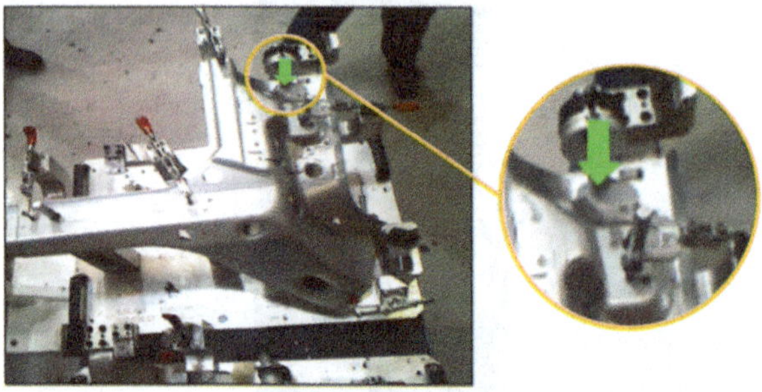

Fig. 4. Virtual content creation through comparator position detection

3 Discussion

At this stage the prototype was tested using recorded data obtained from a traditional human demonstration of the quality control procedure.

On this data, the estimated time needed to perform the complete quality control procedure was 82 s. Using the same sequence and considering only the time required to move the comparator between locations, we estimated the process would require only 36 s while using the AR with validation system triggering the measurements representing a time reduction of 56% per operation. This presents the possibility to process more two pieces in the same amount of time required now for a single one (not considering the time for removing and placing a new part in the gauge). We also argue that the system is flexible and simple enough to be easily transposed to the shop floor, namely because only 3D data is used providing some robustness to lightning conditions changes.

4 Final Remarks and Future Work

Augmented Reality (AR) has great potential for assisting in many industrial tasks (assembly, quality control, maintenance). In this study, we presented a system that shows potential use of this technology in a real quality control procedure. We explored how an action validation mechanism and virtual content authoring in our scenario can speed up the procedure and facilitate the creation of new guides using step-by-step instructions. In the specific case study considered, our tests suggest that is possible to reduce to less than half the task time while guiding the operator through the several repetitive steps, avoiding errors, additional movements and keyboard interactions.

As future work, we plan to perform a more extensive test in real conditions to further evaluate the qualities and limitations of the system and fine tune the system for real use. We also plan to further increase the AR capabilities showing not only information concerning the comparator locations, but also additional information about error location, that might help operators to correct previous mounting steps to reduce the number of faulty parts, resulting in a significant reduction of the discarded pieces along the whole manufacturing process, and thus contributing to a leaner approach in several ways, by reducing time, waist, motion and extra-work [10].

Acknowledgments. The present study was developed in the scope of Produtech – SIF – Solutions for the Industry of the Future [POCI-01- 0247-FEDER-024541]. This study was possible because one of the industry partners presented us the procedure performed in its shop floor allowing us to record one of the procedures performed. This study was also supported by IEETA - Institute of Electronics and Informatics Engineering of Aveiro, funded by National Funds through the FCT - Foundation for Science and Technology, in the context of the project UID/CEC/00127/2019.

References

1. Wu, L.C., Lin, I.C., Tsai, M.H.: Augmented reality instruction for object assembly based on markerless tracking. In: Proceedings of the 20th ACM SIGGRAPH Symposium on Interactive 3D Graphics and Games, pp. 95–102 (2016)
2. Funk, M., Kosch, T., Greenwald, S.W., Schmidt, A.: A benchmark for interactive augmented reality instructions for assembly tasks. In: Proceedings of the 14th International Conference on Mobile and Ubiquitous Multimedia, pp. 253–257 (2015)
3. Michalos, G., Karagiannis, P., Makris, S., Tokçalar, Ö., Chryssolouris, G.: Augmented Reality (AR) applications for supporting human-robot interactive cooperation. Procedia CIRP **41**, 370–375 (2016)
4. Azuma, R.T.: A survey of augmented reality. Presence Teleoper. Virtual Environ. **6**(4), 355–385 (1997)
5. Khuong, B.M., Kiyokawa, K., Miller, A., La Viola, J.J., Mashita, T., Takemura, H.: The effectiveness of an AR-based context-aware assembly support system in object assembly. In: 2014 IEEE Virtual Reality (VR), pp. 57–62 (2014)
6. Borsci, S., Lawson, G., Broome, S.: Empirical evidence, evaluation criteria and challenges for the effectiveness of virtual and mixed reality tools for training operators of car service maintenance. Comput. Ind. **67**, 17–26 (2015)
7. Jetter, J., Eimecke, J., Rese, A.: Augmented reality tools for industrial applications: what are potential key performance indicators and who benefits? Comput. Hum. Behav. **87**, 18–33 (2018)
8. Plopski, A., Fuvattanasilp, V., Poldi, J., Taketomi, T., Sandor, C., Kato, H.: Efficient in-situ creation of augmented reality tutorials. In: 2018 Workshop on Metrology for Industry 4.0 and IoT, pp. 7–11 (2018)
9. Alves, J., Marques, B., Oliveira, M., Araújo, T., Dias, P., Sousa Santos, B.: Comparing spatial and mobile augmented reality for guiding assembling procedures with task validation. In: 19th IEEE International Conference on Autonomous Robot Systems and Competitions (ICARSC) (2019)
10. Lopes, R., Teixeira, L., Ferreira, C.: Lean thinking across the company: successful cases in the manufacturing industry. In: Lean Manufacturing, Implementation, Opportunities and Challenges, Nova Science Publishers, New York (2019)

A Comparison of Three Potato Peeler Designs

Åshild Bøe Drejer[1], Truls Enstad[1], Erik R. M. Fog[1], Thor Oskar F. Viken[1], Fausto Orsi Medola[2], and Frode Eika Sandnes[1,3](✉)

[1] Oslo Metropolitan University, 0130 Oslo, Norway
ashilddrejer@gmail.com, truens@pm.me, erikfog98@hotmail.com, Thor.f.oskar@gmail.com
[2] Sao Paulo State University (UNESP), Bauru, Brazil
fausto.medola@unesp.br
[3] Kristiania University College, 0153 Oslo, Norway

Abstract. This study set out to compare the effectiveness and efficiency of three common potato peeler designs, and whether there were any differences related to gender in terms of use. An experiment was designed involving a timed potato-peeling task using three different potato peeler designs. A balanced group of $N = 20$ males and females was recruited. The results showed that the vertical peeler with flexible blade resulted in the shortest peeling times. This peeler also produced least waste although these differences were not statistically significant. The results did not reveal any statistically significant gender differences.

Keywords: Kitchen utensil ergonomics · Utensil design · Potato peelers

1 Introduction

Potatoes are an important staple for carbohydrates. Although modern life means that individuals increasingly resort to preprocessed food, there are still situations where food is cooked from first principles using the appropriate tools and utensils. There is to the best of our knowledge no previous empirical studies of potato peeler performance. We wanted to investigate if there are measurable differences between different potato peeler designs and any gender related differences in terms of use. We therefore designed a small controlled experiment involving a potato peeling task to compare three commonly available potato peeler designs in terms of peeling speed and amount of waste produced.

2 Related Work

There are seemingly few academic works on potato peelers, although there are several patents issued during the last 70 years that document various designs [1–5]. Some of the academic works are in the domain of agriculture [6, 7] and food production including potato peeling [8] chips making [9].

W. Karwowski et al. (Eds.): IHSED 2020, AISC 1269, pp. 356–362, 2021.
https://doi.org/10.1007/978-3-030-58282-1_56

Within the field of product design kitchen utensils is an active area of research [10, 11]. Key issues include work related injuries caused by certain designs [12] and universal [13] utensil design for the aging population [14]. Kitchen utensils are also an issue in the domain of health and safety [15]. An overview of the evolvement of the domestic kitchen can be found in [16].

3 Method

3.1 Experimental Design

A mixed controlled experiment was designed comprising one within-group independent variable potato peeler with the levels y-peeler, vertical peeler with flexible blade and vertical peeler with fixed blade. Gender was the between-groups independent variable with the levels male and female. Two dependent variables were measured, namely the task completion time, i.e., the time it took to peel each potato, and the amount of peel resulting from each peeling task. The amount of peel is considered a measure of accuracy as removing as little as possible of the non-peel part of the potato is considered more accurate than removing a larger portion of the non-peel part.

3.2 Participants

A total of 20 participants were recruited, comprising 10 males and 10 females. All the participants were recruited at the main campus of the authors' university. Most of the participants were in their early 20s.

3.3 Task

The task comprised peeling potatoes using the three potato peelers.

3.4 Materials

The potatoes used in the experiment was first manually screened to ensure that they were as similar as practically possible both in terms of shape and mass. In total, 60 potatoes were used in the experiment with a mass of (M = 116.0 g, SD = 18.2). The smallest potato had a mass of 80.0 g and the largest 173.0 g.

3.5 Equipment

Figure 1 shows the three common potato peeler designs that were used in the experiment. The y-peeler has a flexible horizontal blade. Typically, the users will move the y-peeler towards themselves to peel. There were two vertical peelers, one with a flexible blade that moves with the contour of the potato and one with a fixed blade that does not move. All the peelers were made in metal, but the vertical peeler with a fixed blade had a hard-plastic handle.

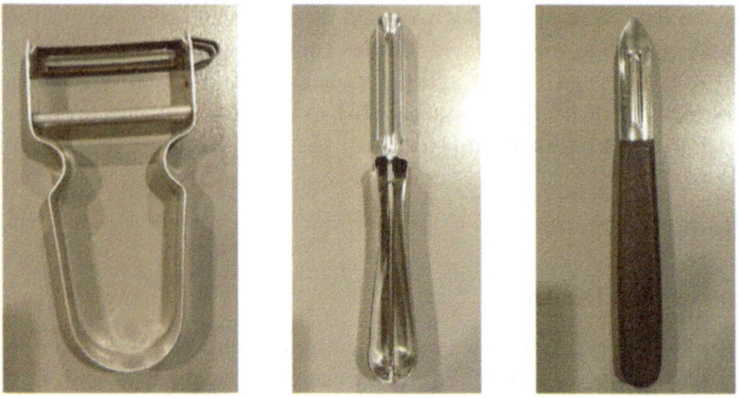

Fig. 1. The three potato peelers used, namely the y-peeler with flexible blade (left), vertical with flexible blade (middle) and vertical with fixed blade (right).

3.6 Procedure

Each participant was first informed about the experiment and then asked to peel three potatoes using the three potato peelers at a comfortable pace in which they were used to. We wanted the experiment to come across as a realistic potato peeling session and not as a competition. The presentation order of the three potato peelers was randomized to minimize any potential bias. For each potato, the total peeling time was measured using a stopwatch. The mass of each potato was also measured before and after peeling process to determine the amount of peel produced in the process. The scales used had a resolution of 1 g.

All the measurements were collected in one session for each participant and participation was voluntary. Participants were therefore anonymous. The data were analyzed using JASP.

4 Results

Figure 2 shows the results of the task completion time measurements. The results show a significant effect on the type of peeler used ($F(2, 36) = 10.974$, $p < .001$), in that the vertical peeler with fixed blade is the slowest ($M = 41.9$, $SD = 28.9$) while Y-peeler ($M = 40.8$, $SD = 21.6$) and the vertical peeler with flexible blade ($M = 38.8$, $SD = 22.7$) yield similar results although the vertical flexible peeler yields the fastest peeling times. Post-hoc tests confirm that the vertical peeler with fixed blade was significantly slower than both the y-peeler ($p < .001$) and the vertical peeler with flexible blade ($p < .001$). There was no significant difference in task completion time between the y-peeler and the vertical peeler with flexible blade.

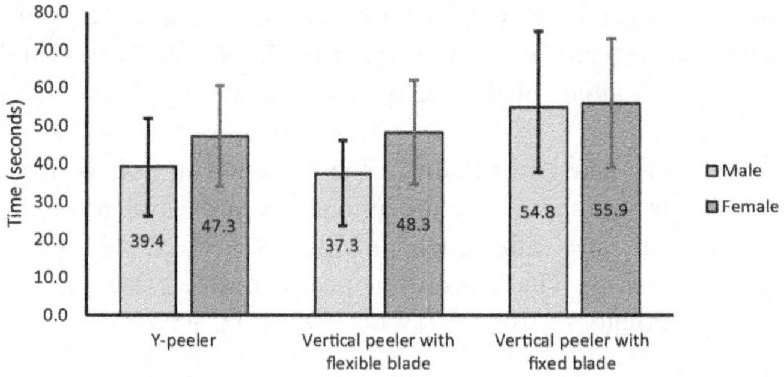

Fig. 2. Task completion times in seconds. Error bars show standard deviation.

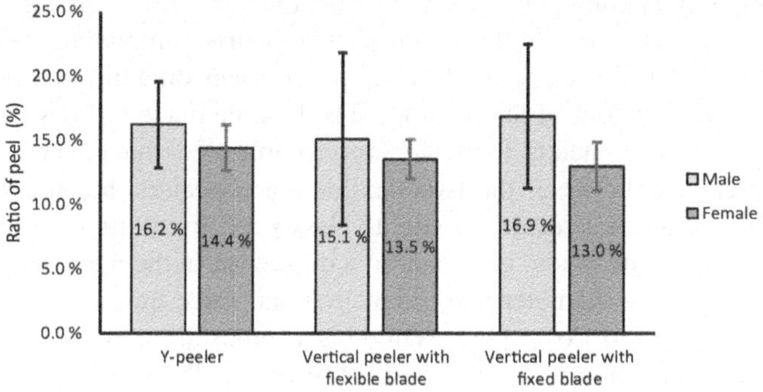

Fig. 3. Accuracy (ratio of peel to potato mass in percentages). Error bars show standard deviation.

There was no significant effect of gender ($F(1, 18) = 0.860$, $p = 0.366$). The results appear to be quite similar for both males and females with the vertical fixed blade peeler. However, although not significant, males seemed to use the two fastest peelers approximately 10 s (or 20%) faster than females. That is, males with y-peeler ($M = 39.4$, $SD = 12.6$) versus females with y-peeler ($M = 47.3$, $SD = 13.6$), and males with vertical flexible peeler ($M = 37.3$, $SD = 8.8$) versus females with vertical flexible peeler ($M = 48.3$, $SD = 13.8$).

Figure 3 shows the results of the mass reduction measurements. No significant effects of either peeler type ($F(2, 36) = 0.739$, $p = .485$) or gender ($F(1, 18) = 1.230$, $p = .282$) could be observed. However, an inspection of the results shows that the females generally produce less peel than the males with all the peelers. Interestingly, females produce the most peel with the y-peeler ($M = 14.4\%$, $5 = 1.8$) and least with the vertical peeler with fixed blade ($M = 13.0\%$, $SD = 1.9$), while males produce the most peel with the vertical peeler with the fixed blade ($M = 16.9\%$, $SD = 5.6$) and least peel with the vertical peeler with the flexible blade ($M = 15.1\%$, $SD = 6.7$). The spread in peel is also much larger for males compared to females. Looking at the results it seems the least peel is achieved with the two vertical peelers. Overall, the spread is much larger with the vertical peeler with fixed blade and the results indicate that the vertical peeler with flexible blade yields more consistent results across all the participants.

Although steps were taken to minimize the variation in potato mass, small variations were unavoidable. To check that the amount of peel is related to the mass of the potato, the reduction in mass after each potato was peeled was correlated with the original mass before peeling. The mass of the peel correlated strongly and significantly with the mass of the potatoes ($r(60) = 0.544$, $p < .001$), 95% CI [0.337, 0.701].

5 Discussion

The results show that potato peelers with flexible blade are faster than the peeler with a fixed blade. One explanation for this could be that the user simply has focus on the side to side motion while peeling and not having to also focus on a turning motion to make the blade follow the contour of the potatoes. The flexible blade follows the shape of the potato. Hence, the user needs to make less complicated maneuvers. There we no significant differences between the two flexible-blade peelers, but the vertical peeler appears slightly faster and results in slightly less peel. In addition to the handle orientation in relation to the blade, there is also a difference in the handgrip diameter. The relation between handle diameter and orientation and force transmission was explored in the study of Kong and Lowe [17]. While big handles are more beneficial when the task involve the application of higher levels of force, smaller handle diameters – which is the case of the two vertical peelers – allow variation and fine adjustments of the objects' position in the hand thus facilitating more accurate movements. The two vertical peelers provided more accuracy than the y-peeler among the females, while only the vertical with flexible blade was more accurate than the y-peeler for the men.

The results did not reveal any significant gender differences. However, the results indicate that some males completed the peeling task faster than females. We did indeed observe that some of the male participants rushed the task, possibly due to have somewhat treated the task as a contest. A possible consequence of rushing the task may have been that some males produced more peel. In fact, using a knife a potato can be quickly peeled with six cuts of a knife if the potato is cut into a cube-like shape. This would clearly be classified as a peeled potato, but also much waste.

The results show that there are large variations in the data. The 20 participants recruited are on the low side. It would have been interesting to have repeated the experiment with more participants in each group. Moreover, we got the impression that most of the participants were already somewhat familiar with the potato peeler designs, but we did not systematically ask the participants about previous experience and familiarity with the peelers. It would probably have been useful to include details about such previous experience and familiarity in the analysis.

6 Conclusions

An experiment was designed to compare three potato peeler designs and to uncover any gender differences related to use. The results showed that the vertical peeler with flexible blade was the fastest. The vertical peeler with the flexible blade also produced least waste although these differences were not statistically significant. The results did not show any significant gender related differences. The results are based on the observation of a small and limited cohort comprising relatively young individuals. It would be interesting to have extended the study with different age groups.

References

1. Dolan, C.H.: U.S. Patent No. 2,407,819: U.S. Patent and Trademark Office, Washington, DC (1946)
2. Fender, F.E.: U.S. Patent No. 2,232,940: U.S. Patent and Trademark Office, Washington, DC (1941)
3. Ness, R.: U.S. Patent No. 3,956,825: U.S. Patent and Trademark Office, Washington, DC (1976)
4. Wang, P.: U.S. Patent No. 6,854,383: U.S. Patent and Trademark Office, Washington, DC (2005)
5. Young, E.C.: U.S. Patent No. 2,958,355: U.S. Patent and Trademark Office, Washington, DC (1960)
6. Pérombelon, M.C.M., Lumb, V.M., Hyman, L.J.: A rapid method to identify and quantify soft rot erwinias on seed potato tubers 1. EPPO Bull. **17**(1), 25–35 (1987)
7. Misener, G.C., Mcleod, C.D., McMillan, L.P.: Evaluation of a prototype potato harvester. Trans. ASAE **27**(1), 24–0028 (1984)
8. Chand, K., Pandey, R.K., Shahi, N.C., Lohani, U.C.: Pedal operated integrated potato peeler and slicer. Agric. Mechanization Asia Afr. Lat. Am. **44**(1), 65–68 (2013)
9. Pradhan, A.K., Dandale, A., Banpurkar, R.: Review Paper on semi automatic chips machine. Int. Res. J. Eng. Technol. **4**(03), 1665–1667 (2017)
10. Branowski, B., Rychlik, M., Sydor, M., Zabłocki, M.: Graphic 3D ergonomic database in evaluation of virtual models of kitchen design/adaptation for needs of handicapped persons, computer-aided design systems. Theory and Practice, Lviv Polytechnic National University, vol. 711, pp. 112–121 (2011)
11. Suhas, M., Prajapathi, P., Jolhe, D.A., Lakhe, R.R.: A study of Indian kitchen from ergonomic consideration. Ind. Eng. J. **12**, 1–19 (2019)
12. Subramaniam, S., Murugen, S.: Investigation of work-related musculoskeletal disorders among male kitchen workers in South India. Int. J. Occup. Saf. Ergon. **21**(4), 524–531 (2015)
13. Sandnes, F.E.: Universell utforming av IKT-systemer, 2nd edn. Universitetsforlaget, Oslo (2018)
14. Maguire, M., Peace, S., Nicolle, C., Marshall, R., Sims, R., Percival, J., Lawton, C.: Kitchen living in later life: exploring ergonomic problems, coping strategies and design solutions. Int. J. Des. **8**(1), 73–91 (2014)
15. Mondal, J.: A review on mechanical and physical hazards at domestic kitchen. Int. J. Occup. Saf. Health **2**(1), 7–10 (2012)

16. Charytonowicz, J., Latala, D.: Evolution of domestic kitchen. In: International Conference on Universal Access in Human-Computer Interaction, pp. 348–357. Springer, Heidelberg (2011)
17. Kong, Y.-K., Lowe, B.D.: Evaluation of handle diameters and orientations in a maximum torque task. Int. J. Ind. Ergon. **35**(12), 1073–1084 (2005)

The Effect of Manual Wheelchair Design on Mobility: A Study with Non-Users and Experienced Wheelchair Users

Sara Raquel Martins Barili[1(✉)], Frode Eika Sandnes[2], Luis Carlos Paschoarelli[1], Galdenoro Botura Junior Botura[1], and Fausto Orsi Medola[1]

[1] UNESP - Sao Paulo State University, Graduate Programme in Design, Bauru 17033-360, Brazil
s.martinsto@gmail.com, {luis.paschoarelli, galdenoro.botura, fausto.medola}@unesp.br
[2] Oslo Metropolitan University, 0130 Oslo, Norway
frodes@oslomet.no

Abstract. The use of inappropriate wheelchairs is believed to limit mobility and reduce the freedom and quality of living for the user. This study therefore set out to investigate the influences of wheelchair design on the performance in a wheelchair agility test. Ten participants performed an agility test involving operating three manual wheelchairs with different designs as fast as possible. The wheelchair designs (independent variable) included a lightweight rigid frame, foldable frame and hospital model. The wheelchairs order was randomized for the agility tests. The time to complete an agility test (dependent variable) was measured with a chronometer. The results show that the use of the rigid frame wheelchair yielded the fastest performance during the agility tests, while the hospital model resulted in longer task-completion times. The findings support the view that active users should be provided with lightweight wheelchairs, as heavy hospital wheelchairs limit mobility.

Keywords: Wheelchairs · Field tests · Equipment design · Mobility

1 Introduction

According to data from the World Health Organization [1] 1 billion of people experience disability worldwide, of which about 1% need a wheelchair. In low-GDP countries, there are many challenges associated with providing people wheelchairs or, more specifically, the most appropriate wheelchairs. As result, user's mobility and satisfaction with the equipment are affected.

Manual propulsion, the main means of moving with a manual wheelchair, has been studied in terms of several biomechanical and physiological outcomes such as: cardiorespiratory and perceived exertion measurement [2–4], the activity of upper limbs muscles [5, 6], upper limb kinematics [7, 8] and hand surface pressure distribution [9]. From the users' point of view, the equipment weight is one of the main aspects considered when selecting and acquiring a wheelchair [10]. A variety of wheelchair

W. Karwowski et al. (Eds.): IHSED 2020, AISC 1269, pp. 363–368, 2021.
https://doi.org/10.1007/978-3-030-58282-1_57

configuration parameters can impact equipment mechanics, the dynamics of a wheelchair in motion and the biomechanical demand during manual propulsion [11]. The tire type, for example, is an important factor that must be considered in the equipment configuration: while solid tires have been claimed as requiring less maintenance, evidence has indicated that these tires increase rolling resistance [12]. In terms of equipment mechanics, wheelchair mass, dimension and mass distribution are important factors affecting the system inertia [13, 14]. Additionally, it has been shown that the influence of the equipment mechanics on the movement dynamics and resistive losses are dependent on the maneuver and trajectory characteristics [15, 16].

Data from the World Health Organization has shown that less than 5% of people who need to use a wheelchair in fact have an adequately adjusted wheelchair [1]. This is most common in low-GDP countries, where many users do not have access to a specialized prescription and provisioning of assistive technology (AT) systems and, as a result, they end up acquiring a wheelchair not matching their characteristics, needs and preferences. In Brazil, it is not uncommon to purchase a wheelchair oneself or have one donated if one is unable to acquire a wheelchair from a specialized rehabilitation center. Consequently, unsuitable wheelchair models are frequently used, such as hospital model wheelchairs intended for short distances or to be driven by a caregiver. Such models may be the only equipment available to the user, since these devices are commonly obtained through donations. However, the use of these devices without the correct prescription can directly influence the performance in daily mobility in addition to causing and aggravating postural problems and upper extremity fatigue, injury, or both. The wheelchair models were selected for this study based on such accounts given by many users and first-hand observations in clinical experience. This study investigated how the wheelchair design can influence the users' ability to move fast and agile on a standard agility test modified for wheelchairs.

2 Materials and Methods

A mixed experimental design was selected for this study with experience level as between groups factor and wheelchair type as a within groups factor. The between groups factor had two levels, namely experienced wheelchair users and novices. The within groups factor had three levels namely the rigid frame, foldable frame and hospital wheelchairs. Mobility performance in terms of task completion time was the dependent variable.

A total of ten male participants, aged between 22 and 43 years voluntarily participated in this study. Half of the participants had mean age of 30.8 + 7.9 years, a mean mass of 72.6 + 9.5 kg and mean height of 1.7 + 0.08 m. These participants were paraplegic due to spinal cord injuries and were active wheelchair users, recruited at SORRI BAURU Rehabilitation Center, with at least 2 years of experience with independent. The other five participants (mean age 25 + 4.6 years, a mean mass of 75.2 + 5.7 kg and a mean height of 1.8 + 0.05 m) did not have any disabilities or experiences with wheelchair usage. Prior to data collection, the participants were informed about the objectives and procedures of the study, read and signed an informed consent form that had been approved by the Ethics Committee of the Faculty of Architecture, Arts and Communication–UNESP (Process. N. 800.500).

Three models of adult manual wheelchairs (Fig. 1) commonly seen in the Brazilian market were used in this study: a rigid-frame wheelchair (RFW) with a mass of 11.55 kg, built with tempered aeronautical aluminum alloy, pneumatic tires in the 24″ rear wheels and foam cushioned seat; a foldable-frame wheelchair (FFW) made of steel with a mass of 21 kg, with backrest tilt and leg/foot support adjustments, pneumatic tires in the 24″ rear wheels, and foam cushioned seat; and a hospital model wheelchair (HMW) with a mass of 15 kg, frame made in iron, seat and backrest in fabric and solid tires in 24″ rear wheels.

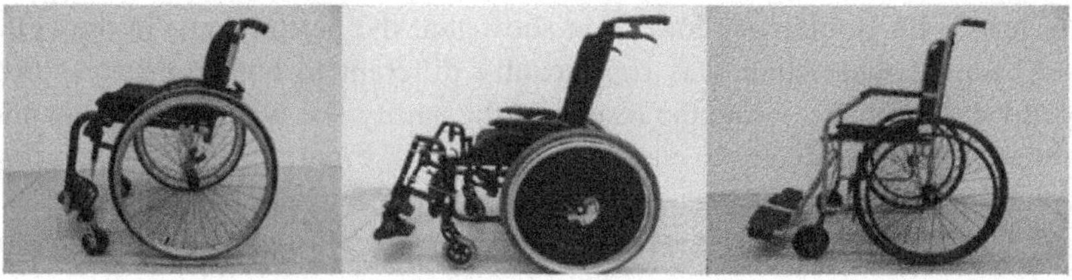

Fig. 1. Manual wheelchair models (from left to right): rigid-frame wheelchair; foldable-frame wheelchair; hospital model wheelchair.

The mobility performance was evaluated with the modified agility test [17]. This test was selected for this study because it comprises straight and turn (to both sides) trajectories, acceleration and breaking (Fig. 2). A video camera and chronometer were used to document study data.

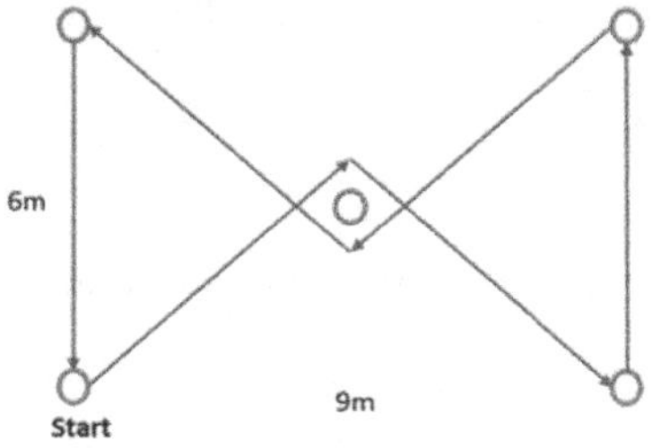

Fig. 2. The modified agility test for wheelchairs.

Participants performed the agility test as fast as possible (repeated three times with each wheelchair type). The sequence of wheelchairs was randomized for all the participants and the time to complete the course was measured using a chronometer. Data is presented descriptively by means of mean and standard deviation (SD). The measurements were transformed using the Aligned Rank Transform (ART). The Aligned Rank Transform is a non-parametric procedure that allows measurements that do not meet the assumptions of parametric tests to be analyzed using traditional repeated measures ANOVA. Post-hoc testing was performed using Bonferroni corrections. Statistical analyses were conducted using JASP version 0.9.1.0 [18] and ArtTool [19].

3 Results

The measurements satisfied the assumptions of sphericity. However, Levene's test revealed that the hospital wheelchair data did not satisfy the assumption of equality of variances ($F(1, 8) = 8.562$, $p = .019$). The data was therefore subjected to the aligned rank transform. A repeated measures ANOVA revealed a significant effect of device type ($F(2, 16) = 9.29$, $p = .002$), but not of participants' experience ($F(1, 8) = 3.95$, $p = .082$), with no interaction between these two factors ($F(2, 16) = 0.73$, $p = .498$). Post hot tests with Bonferroni correction show that the performance in the agility test with the hospital wheelchair was significantly different to both rigid ($p = .005$) and foldable (p = .03) chairs, while the results of the agility test with the foldable and rigid were not statistically different to each other ($p = .45$). Figure 3 shows the performance of users and non-users in the agility test with the three wheelchairs.

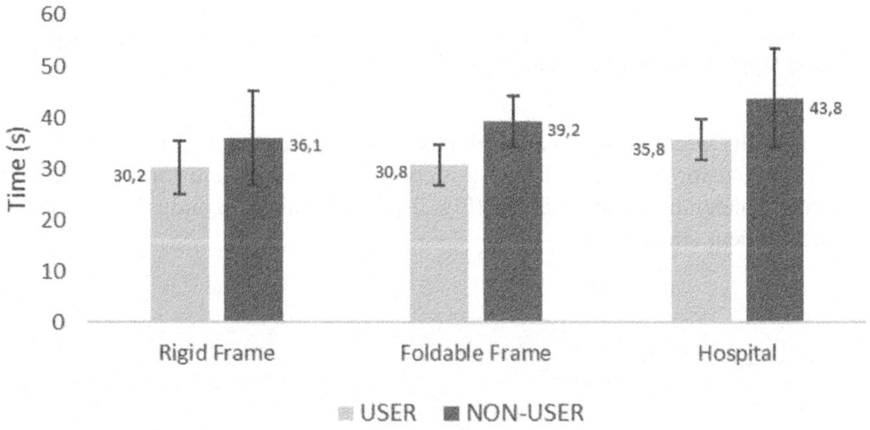

Fig. 3. Time of users and non-users in the agility test with the three wheelchair designs.

4 Discussion

Although it may seem obvious that RFWs allow the fastest performance in comparison to HMWs and FFWs, the magnitude of the time differences in a simple, rapid and controlled test of agility in manual wheelchairs allows to estimate the extent to which the design of manual wheelchairs can affect basic maneuvers in daily mobility. The current findings indicate that the use of the HMW resulted in increased time to complete the agility test in comparison to the other wheelchair models. The equipment configuration and the use of solid tires are probably two factors that contributed to this result. The study of Kwarciak et al. [12] showed that solid tires increase wheels' rolling resistance.

The findings support the claims that users should be provided with a lightweight wheelchair, as well as the risk of mobility limitation that the user might be exposed when using heavy manual wheelchairs or with configuration and accessories that do not target optimal performance. A previous study reported benefits of lighter wheelchairs [20]. In this context, designers, engineers, rehabilitators and other professionals may benefit from the information presented herein. However, this study has limitations that should be noted. The small sample size and the fact that half of the participants' sample being participants without disabilities and with no experience in manual wheelchair usage may limit power of the current findings, which, thus, may be not fully representative of the wheelchair user population. Future studies should address these issues, as well as measuring inertial properties of the different wheelchair designs so that it can be correlated with the data from the participants. Also, from the perspective of the users' efforts, measuring physiological parameters such as oxygen consumption may provide objective data in terms of energy costs for the users when moving with manual wheelchairs of different designs.

5 Conclusion

This study addressed how the use of different designs of manual wheelchair influence the performance in an agility test. The results indicate that lightweight rigid frame wheelchairs are more efficient, while hospital model wheelchair with solid tires demand more time to complete the course trajectory, therefore affecting the mobility performance negatively. The results support the view that specialized prescriptions provide the most appropriate wheelchair for the user.

Acknowledgments. The authors would like to thank FAPESP (Process No. 16/05026-6) and the Norwegian Agency for International Cooperation and Quality Enhancement in Higher Education–DIKU (Project No. UTF-2016-long-term/10053) for the financial support.

References

1. World Health Organization. Health statistics and information systems: Global Health Estimates (GHE) (2011). http://www.who.int/mediacentre/news/releases/2011/disabilities_20110609/en/
2. Gauthier, C., Grangeon, M., Ananos, L., Brosseau, R., Gagnon, D.H.: Quantifying cardiorespiratory responses resulting from speed and slope increments during motorized treadmill propulsion among manual wheelchair users. Ann. Phys. Rehabil. Med. **60**(5), 281–288 (2017)
3. Qi, L., Ferguson-Pell, M., Salimi, Z., Haennel, R., Ramadi, A.: Wheelchair users' perceived exertion during typical mobility activities. Spinal Cord **53**(9), 687–691 (2015)
4. Paulson, T.A., Bishop, N.C., Eston, R.G., Goosey-Tolfrey, V.L.: Differentiated perceived exertion and self-regulated wheelchair exercise. Arch. Phys. Med. Rehabil. **94**(11), 2269–2276 (2013)

5. Bertolaccini, G.S., Filho, I.F.P.C., Christofoletti, G., Paschoarelli, L.C., Medola, F.O.: The influence of axle position and the use of accessories on the activity of upper limb muscles during manual wheelchair propulsion. Int. J. Occup. Saf. Ergon. **24**(2), 311–315 (2018)
6. Louis, N., Gorce, P.: Surface electromyography activity of upper limb muscle during wheelchair propulsion: influence of wheelchair configuration. Clin. Biomech. (Briston, Avon) **25**(9), 879–885 (2010)
7. Bertolaccini, G.S., Sandnes, F.E., Paschoarelli, L.C., Medola, F.O.: A descriptive study on the influence of wheelchair design and movement trajectory on the upper limbs' joint angles. In: Rebelo, F., Soares, M. (eds.) Advances in Ergonomics in Design. (AHFE 2017). Advances in Intelligent Systems and Computing, vol. 288, pp. 645–651. Springer, Cham, Amsterdam (2018)
8. Gorce, P., Louis, N.: Wheelchair propulsion kinematics in beginners and expert users: influence of wheelchair settings. Clin. Biomech. (Bristol, Avon) **27**, 7–15 (2012)
9. Medola, F.O., Paschoarelli, L.C., Silva, D.C., Elui, V.M.C., Fortulan, C.A.: Pressure on hands during manual wheelchair propulsion: a comparative study with two types of handrim. In: European Seating Symposium, pp. 63–65 (2011)
10. Medola, F.O., Elui, V.M.C., Santana, C.S.: La selección de la silla de ruedas y la satisfacción de indivíduos con lesion medular. Rev Iberoam de Fisioterapia y Kinesiología **13**(1), 17–21 (2010)
11. Medola, F.O., Elui, V.M.C., Santana, C.S., Fortulan, C.A.: Aspects of manual wheelchair configuration affecting mobility: a review. J. Phys. Ther. Sci. **26**(2), 313–318 (2014)
12. Kwarciak, A.M., Yarossi, M., Ramanujam, A., Dyson-Hudson, T.A., Sisto, S.A.: Evaluation of wheelchair tire rolling resistance using dynamometer-based coast-down tests. J. Rehabil. Res. Dev. **46**(7), 931–938 (2009)
13. Sprigle, S., Huang, M.: Impact of mass and weight distribution on manual wheelchair propulsion torque. Assist. Technol. **27**(4), 226–235 (2015)
14. Caspall, J.J., Seligsohn, E., Dao, P.V., Sprigle, S.: Changes in inertia and effect on turning effort across different wheelchair configurations. J. Rehabil. Res. Dev. **50**(10), 1353–1362 (2013)
15. Lin, J.T., Huang, M., Sprigle, S.: Evaluation of wheelchair resistive forces during straight and turning trajectories across different wheelchair configurations using free-wheeling coast-down test. J. Rehabil. Res. Dev. **52**(7), 763–774 (2015)
16. Medola, F.O., Dao, P.V., Caspall, J.J., Sprigle, S.: Partitioning kinetic energy during freewheeling wheelchair maneuvers. IEEE Trans. Neural Syst. Rehabil. Eng. **22**(2), 326–333 (2014)
17. Belasco Jr., D., Silva, A.C.: Consistência dos resultados do teste de corrida em ziguezague de Barrow (modificado) em jogadores de basquetebol em cadeira de rodas. International Congress of Motor Rehabilitation, Águas de Lindóia (1998)
18. JASP Team.: JASP (Version 0.9) Computer software (2018)
19. Wobbrock, J.O., Findlater, L., Gergle, D., Higgins, J.J.: The aligned rank transform for nonparametric factorial analyses using only anova procedures. In: Proceedings of the SIGCHI Conference on Human Factors in Computing Systems 2011, pp. 143–146. ACM, Vancouver (2011)
20. Oliveira, N., Blochlinger, S., Ehrenberg, N., Defosse, T., Forrest, G., Dyson-Hudson, T., Barrance, P.: Kinematics and pushrim kinetics in adolescents propelling high-strength lightweight and ultra-lightweight manual wheelchairs. Disabil. Rehabil. Assist. Technol. **14**(3), 209–216 (2019)

Rehabilitation of Children Affected by Attention Deficit Disorder

Carlos Ramos-Galarza[1,2(✉)], Pamela Acosta-Rodas[1], Janio Jadán-Guerrero[2], Mónica Bolaños-Pasquel[2], and Fabiola Saez-Delgado[3]

[1] Facultad de Psicología, Pontificia Universidad Católica del Ecuador, Av. 12 de Octubre y Roca, Quito, Ecuador
caramos@puce.edu.ec

[2] Centro de Investigación en Mecatrónica y Sistemas Interactivos MIST, Carrera de Ingeniería en Ciencias de la Computación y Carrera de Psicología, Universidad Tecnológica Indoamérica, Av. Machala y Sabanilla, Quito, Ecuador
janiojadan@uti.edu.ec

[3] Departamento de Fundamentos de la Pedagogía, Facultad de Educación, Universidad Católica de La Santísima Concepción, Concepción, Chile
fsaez@ucsc.cl

Abstract. This paper presents the conceptual development of a technological application for the treatment of children with attention deficit disorder (ADD). This technological application will consist on serious-gaming to stimulate the different levels of attention. The first level is focused in attention and stimulates it by identifying stimuli through tactile interaction with the user. The second level is selective attention, for which, the child must select a stimulus leaving the irrelevant stimuli aside. The third level will work on sustained attention, where the child must fulfill a game with a longer duration. The next level will contain an activity to stimulate alternating attention, where the child must solve two tasks by alternating the focus of attention. Finally, for the divided attention, it must be resolved two tasks at the same time. With this application it is expected to improve the child's motivation in the treatment and provide an intervention tool for the clinician.

Keywords: Technological applications · Attention deficit disorder · Neuropsychological rehabilitation · Neurodevelopmental disorders

1 Introduction

Attention deficit disorder (ADD) is one of the most common neuro-developmental disorder that affects around 5% of children population [1]. This disorder is characterized for presenting a pattern of difficulties for attending external and internal stimuli, producing in the child a clinical alteration in the majority of areas where those take place [2].

The American Psychiatric Association describes some of the signs of impaired attention such as difficulties attending to details, presenting mistakes when doing homework, problems maintaining attention in tasks, homework o game activities,

W. Karwowski et al. (Eds.): IHSED 2020, AISC 1269, pp. 369–373, 2021.
https://doi.org/10.1007/978-3-030-58282-1_58

appearance of not attending when someone speaks directly to him or her, difficulties when following instructions, not ending homework or home tasks, difficulties to organize tasks and activities, dislike or avoiding tasks requiring cognitive effort, become easily distracted by irrelevant stimuli and neglect in daily activities [3].

Various authors have stated that the core problem in ADD is the immaturity of the attention mechanism associated to mental skills such as the inhibitory control or cerebral structures as the frontal lobe, which would explain the child's difficulties sustaining the attention in classroom, when doing homework or any other activity [4, 5, 9].

2 Brain Systems of Attention

The attention process includes three brain systems (see Fig. 1) that allow maintaining an adequate state of alert level when receiving information from sensorial receptors and, as consequence, cognitive regulation responses are produced [6].

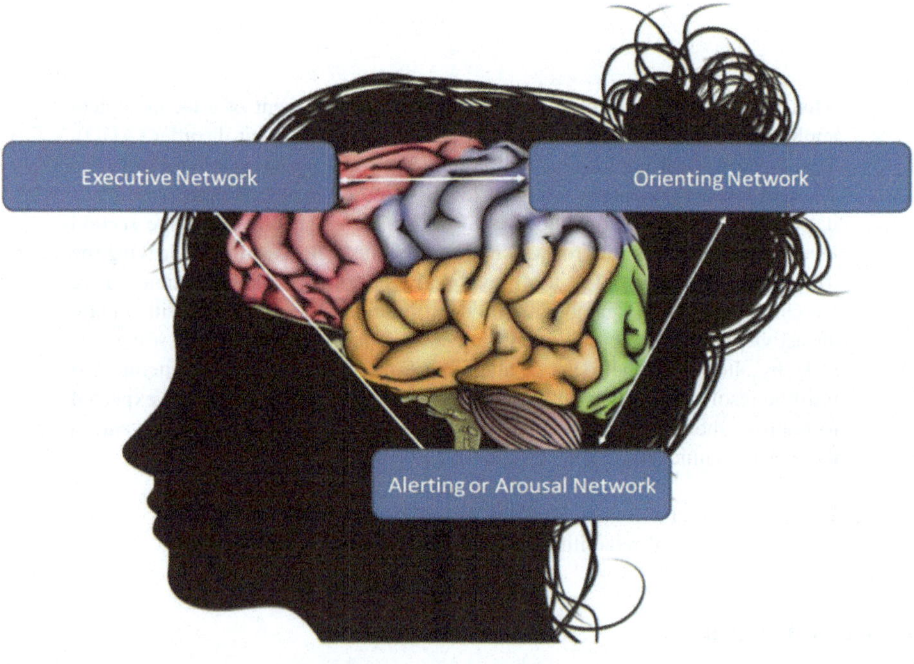

Fig. 1. Model of three brain systems of attention [7]

Attention process has five levels of complexity that matures as development and cognitive skills advance. The first level is focused attention which refers to the capacity to focus attention on a specific stimulus. The second level is sustained attention which is the ability to attend over a prolonged period of time. The third level is selective attention which allows attending to a specific stimulus inhibiting irrelevant elements in tasks. The fourth level is alternating attention, which is the ability for changing focus attention between two or more stimuli. The last level is divided attention which allows attending different stimuli or attention at the same time [8, 10].

Previous research has reported that children with ADD have serious difficulties in the majority of attention levels [11, 12], presenting problems in their neuropsychological treatment, since most of the applied techniques are based on paper and pencil materials, diminishing the motivation of children with ADD in their rehabilitation process [13].

Because of all the descriptions made above, we think it is necessary developing technological innovations that help the different levels of attention process of children with ADD. The conceptual content of a technological application for the rehabilitation of the attention process of children with ADD is presented as following.

3 Technological Application to Train Attention Systems in ADD

As it was mentioned previously, one of the main problems in the rehabilitation of children with ADD is their lack of motivation for doing the exercises based on paper and pencil materials. To solve this issue, it is proposed a technological application based on serious-gaming for tablet with exercises for the five levels of the attention process (see Fig. 2).

Fig. 2. Graphical representation of child with ADD doing the exercises for training the attention process

- Focused Attention: the game that will be implemented will show figures to be paint, to find matching figures, to organize stimuli, to complete sequences and by doing so, identifying stimuli through tactile interaction with the user.
- Sustained Attention: where the child must fulfill a game with a longer duration and the task will consist on solving problems in different stories proposed.
- Selective Attention: the scenario that will be proposed will ask the child to select a stimulus leaving the irrelevant stimuli aside.

- Alternating Attention: where the child will have to solve two tasks by alternating the focus of attention
- Divided Attention: the procedure that will be present will have some activities about solving two tasks at the same time.

4 Clinician Feedback

The clinician will have the opportunity of supervising goals accomplished, and the progress of children with ADD that use this technological application through the daily reception of the practice carried out by the child, as well as his/her performance in each exercise. In this way, the clinician will be able to assess whether the results obtained by the patient are allowing the improvement of their attention skills.

5 Conclusions

In this paper, the conceptual development of a technological application to improve the attention skills of children with ADD has been reported. Its content allows working in the different levels involved in the attention process: focused attention, sustained attention, selective attention, alternating attention and divided attention.

There are some benefits of this technological resource, for example, children with ADD will be motivated to continue from beginning to the end with the therapy, as well as compromised with the exercises proposed within the therapeutical setting, which will be realized independently and by their own, and the clinician will receive feedback of the children with ADD performance.

The challenge for the future implied in this type of intervention lies on the possibility of generalizing the improvement achieved in mental ability. The main objective is that these kind of advances could be evidenced not solely on the application, but in daily life.

As future research, it is of great interest for the research team to develop and apply this technological device through an experiment to children with ADD and compare them with a control group, analyzing the benefits of this intervention.

References

1. Ramos-Galarza, C., Bolaños, M., Ramos, D.: Prevalence of attention deficit hyperactivity disorder in ecuadorian students. Revista Científica y Tecnológica **3**(1), 13–19 (2015). https://doi.org/10.26423/rctu.v3i1.72
2. Ramos-Galarza, C., Pérez-Salas, C.: Relationship between the hybrid model of executive functions and attention deficit hyperactivity disorder. Revista Psicología desde el Caribe **32** (2), 299–314 (2015). http://dx.doi.org/10.14482/psdc.32.2.5986
3. American Psychiatric Association. Diagnostic and Statistical Manual of Mental Disorders (5th edn.). American Psychiatric Publishing, Washington, DC (2013). https://doi.org/10.1176/appi.books.9780890425596

4. Ramos-Galarza, C., Acosta-Rodas, P., Perez-Salas, C., Ramos, V.: Inhibitory control and symptomatology of attention deficit hyperactivity disorder. Revista Ecuatoriana de Neurología **28**(3), 41–46 (2019). http://revecuatneurol.com/magazine_issue_article/inhibitory-control-symptomatology-attention-deficit-hyperactivity-disorder-control-inhibitorio-sintomatologia-trastorno-deficit-atencion-hiperactividad/
5. Barkley, R.: Behavioral inhibition, sustained attention, and executive functions: constructing a unifying theory of ADHD. Psychol. Bull. **121**(1), 65–94 (1997). https://doi.org/10.1037/0033-2909.121.1.65
6. Posner, M., Rothbart, M., Sheese, B., Voelker, P.: Developing attention: behavioral and brain mechanism. Adv. Neurosci. **1**, 1–9 (2014). https://www.researchgate.net/deref/http%3A%2F%2Fdx.doi.org%2F10.1155%2F2014%2F405094
7. Ramos-Galarza, C.: Attention systems of university students from Quito-Ecuador. Revista Ecuatoriana de Neurología **25**(1–3), 34-38. http://revecuatneurol.com/magazine_issue_article/sistemas-atencion-focalizada-sostenida-selectiva-universitarios-quito-ecuador/
8. Sohlberg, M., Mateer, C.: Cognitive Rehabilitation an Integrative Neuropsychological Approach. The Guilford Press, New York (2001)
9. Salum, G., Sergeant, E., Sonuga-Barke, E., Vandekerckhove, J., Gadelha, A., Pan, P., Moriyama, T., Graeff-Martins, A., Gomes de Alvarenga, P., Do Rosário, M., Manfro, G., Polanczyk, G., Rohde, L.: Specify of basic information processing and inhibitory control in attention deficit hyperactivity disorder. Psy. Med. **44**, 617–631 (2014). https://doi.org/10.1017/S0033291713000639
10. Cognifit. Attention One of our Cognitive Domains. https://www.cognifit.com/attention
11. Tucha, O., Tucha, L., Kaumann, G., König, S., Lange, K., Dorota, S., Streather, Z., Engelschalk, T., Lange, K.: Training of attention functions in children with attention deficit hyperactivity disorder. Attention Deficit Hyperactivity Disord. **3**(3), 271–283 (2011). https://doi.org/10.1007/s12402-011-0059-x
12. Tamm, L., Hughes, C., Ames, L., Pickering, J., et al.: Attention training for school-aged children with ADHD: results of an open trial. J. Attention Disord. **14**(1), 86–94 (2010). https://doi.org/10.1177/1087054709347446
13. Ramos-Galarza, C., Acosta-Rodas, M., Acosta-Vargas, P., Salvador-Ullauri, L.: Technological Innovation to Assess Cognitive Functions in Attention Deficit Hyperactivity Disorder (2020, in press)

Author Index

W. Karwowski et al. (Eds.): IHSED 2020, AISC 1269, pp. 375–377, 2021.
https://doi.org/10.1007/978-3-030-58282-1

The manufacturer's authorised representative in the EU is Springer Nature Customer Service Centre GmbH, Europaplatz 3, 69115 Heidelberg, Germany. If you have any concerns regarding our products, please contact ProductSafety@springernature.com

Printed and bound by CPI Group (UK) Ltd, Croydon, CR0 4YY
30/09/2026
02225495-0011